PRIMARY PRODUCTS
OF METABOLISM

ECONOMIC MICROBIOLOGY

Series Editor

A. H. ROSE

Volume 1. Alcoholic Beverages

Volume 2. Primary Products of Metabolism

In preparation

Volume 3. Secondary Products of Metabolism

Volume 4. Microbial Cell Material: Biomass

Volume 5. Microbial Enzymes and Transformations

Volume 6. Microbial Biodegradation

ECONOMIC MICROBIOLOGY
Volume 2

PRIMARY PRODUCTS OF METABOLISM

edited by

A. H. ROSE

School of Biological Sciences
University of Bath,
Bath, England

1978

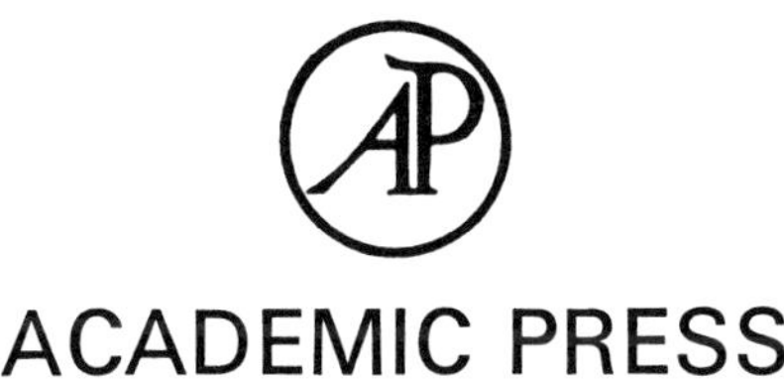

ACADEMIC PRESS

LONDON NEW YORK SAN FRANCISCO

A Subsidiary of Harcourt Brace Jovanovich, Publishers

ACADEMIC PRESS INC. (LONDON) LTD.
24/28 Oval Road
London NW1

United States edition published by
ACADEMIC PRESS INC.
111 Fifth Avenue
New York, New York 10003

Library of Congress Catalog Card Number: 77 77 361
ISBN: 0-12-596552-4

*Printed in Great Britain at The Spottiswoode Ballantyne Press
by William Clowes & Sons Limited, London, Colchester and Beccles*

CONTRIBUTORS

ARNOLD L. DEMAIN, Department of Nutrition and Food Science, Massachusetts Institute of Technology, Cambridge, Massachusetts 02139, U.S.A.

R. N. GREENSHIELDS, Department of Biological Sciences, University of Aston in Birmingham, Birmingham, England.

JOHN J. HASTINGS, Formerly of Commercial Solvents (Great Britain) Limited.

S. KINOSHITA, Tokyo Research Laboratory, Kyowa Hakko Kogyo Co. Ltd., Tokyo, Japan.

C. J. LAWSON, Tate and Lyle Ltd., Group Research Laboratory, Reading, Berks, England.

L. M. MIALL, Pfizer Central Research, Sandwich, Kent, England.

K. NAKAYAMA, Tokyo Research Laboratory, Kyowa Hakko Kogyo Co. Ltd., Tokyo, Japan.

D. PERLMAN, School of Pharmacy, University of Wisconsin, Madison, Wisconsin, U.S.A.

COLIN RATLEDGE, Department of Biochemistry, The University of Hull, Hull, England.

ANTHONY H. ROSE, Zymology Laboratory, School of Biological Sciences, University of Bath, Bath, Avon, England.

DOROTHY M. SPENCER, Biology Department, Goldsmith's College, University of London, England.

J. F. T. SPENCER, Biology Department, Goldsmith's College, University of London, England.

I. W. SUTHERLAND, Department of Microbiology, University of Edinburgh, Edinburgh, Scotland.

PREFACE TO THE SERIES

Controlling and exploiting the World's flora and fauna have been fundamental to Man's colonization of this planet. His ability to regulate the activities, both pathogenic and saprophytic, of micro-organisms, and to go on to harness microbial activity in the manufacture of foods and chemicals represents a truly outstanding achievement especially when one remembers that microbes represented an invisible activity or agent until microbiology became established as a science during the latter half of the last Century. Only then did it become apparent that Man's very existence depends on microbial activity.

This multi-volume series aims to provide authoritative accounts of the many facets of exploitation and control of microbial activity. The first volume describes production of alcoholic beverages, and in the second and third volumes there are accounts of the microbiological production of commercially important chemicals. Production of microbial biomass is the subject of the fourth volume, and in the fifth there are accounts of production of enzymes from micro-organisms and of industrially-important chemical conversions or reactions mediated by microbes. Later volumes will deal with biodeterioration caused by microbes, sewage purification and the microbiology of foods. Throughout the volumes, emphasis is placed on the chemical activities of micro-organisms for it is these activities which affect with such impact the activities of man. It is hoped that the series will provide an adequate testimony to the unique relationship which Man has forged with his smallest servants.

January, 1977   ANTHONY H. ROSE

PREFACE TO VOLUME 2

Notwithstanding the achievements of the industrial chemist over the past century and a half, many commercially important chemicals are manufactured on an industrial scale using micro-organisms; these processes are known collectively as fermentations. The reasons for the continued viability of these microbiological processes, which inherently must be less efficient than the strictly chemical production, are twofold.

First, manufacture of a few chemicals is only legally permitted by microbiological processes; examples are the production of ethanol in alcoholic beverages, which is described in the first volume of this series, and the wide use of acetic acid in vinegar, an account of which appears in the present volume. A more important consideration, however, is that the cost of producing several industrially important chemicals by microbiological processes is far cheaper than the cost of synthesizing them. This is easily appreciated when one attempts to estimate the cost of producing large quantities of a cobamide or a macrolide antibiotic by purely chemical means, but it is also relevant to such relatively simple molecules as acetic acid.

For convenience, production of industrially important chemicals by microbiological processes is divided into two volumes. The present one deals with the production of primary products of metabolism, the subsequent volume with secondary products of metabolism.

This volume includes accounts of the production of organic acids, nucleotides and amino acids which form large and stable sectors of the microbiological industries, and also includes information on polysaccharide fermentations which are currently undergoing extensive development. Further, there are accounts of the production of lipids and polyhydroxy alcohols which as yet have not been introduced on a commercial scale but which could well become economically viable in the near future. Finally, there is also an account of the production of acetone and butanol by bacteria. This fermentation process featured significantly in the career of Chaim Weizmann, the first President of the State of Israel, and it is still operated in some countries. The reason for its demise elsewhere is that these two solvents can be made cheaply from petrochemicals. However, with the increasing cost of petroleum, this fermentation process could well stage a comeback in the foreseeable future.

November, 1977   ANTHONY H. ROSE

CONTENTS

1. Production and Industrial Importance of Primary Products of Microbial Metabolism

A. H. ROSE

2. Acetone–Butyl Alcohol Fermentation

JOHN J. H. HASTINGS

3. Organic Acids

L. M. MIALL

4. Acetic Acid: Vinegar

R. N. GREENSHIELDS

5. Production of Nucleotides by Micro-Organisms

ARNOLD L. DEMAIN

6. Amino Acids

S. KINOSHITA and K. NAKAYAMA

7. Lipids and Fatty Acids

COLIN RATLEDGE

8. Vitamins

D. PERLMAN

9. Polysaccharides

C. J. LAWSON and I. W. SUTHERLAND

10. Production of Polyhydroxy Alcohols by Osmotolerant Yeasts
J. F. T. SPENCER and DOROTHY M. SPENCER

NOTES

Abbreviations

The abbreviations used for chemical and biochemical compounds in this book are those recommended by the International Union of Pure and Applied Chemistry—International Union of Biochemistry Commission on Biochemical Nomenclature, and summarized in the *Biochemical Journal* (1976; **153**, 1–24).

Names of Micro-Organisms

In general, the names of bacteria used are those recommended in *Bergey's Manual of Determinative Bacteriology* (8th edition, 1974, edited by R. E. Buchanan and N. E. Gibbons, and published by Williams and Wilkins Co. of Baltimore) and those of filamentous fungi which were adopted in the *Dictionary of the Fungi* (6th edition, written by G. C. Ainsworth and W. Bisby, and published in 1971 by the British Commonwealth Mycological Institute at Kew). Names of yeasts are those recommended in *The Yeasts, a Taxonomic Study* (2nd edition, 1970, edited by J. Lodder, and published by the North-Holland Publishing Co. of Amsterdam, Holland).

1. Production and Industrial Importance of Primary Products of Microbial Metabolism

ANTHONY H. ROSE

*Zymology Laboratory, School of Biological Sciences,
University of Bath, Bath, Avon, England*

I. NATURE OF PRIMARY PRODUCTS OF MICROBIAL METABOLISM

A. Growth Phases in Batch Culture

After Robert Koch (1843–1910) had devised techniques for growing bacteria in pure culture, an achievement which made the study of micro-organisms an infinitely more attractive discipline, it was to be expected that students of microbiology would go on to describe bacterial growth in pure culture more fully and in quantitative terms. To begin with, many of them endeavoured to estimate the rate of growth of bacteria which Nägeli (1887), for example, did by determining the amount of acid produced in a culture whilst others (e.g. Buchner *et al.*, 1887) used the plating-out method of Koch in a quantitative manner. These early studies showed that bacteria, when freshly inoculated into a suitable medium, do not immediately begin to multiply, but do so only after a short interval of time. Observations along these lines, which Lane-Claypon (1909) and later Slator (1917) did much to develop, led ultimately to the recognition of the phases of growth in batch cultures of micro-organisms which form a cornerstone in the history of microbiology. Pride of place must go to Lane-Claypon (1909) for delineating the lag, exponential and station-

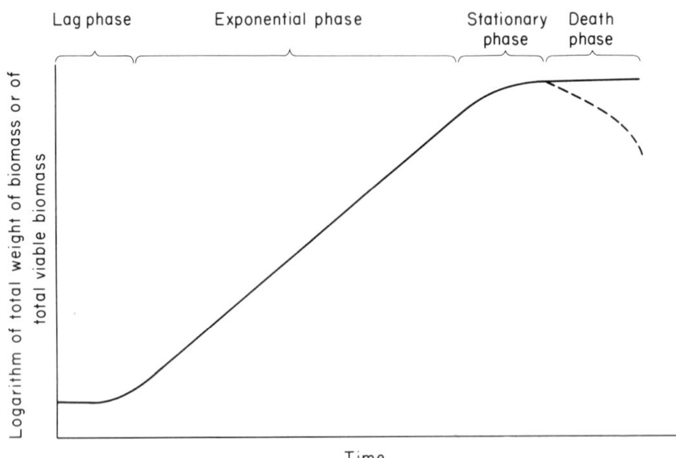

Fig. 1. Generalized growth curve for a unicellular micro-organism.

ary phases of growth (Fig. 1), although the paper by Buchanen (1918) is a major landmark in the description of growth phases in batch culture, despite the fact that he was sufficiently ambitious to recognize no fewer than seven such phases of growth.

B. Primary and Secondary Products of Metabolism

While the early microbial physiologists appreciated that, in the exponential phase of growth, there is extensive metabolism with the microbes rapidly replicating their cell components as a prerequisite to growth and cell division, it was at first assumed that the stationary phase of growth represented virtually total metabolic inactivity on the part of the micro-organisms. It was not the microbiologist but the natural-product chemist who first realized the falseness of this assumption, for, during the 1920s and 1930s, organic chemists discovered, particularly in stationary-phase cultures of filamentous fungi, a rich—almost never-ending—source of complex organic compounds elucidation of whose chemical structures offered a formidable challenge. As the structures of these excreted compounds were described, it became apparent that they were not compounds which play an important role during exponential growth of micro-organisms. Plant physiologists had some years earlier recognized two similar classes of compounds produced by higher plants. There were compounds such as chlorophylls which are synthesized by almost all plants; these were termed *primary products* of metabolism, and they contrasted with compounds such as camphor and tannins which can be obtained only from particular plant species and which could not be assigned a general metabolic function. These latter compounds were dubbed *secondary products* of metabolism. Recognizing the differences between primary and secondary products of microbial metabolism was largely pioneered by the British microbial biochemist John D. Bu'Lock, who has written extensively on the subject (Bu'Lock, 1961, 1965, 1967, 1975). He has also introduced new terms to describe the phases of primary microbial metabolism, namely τρφε (tropophase from the Greek for nutrient), and of secondary metabolism, ιδοφε (idiophase, Gr. for peculiar; indicating that the phase is characterized by metabolic idiosyncracies).

A brief account of primary products of microbial metabolism and their biosynthesis is given in Section IIA (pg. 5) of this chapter. This volume is concerned with those primary products of microbial metabolism that have actual or potential economic importance. Such products include microbial biomass, which forms the subject of a separate volume (Volume 4) in the series. The present volume describes production of cell constituents, including polysaccharides, lipids and vitamins, and intermediates in the biosynthesis of cell constituents such as nucleotides and amino acids, which have proven or potential economic importance. All of these are products of anabolic primary metabolism. But some products of microbial catabolism, particularly fermentation products that include ethanol, acetone, butanol, acetic acid and even carbon dioxide, are also of commercial importance and are produced industrially from micro-organisms. Synthesis of ATP by micro-organisms has also been exploited in fuel cells as a source of energy useful for Man.

Production of commercially important secondary products of microbial metabolism is described in Volume 3 of this series. Undoubtedly the most important group of these secondary products of metabolism are antibiotics, but they also include alkaloids and toxins. These compounds characteristically have a much greater molecular complexity than primary products of metabolism, which explains their interest to the natural-product chemist. The geneticist, too, is interested to learn of the regulatory processes that cause genes concerned with production of primary products of metabolism to become largely if not entirely inactive in transcription, and induce expression of those genes concerned with production of secondary products of metabolism.

Interest in the expression of genetic information in micro-organisms to produce commercially important chemical compounds has recently taken a dramatic turn with the discovery that, through the agency of phage or plasmid vectors, genetic information from conceivably any organism can be introduced into bacteria. This is the practice of *genetic engineering*, and theoretically it introduces the possibility of using bacteria to produce, for example, insulin. There are formidable biochemical and genetical problems to be overcome before such exploitation of genetic engineering can be realized. Meanwhile, the problems that could arise as a result of an inadequately monitored programme of genetic engineering, problems such as the introduction, inadvertent or otherwise, into the environment of bacteria with genomes that contain DNA from an oncogenic virus, have been discussed World-wide in what has become an international debate amongst microbial geneticists (Cohen, 1975; Curtiss, 1976).

II. MICROBIAL METABOLISM AND ITS REGULATION

A. Principles of Microbial Metabolism

Growth and reproduction of a microbe is the result of an exquisitely ordered and intricate series of reactions. For simplicity, the microbial biochemist finds it convenient to recognize three separate unit processes in microbial growth and reproduction. The first of these is entry of chemical compounds from the environment or medium into the organism. The cell wall and extramural layers of a micro-organism are freely permeable to the majority of compounds present in the environment, and the main barrier between a chemical compound and the metabolic machinery inside a micro-organism is the plasma membrane. The mechanisms involved in passage of compounds into and out of a micro-organism have only recently begun to be elucidated, mainly because this is a vectorial process, involving movement of molecules in space (a distance of about 7.5 nm), which is not as amenable to investigation as the scalar or metabolic reactions that comprise the other two unit processes in microbial growth and reproduction. Accounts of the mechanisms that operate during passage of compounds across the microbial plasma membrane can be found in many of the basic texts in microbial physiology (A. H. Rose, 1976; Dawes and Sutherland, 1976). Many solute molecules that pass across the microbial plasma membrane do so through the mediation of carrier or transport mechanisms that involve specific proteins. The solute may arrive on the inside of the plasma membrane in a chemically unmodified form or, as with some sugar-transport processes, as a phosphate. Other solutes may pass across the membrane by diffusion, these solutes being soluble in membrane lipids. Ironically, the microbial physiologist is still largely ignorant of the manner in which the most ancient of industrially important primary products of microbial metabolism and one which still forms the basis of a large group of industries, namely ethanol, passes across the yeast plasma membrane.

When inside a micro-organism, a nutrient is subjected to a series of chemical modifications, that is, it is *metabolized*. The sequence of reactions by which a compound is metabolized is referred to as a *metabolic pathway* (Fig. 2). On entering a pathway, a nutrient is converted into one or more different compounds, known as *intermediates*, each of which is further metabolized to give ultimately the

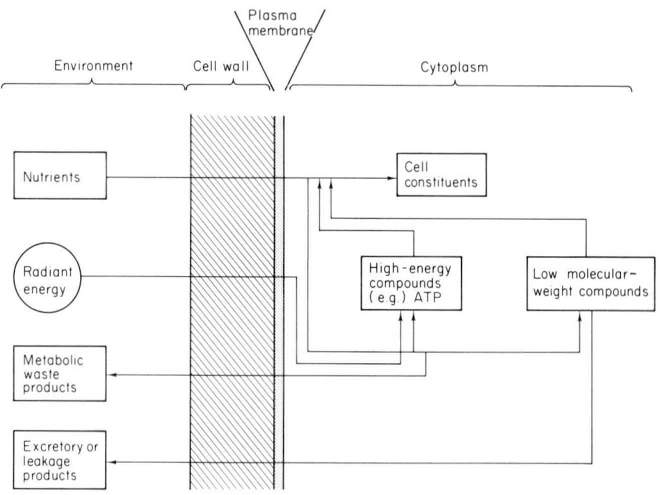

Fig. 2. Flow diagram summarizing the salient features of microbial metabolism.

end product of the pathway. Each of the reactions on a metabolic pathway is catalysed by a specific enzyme. Another term which is often used when discussing cell metabolism is *precursor*; this is defined as any compound which is formed within the cell, or supplied in the medium, and which is metabolized to give some end product.

When a nutrient enters a metabolic pathway, it is subjected either to a decrease or an increase in molecular complexity. Those pathways that effect a decrease in molecular complexity are referred to as *catabolic* pathways to distinguish them from *anabolic* pathways that lead to an increase in molecular complexity of the compounds being metabolized. Catabolic pathways furnish the cell with a supply of energy (ATP) and low molecular-weight (C_2, C_3 and C_4) compounds some of which then enter anabolic pathways to be used in synthesis of cell components. Other low molecular-weight compounds are waste products of metabolism and are excreted. Reactions on both catabolic and anabolic pathways are subject to a variety of control or regulatory mechanisms the aim of which is to ensure that just sufficient ATP and low molecular-weight compounds are generated on catabolic pathways, while anabolic pathways do not lead to overproduction of cell constituents. These regulatory processes are hardly ever completely efficient, and production of industrially important primary products of microbial metabolism exploits these inefficiencies, which are often exaggerated by growing the microbe under conditions that further derange cell metabolism.

B. Primary Metabolic Pathways

1. Catabolic Pathways

Energy is obtained by micro-organisms from one or more of three sources, namely organic compounds, inorganic compounds and visible radiation. This chapter makes no attempt to describe all of the catabolic pathways which have been charted in micro-organisms; again, basic texts in microbial physiology can be consulted for this information (Mandelstam and McQuillen, 1973; A. H. Rose, 1976). Figure 3 summarizes the main catabolic highways that operate in organotrophic microbes which obtain their energy from organic compounds. Organic substrates, usually sugars, are catabolized along one or more of a small number of glycolytic pathways (Embden–Meyerhof–Parnas, hexose monophosphate, phosphoketolase and Entner–Doudoroff), via various C_3 and C_2 compounds, to yield pyruvate. Three-carbon compounds, such as glycerol, can also enter these glycolytic pathways after being converted into an intermediate. During glycolysis, ATP is formed by substrate-level phosphorylation. Under aerobic conditions, pyruvate is oxidatively decarboxylated to acetyl-CoA, the acetyl residue in which is then completely oxidized in the tricarboxylic acid cycle. Oxidation of the reduced nicotinamide and flavin nucleotides, which are produced during operation of

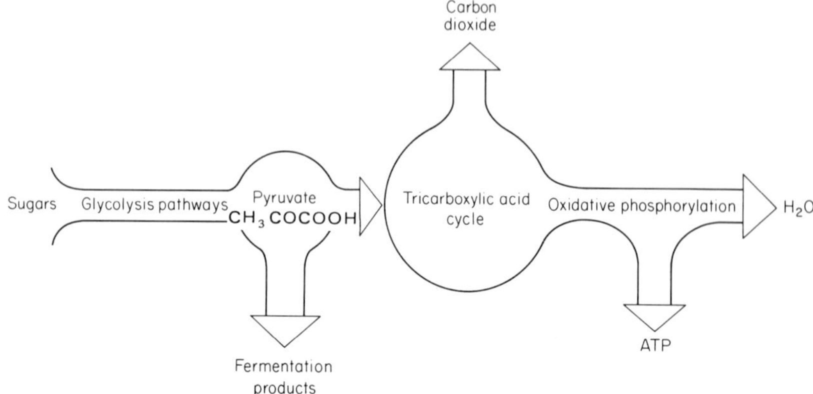

Fig. 3. Diagram showing the main highways of catabolism in organotrophic micro-organisms.

the tricarboxylic acid cycle, takes place in an oxidative phosphorylation chain. This is the process of *aerobic respiration* with molecular
oxygen acting as the electron acceptor. Compounds such as nitrate
and sulphate can, with certain micro-organisms, take over the
electron-accepting role of oxygen in aerobic respiration, when the
process becomes *anaerobic respiration*. However, many organotrophic micro-organisms when they encounter anaerobic conditions
carry out a *fermentation* which is defined as an energy-yielding
process in which an organic compound functions as both the electron
donor and the electron acceptor. The fate of pyruvate during a
fermentation varies in different micro-organisms, although its metabolism always involves regeneration of NAD^+ in order that further
glycolysis can take place. Many fermentation end products are of
actual or potential industrial importance; they include ethanol
(which forms the basis of the alcoholic beverage industries described
in Volume 1 of this series) as well as acetone, butanol and lactic acid
which are dealt with in the present volume.

Strictly speaking, a fermentation must be an anaerobic process,
since an organic compound rather than molecular oxygen is acting as
the electron acceptor. Nevertheless, the industrial microbiologist
regularly refers to any medium—or large-scale culture of a micro-
organism as a fermentation, and the vessel in which it is held as a
fermenter, although, quite often, massive volumes of air are being
pumped into the vessel. This use of the term is a legacy from the

early days of industrial microbiology when the processes—production of beer, wines and spirits (see Volume 1), and of acetone and butanol (described in Chapter 2)—were genuine fermentation processes. In a basic text written some years ago (Rose, 1961), I made strenuous efforts to use the term fermentation in its correct physiological sense, but my example completely failed to make any impression on industrial microbiologists.

Compared with a microbe such as *Escherichia coli*, which when grown aerobically oxidizes its energy-providing substrates very efficiently to carbon dioxide and water, and moreover excretes very little organic material into the medium, microbes that carry out anaerobic fermentations are grossly inefficient, if only because the energy yield from catabolism of the substrate is very much lower. Many micro-organisms, especially filamentous moulds, when growing aerobically excrete small quantities of some of the organic acids that act as intermediates in the tricarboxylic acid cycle or of compounds, such as itaconic acid, that are formed from these acids. Commercial production of citric acid, using *Aspergillus niger*, which is described in Chapter 3 of this volume, exploits this inefficiency, although the process also involves perturbation of the mould metabolism to increase the amounts of citric acid excreted. A few micro-organisms, notably the acetic-acid bacteria, even when grown aerobically, carry out only a partial oxidation of their organic substrate, and produce from organic substrates compounds such as acetic acid and oxogluconic acids. Chapter 3 of the volume gives an account of production of oxogluconic acids, while manufacture of acetic acid and vinegar is described in Chapter 4.

2. *Anabolic Pathways*

The nature of intermediates on pathways that lead to synthesis of cell constituents, namely anabolic pathways, was discovered to a large extent some time after catabolic pathways had been charted, and while not yet complete this body of information is very nearly so. As data on anabolic pathways accumulated, the unity which had been encountered in the biochemistry of catabolic pathways, and which had been emphasized from early on by the Dutch micro-biologist Albert J. Kluyver (Kluyver, 1931; Kluyver and Van Niel, 1956), was found to exist also on biosynthetic pathways. Since pathways leading to biosynthetic end-products are very similar in all micro-organisms, it might be expected that the industrial micro-biologist would use a wide range of different microbes in the large-scale production of commercially useful compounds. However, this is not so, for the majority of industrially useful micro-organisms are organotrophs. The reasons for this are mainly economic. Large-scale cultivation of algae is very costly with the need to provide artificial light and increased partial pressures of carbon dioxide (Vincent, 1971), while cultivation of chemolithotrophic bacteria has many attendant difficulties and invariably produces poor cell yields.

Details of biosynthetic pathways that operate in micro-organisms can be found in basic texts in biochemistry (Lehninger, 1970; Dagley and Nicholson, 1970) and microbial physiology (A. H. Rose, 1976). There follows a brief summary of these pathways to allow the reader to appreciate the background to microbial production of primary products of metabolism.

a. *Nucleic acids and proteins.* Synthesis of nucleic acids (DNA and RNA) and proteins is at the basis of cell metabolism because the majority of the proteins produced, in which are translated information encoded on DNA, have an enzymic or regulatory function. The ammonium ion, which is the form in which nitrogen is made available in many fermentation processes, is incorporated into one or more of three amino acids, namely α-oxoglutarate, aspartate or alanine. These amino acids are then used as sources of the amino group for synthesis of 15 other amino acids; asparagine and gluta-mine are formed by amidation of, respectively, aspartate and gluta-mate. There exists in all micro-organisms an intracellular pool of amino acids and, with some microbes, this is sufficiently large or unbalanced to allow the organism to excrete amino acids. Techniques for increasing the extent to which amino acids are excreted by micro-organisms are referred to in Section IIIB of this chapter (pg. 22), and in more detail in Chapter 8. Amino-acid residues, in the form of their tRNA derivatives, are polymerized into polypeptides, on polyribosomes.

The purine ribonucleotides, adenylic acid and guanylic acid, are synthesized from inosinic acid, which in turn arises from a pathway on which phosphoribosyl pyrophosphate and glutamine are pre-cursors. Uridylic acid, a pyrimidine nucleotide, is formed on a pathway in which carbamoyl phosphate and aspartate are precursors; CTP arises from UTP in a reaction which involves ATP and NH_4^+. Deoxyribonucleoside diphosphates are formed from the corresponding ribonucleoside diphosphates in reactions catalysed by a ribonucleoside diphosphate reductase. Microbes contain a pool of ribonucleosides and deoxyribonucleosides which, as with the amino-acid pool, can become excessively large and imbalanced leading to excretion of some of the pool constituents. These microbes are a source of industrially important nucleotides, as described in Chapter 5. Deoxyribonucleic acid and ribonucleic acids are synthesized in reactions catalysed by the appropriate polymerases and involving nucleoside triphosphates.

b. *Polysaccharides.* Microbial walls contain a variety of polysaccharides, often heteropolysaccharides, added to which some microbes have the capacity to produce extramural layers (macrocapsules, microcapsules and slime layers) which with a few exceptions are composed of polysaccharides. The sugars used in media for commercially important microbes are hydrolysed to monomers, if necessary, and converted into other sugar residues usually at the level of the nucleoside diphosphate derivative. Polymerization of sugar residues into polysaccharides involves a variety of transferases, and the structure of the polysaccharide is determined by the specificities of these enzymes. However, synthesis of a few wall polysaccharides involves a lipid intermediate (an undecaprenyl phosphate or dolichol phosphate) which conveys preformed repeating units to the growing polysaccharide chain.

Regulation of polysaccharide synthesis is not under very strict regulation and many microbes, particularly those that synthesize extramural layers, tend to overproduce polysaccharides, especially when the medium is rich in carbon source and relatively depleted of a nitrogen source. Recent years have witnessed a steady increase in exploitation of polysaccharide overproduction by microbes, a development which is covered in Chapter 10.

c. *Lipids.* Micro-organisms synthesize two types of lipid, namely polar and neutral lipids. The principal polar lipids are phospholipids and glycolipids, which are located in membranes and because of their amphipathic nature confer on a membrane (in collaboration with membrane proteins) barrier properties. Neutral lipids are also synthesized by eukaryotic microbes; these are principally triacylglycerols. Phospholipid synthesis in micro-organisms is under very rigorous control, and there are very few if any documented examples of overproduction of phospholipids by micro-organisms. Not so as regards triacylglycerols, for these neutral lipids appear to be oversynthesized in appreciable quantities by many eukaryotic micro-organisms (Bartlett and Mercer, 1974; Hossack *et al.*, 1977), although the excess lipid remains intracellular. Other types of lipid are, however, excreted by a few microbes, which include yeasts (species of *Candida, Hansenula* and *Torulopsis*) and filamentous fungi (Hunter and Rose, 1971). These lipids include sphingolipids, polyol fatty-acyl esters, sophorosides of fatty acids and substituted acids.

Although as yet it has not been established as a industrially viable project, there is some commercial interest in the fatty-acyl residues present in microbial phospholipids and triacylglycerols, as indicated in Chapter 7. Polyols which are elaborated extracellularly by some yeasts also have a potential commercial value, as described in Chapter 10.

Lipid synthesis by micro-organisms involves, firstly, production of the coenzyme-A ester of a fatty acid, followed by esterification of the hydroxyl groups on a glycerol residue in glycerol 3-phosphate. Details of lipid synthesis in micro-organisms can be found in the text by Weete (1974) and those edited by Erwin (1973) and Wakil (1970).

d. *Tetrapyrroles and terpenes.* The two principal classes of tetrapyrroles which are synthesized by micro-organisms are haems and chlorophylls. Their biosynthetic precursors are succinyl-CoA and glycine, and there is a very tight regulation on the pathway since neither class of end product is overproduced to any appreciable extent. The two main terpene classes in micro-organisms, namely carotenoids and sterols, which are synthesized from acetyl-CoA and leucine, are by contrast overproduced by many microbes, although they are still retained inside the cell. Reference to carotenoid production by microbes appears in Chapter 8 (pg. 317), β-carotene being a provitamin which is converted into vitamin A in the liver. Interest has been shown in *Sacch. cerevisiae* and other yeasts and filamentous fungi as sources of sterols in the manufacture of contraceptive drugs. By suitably adjusting cultural conditions, it is possible to obtain *Sacch. cerevisiae* with up to 10% of its dry weight as ergosterol, probably as a fatty-acyl ester (Dulaney *et al.*, 1954). A possible shortage of plant sterols which are used in the manufacture of contraceptive drugs explains the continued interest in yeast as a source of sterols.

e. *Vitamins.* Vitamins, principally those in the B group, are required by micro-organisms, in common with all other living cells, mainly as coenzymes. The biosynthetic pathways leading to synthesis of vitamins are varied, and an account of these can be found in the text by Goodwin (1963). Because they act metabolically as parts of enzymes, the amounts of vitamins synthesized by micro-organisms are very small. At the same time, metabolic regulation on these pathways is relatively inefficient, with many microbes being able to excrete appreciable amounts of B-group vitamins. This lack of efficient control over vitamin synthesis does not place most micro-organisms at any great disadvantage, for in terms of the amount of vitamin excreted, the energy wastage is clearly small. However, not so with some microbes which are used as commercial sources of vitamins, when the concentration in the culture fluid may be as high as 10 mg/ml, as described in Chapter 8.

C. Regulation of Microbial Metabolism

While the microbial physiologist is still far from understanding fully the manner in which microbial metabolism is regulated, it is clear that, broadly, this control is effected in three ways, namely by control of genetic information, regulation of enzyme action, and regulation of enzyme synthesis.

1. *Genetic Control*

The genetic material in cells, which is collectively referred to as the genome, is made up of DNA which has encoded on it instructions for the 'running' of the cell. Synthesis of each enzyme, or strictly speaking each polypeptide, is regulated by a separate gene or sequence of genetic information, made up of approximately 600 base pairs. If the gene for a particular enzyme is not present in a microbial genome, then clearly that microbe cannot synthesize the corresponding enzyme. But not all genes are structural genes, that is genes which code for enzymically active proteins. The products of other genes have a regulatory role, and are proteins which stimulate or inhibit various metabolic processes. Details of the genetic code can be found in any of the basic texts in biochemistry and microbiology.

2. *Regulation of Enzyme Action*

Regulation of enzyme action is effected in at least four different ways.

a. *Access of substrates to enzymes.* A basic, but often overlooked, way in which activity of an enzyme can be curbed is to restrict access of the substrate to the enzyme. This is brought about in several different ways, ranging from regulation of the rate at which a substrate is transported across the plasma membrane into the cell to the formation, in eukaryotic micro-organisms, of organelles such as mitochondria that are bounded by membranes.

b. *Feedback inhibition of enzyme action.* The most extensively investigated of the mechanisms by which microbes regulate enzyme action is feedback inhibition. If a metabolic product, E, is synthesized in a series of reactions in which A is converted into B, B to C and so on, feedback inhibition can operate by the end product E inhibiting the action of one or more of the enzymes that catalyse reactions earlier in the sequence, thereby restricting the amount of E that is produced. Low molecular-weight end products which interact with enzymes are referred to as *effectors*, and they bring about their action by altering the conformation of the enzyme. They do so by combining with the enzyme at the allosteric site, which differs from the active site. Enzymes which combine with effectors in this way are known as allosteric enzymes.

Many variations on the basic theme of feedback inhibition of enzyme action have been found to operate in micro-organisms. Immediately the phenomenon was shown to operate on linear metabolic pathways, it was clear that a modified type of mechanism must be imposed upon branched pathways which lead to the production of two or more end products but in which certain reactions are shared. Clearly, overproduction of one of the end products could, by feedback inhibition of the first enzyme on the pathway, restrict synthesis of the other end product although the cell may not have synthesized sufficient of it to meet its immediate metabolic demands. One way in which this problem is overcome is for the end product of a branch on a pathway to inhibit action only of the enzyme at the base of the branch. Flow through the pathway can also be retarded by the end product inhibiting action of one of a number of isoenzymes, which are proteins each with the same enzymic activity, catalysing the first reaction on a pathway.

The complexity of operation of feedback mechanisms is exemplified by two other phenomena. One has been termed *concerted feedback* which operates, for example, on a pathway in bacteria that leads to synthesis of lysine, methionine and threonine. The first step on this pathway is catalysed by aspartate kinase (A. H. Rose, 1976), but there is only one enzyme species produced even though this is a branched pathway. This enzyme is inhibited little or not at all by individual amino-acid end products, but strongly by two of the amino-acid end products.

Glutamine synthetase is an important enzyme in metabolic regulation because the end product of its action is involved in many anabolic reactions. Its activity is regulated by *cumulative feedback inhibition*. *In vitro* experiments with extracts of prokaryotic and eukaryotic microbes have shown that the total residual activity of glutamine synthetase in reaction mixtures containing several different feedback inhibitors is equal to the product of the fractional activities manifest when each is present separately in the reaction mixture.

Feedback inhibition also operates on catabolic pathways the main products from which are ATP and reduced nicotinamide nucleotides. Not much is known of the manner in which reduced nicotinamide nucleotides regulate enzyme activity, although there is some understanding of the way in which cellular concentrations of AMP, ADP and ATP affect enzyme activity. The relative concentrations of these nucleotides in a cell can be used to calculate the *energy charge* or *adenylate charge* of the cells, using the equation:

$$\text{Energy charge} = \frac{0.5\,\{[\text{ADP}] + 2[\text{ATP}]\}}{[\text{AMP}] + [\text{ADP}] + [\text{ATP}]}$$

Using this equation, values for the energy charge of a micro-organism fall in the range 0–1.0 (Chapman and Atkinson, 1977). *In vitro* experiments have shown that the activity of many enzymes associated with ATP generation, such as isocitrate dehydrogenase, is inhibited when the energy charge in the reaction mixture is adjusted to a high value (greater than about 0.8). Conversely, the rate of action of enzymes which catalyse reactions on anabolic pathways and which consume ATP—aspartate kinase is a case in point—is increased when the energy charge in the reaction mixture is high.

Feedback inhibition also operates on transport proteins. For example, the apparent K_m value for uptake of glucose by *Sacch. cerevisiae* is greater for cells grown aerobically than anaerobically. This lower affinity of the glucose-transport proteins for solute in aerobically-grown yeast appears to explain at least in part Pasteur's observation that, when *Sacch. cerevisiae* is transferred from an anaerobic to an aerobic environment, growth is accelerated and uptake of sugar decreased. This was one of the first observations made on regulation of microbial metabolism, and is often referred to as the *Pasteur effect*.

c. *Modification to enzyme structure.* The activity of certain enzymes in microbes can be curbed as a result of their undergoing reversibly either a chemical or a physical modification. For example, glycogen synthetase in *Sacch. cerevisiae* and *Neurospora crassa* can be altered as a result of its being phosphorylated. Other enzymes are specifically modified, and rendered inactive, following adenylylation. During physical modification of enzymes, the conformation of the protein is changed, so as to render it enzymically inactive, usually the result of the protein combining with cations.

d. *Degradation of individual enzymes.* Undoubtedly the most sophisticated mechanism yet shown to operate during control of enzyme activity in microbes is that in which highly specific proteolytic enzymes degrade individual enzymes which are no longer required in metabolism. Reflecting again the popularity of the microbe in basic cell research, one of the best documented examples has come from work on *Sacch. cerevisiae*. In this yeast, the enzyme tryptophan synthase is specifically inactivated in cells when cultures enter the stationary phase of growth, that is when the environmental conditions are such that synthesis of tryptophan is no longer necessary.

3. *Regulation of Enzyme Synthesis*

Inhibition of enzyme action, in its many forms, provides the microbe with a rapid and sensitive method for preventing overproduction of low molecular-weight end products. But it is basically inefficient in that enzymes are synthesized only to be prevented from functioning, or indeed broken down. Microbes also employ other more efficient mechanisms which act by controlling enzyme synthesis, thereby conserving cellular protein by ensuring that the microbe stops synthesizing those enzymes that are not required.

a. *Induction of enzyme synthesis.* Some enzymes are synthesized by a micro-organism irrespective of the composition of the medium in which it is grown. These are referred to as *constitutive enzymes* to distinguish them from *induced enzymes* which are synthesized only in response to the presence in the medium of an inducer which is usually the substrate for the enzyme or some structurally related compound. The product formed by action of the induced enzyme on the inducer can frequently itself induce synthesis of a further enzyme, and so on, a process known as *sequential induction.* Induction in *Escherichia coli* of β-galactosidase, an enzyme which catalyses hydrolysis of lactose to glucose and galactose, is without doubt the most intensively studied example of microbial enzyme induction.

b. *Repression of enzyme synthesis.* Just as microbes conserve cellular protein by synthesizing certain enzymes only when the substrates for these enzymes are available, so they can stop synthesis of other enzymes when the end-product of the pathway on which that enzyme operates is no longer required. The latter process is known as repression of enzyme synthesis. End-product repression of enzyme synthesis on a pathway differs from feedback inhibition of enzyme activity in that all of the enzymes on the pathway are usually affected. On some pathways, such as that leading to histidine synthesis in *Salmonella typhimurium*, synthesis of each of the enzymes is repressed by the end product to the same extent, a phenomenon known as *co-ordinate repression.*

An interesting example of enzyme repression operates on pathways that lead to synthesis of the branched-chain amino acids leucine, valine and isoleucine. These two quite separate pathways involve a homologous series of reactions, and enzymes catalyse reactions on both pathways. Studies with several bacteria have shown that there is no repression of synthesis of the common enzymes until all of the end products, namely valine, leucine and isoleucine, are simultaneously present in excessive concentrations. This type of concerted action has been termed *multivalent repression.*

Regulation of microbial metabolism is brought about by the co-ordinated action of the various mechanisms for modifying enzyme action and of those that control enzyme synthesis. With some pathways, there has been an intensive study of the contribution which the different mechanisms make to overall metabolic regulation. One of the best studied is the β-keto-adipate pathway for catabolism of aromatic compounds in pseudomonads (Stanier and Ornston, 1973). Several of the pathways for amino-acid synthesis in bacteria have also been intensively studied (see Chapter 6, pg. 210).

   c. *Molecular mechanisms in regulation of enzyme synthesis.* One of the greatest achievements in molecular biology has been the discovery over the past 15 years of the molecular mechanisms that operate during regulation of gene expression, in other words how genes are switched on and off. These mechanisms involve both negative and positive control of protein synthesis.

The classic work of Francois Jacob and the late Jacques Monod (1961) first focused attention on negative control of protein synthesis. They proposed that synthesis of enzymic proteins—or polypeptides—is controlled by *structural genes* (Fig. 4), groups of which lie closely together on the genome. They further suggested that separate *regulator genes* are responsible for producing compounds, which are proteins, that act by controlling expression of a group of structural genes. These proteins are known as *repressor proteins,* and they act not on individual structural genes but on a separate region of the genome where expression of, that is mRNA synthesis on, is controlled. This site is known as the *operator gene* (Fig. 4). One other genetic element, existence of which was not proposed by Jacob

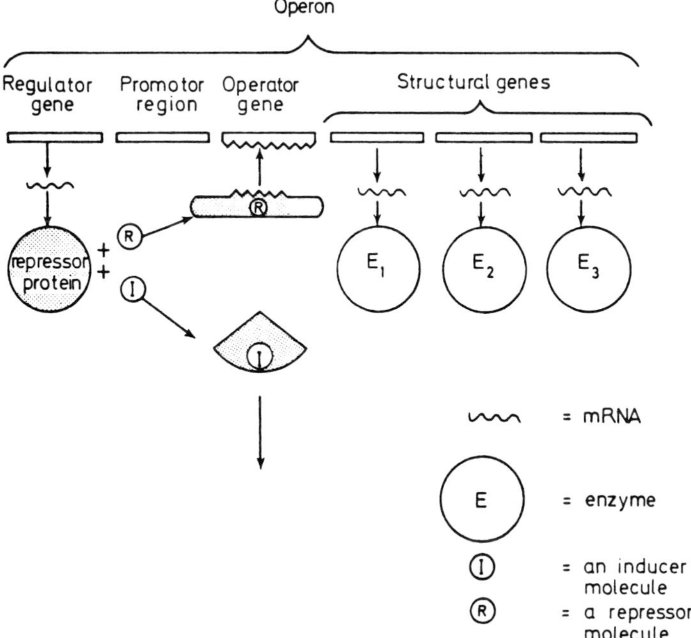

Fig. 4. Diagram of the Jacob and Monod (1961) scheme for negative control of protein synthesis in micro-organisms. See text for an explanation.

and Monod but whose function came to light from later experiments, is the *promotor region* which is the site on the genome where RNA polymerase becomes attached. A group of structural genes, together with associated regulator and operator genes and promotor region, is known as an *operon*. It must be remembered that, while evidence that the Jacob and Monod (1961) mechanism operates during gene expression in prokaryotic microbes is now very extensive, there is only fragmentary evidence for its operation on the eukaryotic genome.

Jacob and Monod (1961) proposed that induction of protein synthesis can only take place when the repressor protein is not in contact with the operator gene. In the absence of an inducer, either supplied in the medium or synthesized in the cell, the repressor protein remains in contact with the operator gene, thereby preventing expression of the associated structural genes. The repressor protein is thought to act by occluding part of the promotor region so preventing RNA polymerase from making contact with the promotor region. Repression of enzyme synthesis was explained by postulating that, in the absence of the effector molecule, the repressor protein has no affinity for the operator gene, but that it acquires an affinity after undergoing a conformational change by interacting with the effector. The common feature to these mechanisms is that gene expression can only occur when the operator gene is free, and for this reason the mechanisms are said to be negative in nature.

Positive control of protein synthesis, in which molecules need to be present at certain loci on the genome, has been less well researched. It does, however, operate during arabinose utilization in *E. coli* (A. H. Rose, 1976). On this operon, the product of one gene (*ara* C) is a protein which combines with the inducer, arabinose, to form an activator molecule which acts at an initiator site on the operon. This activator molecule is therefore responsible for positive control of expression of the *ara* structural genes. Positive control of protein synthesis also operates during catabolite repression. Relief from catabolite repression, that is initiation of gene expression, is brought about by cyclic adenosine monophosphate (cAMP) which combines with a protein, known as the catabolite gene activator, to form a complex that facilitates binding of RNA polymerase to the promotor region on the genome. Catabolites act by controlling the intracellular concentration of cAMP.

III. BIOCHEMICAL BASIS FOR INDUSTRIAL PRODUCTION OF PRIMARY PRODUCTS OF METABOLISM

The previous section of this chapter described briefly the primary pathways of microbial metabolism, and the manner in which activity on these pathways is regulated. Stress was placed on the extent to which operation of the various pathways is inefficient, as witnessed by excretion or overproduction by the microbe of an end product of primary metabolism or of an intermediate on a pathway leading to synthesis of one of these products. *The overall objective of the industrial microbiologist concerned with production of primary products of metabolism is to select strains of the appropriate microbe, and adjust the medium and environmental conditions, to maximize production of the primary product.* The task is far more difficult than with production of secondary products of metabolism which, in general, are excreted to a much greater extent than primary products. In selecting the most suitable strain and cultural conditions—that is in manipulating the metabolism of the microbe—the industrial microbiologist is helped very considerably by a knowledge of the metabolic pathway along which the primary product is synthesized. However, as indicated in several of the chapters in this volume, knowledge of the biosynthetic pathway is not essential, it being possible to optimize a production process to a not inconsiderable extent using a purely pragmatic approach.

A. Manipulation of Microbial Metabolism

The ability of certain chemical compounds to depress or inhibit microbial growth has been known for many years, knowledge which antedates even the development of microbiology as a science. These compounds may be said to perturb microbial metabolism to the extent that growth is affected. But the industrial microbiologist is interested in a rather different type of manipulation in which metabolism of a microbe is affected such that greater quantities of compounds are excreted, or a different end product of metabolism is produced, without growth of the organism being materially influenced. Many examples of this type of manipulation are documented in the literature, and they fall into two classes, namely those in which metabolism is manipulated by altering the composition of the medium, and those in which it is effected by altering the composition of the genome.

The first example of a manipulation of microbial metabolism to be reported came as early as 1918 at a time when the very first metabolic pathway, the Embden–Meyerhof–Parnas pathway was being charted. Neuberg and Reinfurth (1918) discovered that, when sulphite was included in a medium in which *Sacch. cerevisiae* was fermenting sugar, the main product of the fermentation was not a mixture of ethanol and carbon dioxide but glycerol. This was dubbed Neuberg's 'second form of fermentation', the first being that which leads to production of ethanol and carbon dioxide. The second form of fermentation is based upon the ability of sulphite to bind acetyldehyde thereby blocking reactions by which NAD^+ is regenerated. Instead, the oxidized carrier is regenerated during reduction of dihydroxyacetone phosphate to glycerol 3-phosphate, which in turn is hydrolysed to yield glycerol (Sols *et al.*, 1971). Neuberg's second form of fermentation was exploited briefly in commercial production of glycerol. A third form of fermentation by *Sacch. cerevisiae* was later reported brought about by adjusting the yeast medium to an alkaline pH value. Since these early reports, numerous other instances of environmentally-induced perturbation of microbial metabolism, too numerous to document, leading to either an increased excretion of a product or excretion of a quite different compound, have been reported. Many of these involve stressing a microbe by depriving it of a sufficient supply of an essential nutrient.

Techniques for selecting mutant strains of micro-organisms were developed in the early 1940s by George Beadle and his colleagues. Their mutants of *Neurospora crassa* made a valuable contribution in charting the first biosynthetic pathways, and showed the immense value of mutants in studies in microbial physiology. Hence the dictum that the microbial physiologist is better off if he has mutant strains of a microbe than if he has not. Techniques for selecting mutant strains of microbes were rapidly developed, not only for fungi but also for bacteria. A more positive direction in this research was clearly discernible following publication of the Jacob and Monod (1961) theory of regulation of gene expression since it was then possible to investigate the molecular basis of the behaviour of a mutant (Drake, 1970). Today, microbial mutants have a paramount position in research into the physiology of micro-organisms.

B. Application of Metabolic Manipulations in Industrial Fermentations

1. Environmentally-Controlled Perturbations

Once it has been established that a particular chemical compound excreted by, or in the cell mass of, a microbe has commercial importance because it may be manufactured by a microbiological process more cheaply than by chemical synthesis, the industrial microbiologist immediately mounts a research programme in which environmental conditions for maximum excretion (or occasionally optimum excretion taking into consideration questions of cost) are sought. More often than not, this is a pragmatic exercise, although increasingly a knowledge of the nature of precursors and intermediates on metabolic pathways, when these are known, is used to direct this experimental approach. At the beginning of this exercise, considerable increase in productivity can usually be obtained by optimizing the cultural conditions, such as temperature, oxygen-transfer rate and pH value. For example, production of vitamin B_{12} by *Propionibacterium* spp. was considerably enhanced when the bacterium was grown anaerobically at first followed by an aerobic phase (see pg. 304).

Imposition of anaerobic conditions on *Sacch. cerevisiae*, leading to formation of ethanol and carbon dioxide as fermentation end products, may be considered a form of metabolic manipulation. Production of many secondary products of metabolism (see Volume 3 of this series) is enhanced by including a precursor of the product in the medium. There are fewer examples of this type of 'directed fermentation' (Perlman, 1973) in production of primary products of metabolism, although cobalt salts are used to increase vitamin B_{12} production by *Propionibacterium* spp. and *Streptomyces* spp. (see pg. 304). There are many more examples in large-scale production of primary products of metabolism when a metabolic inhibitor or poison is used to manipulate the metabolism of the microbe in order to increase productivity. Inclusion of ferrocyanide in beet-molasses medium used in production of citric acid by *Aspergullus niger* increases productivity largely, it is suggested, because ferrocyanide sequesters cations (see pg. 52). A similar explanation has been adduced to explain the stimulatory effects of methanol and other short-chain aliphatic alcohols on citric acid production by *A. niger* in a blackstrap-molasses medium. Riboflavin production by clostridia is stimulated by α,α-dipyridyl, although no satisfactory explanation of the effect of this inhibitor has been forthcoming (Perlman, 1973). Some of the most interesting effects of inhibitors on production of primary products of metabolism are seen in the amino-acid fermentations (see pg. 210). For example, glutamate excretion by bacteria is enchanced when the permeability of the ccll envelope is increased, and this can be brought about by biotin, oleic acid or glycerol deficiencies being imposed on mutant strains of bacteria that are auxotrophic for these compounds, or by the action of penicillin. The last of these compounds is used in the industrial production process.

C. Strain Selection

Once it has been established that a particular strain or species of microbe is capable of producing a commercially useful compound on a scale that can make the microbiological process compete with chemical synthesis of the compound, examination of further strains of the microbe is a natural first step in process development. This was realized in one of the oldest fermentation processes, production of acetone and butanol by clostridia (see, pg. 31), when strains were isolated which produced more butanol and less acetone, the former compound then being the desired product. In recent years, strain selection, usually involving production of mutants, has become an integral part of process development in all industrial microbiological processes. When coupled with a detailed knowledge of the bio-synthetic pathway involved, use of mutants can be extremely rewarding. There is probably no better example of the use of mutant strains of microbes in commercial production of primary products of metabolism than that afforded by the amino-acid fermentations (see pg. 210). Some idea of the impact which genetics is making in industrial microbiological processes can be gathered by perusing the proceedings of the first two International Symposia on the Genetics of Industrial Micro-organisms (Vanek *et al.*, 1973; MacDonald, 1976).

IV. INDUSTRIAL IMPORTANCE OF PRIMARY PRODUCTS OF MICROBIAL METABOLISM

The number of primary products of microbial metabolism is vast, and those which are produced on a commercial scale are industrially important chemical compounds whose synthesis by purely chemical means is more expensive than their production by a microbiological process. Perlman (1973b) compiled a valuable check list of the major industrial fermentations then in operation, which he has recently up-dated and extended (Perlman, 1977). Brown (1976) has also compiled a review of the World-wide output of fermentation products.

A. Fermentation Products

The number of true fermentation products—that is compounds which are end products of catabolism in which organic compounds are both the electron donor and the electron acceptor (see pg. 8)—which are manufactured on a commercial scale is quite small.

Historically, and in terms of the amount produced, the most important of these is ethanol, which is manufactured by fermentation of sugars by strains of *Saccharomyces cerevisiae* and distillation of the fermented medium, using processes very similar to those described in Volume 1 of this series for production of neutral potable spirit. (Simpson, 1977). Ethanol production by fermentation was a large industry until the availability in the 1930s of cheap petroleum allowed this important industrial solvent to be manufactured much more cheaply by purely chemical means. Some users of ethanol, notably the pharmaceutical industry and less industrialized countries not blessed by abundant local petroleum resources, continued to use fermentation ethanol, so that the fermentation industry was kept alive (World Wide Survey of Fermentation Industries, 1971). The explosive increase in the cost of petroleum World-wide over the past three years has prompted many industrial microbiological firms to re-examine the feasibility of increasing the production of fermentation ethanol. Already several organizations have done so, usually using molasses as a substrate (D. Rose, 1976). A number of petroleum producers now dilute their commercial gasoline with up to 20% ethanol; Brazil, which is the World's largest cane-sugar grower and therefore has readily available supplies of molasses, is well to the fore (Jackson, 1976).

Two other industrially important solvents, acetone and butanol, were once manufactured on a large scale using clostridia, as described in Chapter 2 by Hastings (pg. 31). This fermentation, which has an important political significance through its connection with Chaim Weizmann (Bunker, 1957; Hastings, 1971), also suffered a setback with the arrival of cheap petroleum. Nevertheless, production of these solvents by fermentation continued to be fairly active in less industrialized countries, notably Argentina and Spain (World Wide

Survey of Fermentation Industries, 1971), and could experience a renaissance with the recent increased costs of petroleum.

Lactic acid is produced both synthetically and by fermentation, as described by Miall in Chapter 3 (pg. 95). Perlman (1977) lists four producing companies in the World, and the World Wide Survey of Fermentation Industries (1971), quoting data collected for 1967, gave production figures for Japan and Spain. Another organic acid which is a valuable industrial chemical, is acetic acid. This too is made entirely by chemical synthesis, although manufacture of vinegar by acetification of fermented worts or juices involves production of acetic acid by microbes. In many countries, production of synthetic vinegar is either legally banned or is a small industry. Vinegar production is described by Greenshields in Chapter 4.

An industrial chemical whose production by microbiological means was developed largely because of the lack of natural rubber during World War II is 2,3-butanediol. Synthetic rubber, production of which during that war increased from 5,000 to nearly a million tons per annum in the United States of America, uses 1,3-butadiene and chloroprene as raw materials. Butadiene is not produced in large quantities by micro-organisms, but 2,3-butanediol, which can be converted chemically into 1,3-butadiene, is. The main organisms examined as a source of 2,3-butanediol were strains of *Aerobacter, Bacillus* and *Serratia,* and the pioneering work was carried out largely at the National Research Council of Canada laboratories in Ottawa (Haskins, 1972) and in the U.S. Department of Agriculture Laboratories at Peoria, Illinois (Ward, 1970). Present day microbiological production of 2,3-butanediol is small—Perlman (1977) lists just one producer—and it is not described in this volume. It could however grow again in importance because of the cost of the alternative source of the synthetic rubber raw material, namely petroleum. Long and Patrick (1963) reviewed post-war progress with this fermentation.

Kluyver (1931) pointed out many years ago that microbial fermentations involve the partitioning of the energy of a substrate into the various fermentation products, and the World-wide concern over energy shortage has prompted microbiologists to consider the possibility of using yet other fermentation products as sources of energy. Hungate (1974), in a thoughtful article, considers methane and molecular hydrogen in this role, and concludes that methane could well be produced on a much greater scale as an energy source. Hydrogen he considers to be less attractive, since it contains less energy, volume for volume, compared with methane. Methane is already produced in many sewage works as a source of energy, while cattle are mobile, automated methanogenic fermenters! There is one problem associated with an increased use of microbiologically produced methane as a source of energy, and that is its storage. This might be overcome by converting it into methanol.

B. Metabolic Intermediates

Several of the organic acids which are intermediates on the tricarboxylic acid cycle, or acids derived from them, are important industrial chemicals which are manufactured microbiologically. These processes are described by Miall in Chapter 3 (pg. 47). Large-scale production of citric acid by moulds, which was pioneered in the United States of America in the 1930s, led to a severe cut-back in production of this food acidulent from lemons. Italy was the country most affected by this development, although that country still produces sizeable quantities of citric acid from fruits (World Wide Survey of Fermentation Industries, 1971). Perlman (1977) lists six other organic acids, namely erythorbic acid, gluconic acid, itaconic acid, 2-oxogluconic acid, α-oxoglutaric acid and malic acid, as being produced by fermentation. Production of these acids is described in Chapter 3.

Although glutamic acid has been manufactured microbiologically from the early 1950s, and marketed as a condiment or flavouring agent in the form of monosodium glutamate, microbiological production of the nutritionally essential amino acids came later, and then very largely from laboratories in Japan. These are among the most sophisticated of industrial microbiological processes, particularly as regards the use of mutant strains of bacteria. They are described by Kinoshita and Nakayama in Chapter 6 (pg. 210). A comparable level of sophistication can be seen in processes leading to production of nucleotides by micro-organisms. These were developed after the amino-acid fermentations, and again in Japan. They are described in Chapter 5 by Demain. The main use for microbiologically produced nucleotides is as flavour-enhancing agents in foodstuffs.

Microbiologically produced vitamins were first obtained in the late 1930s when certain American firms installed plant for the recovery of riboflavin from the residues remaining in the acetone-butanol fermentation. Some riboflavin is still produced microbiologically, although the process faces severe competition from chemical synthesis (Perlman, 1977). The other vitamin produced commercially using micro-organisms is vitamin B_{12} (Perlman, 1977). Both production processes are described by Perlman in Chapter 8.

C. Cell Components

In commercial terms, undoubtedly the most important microbial cell component is protein. Several microbes are grown on a large scale as a dietary source of protein—usually referred to as 'single-cell protein'. These processes are described in Volume 4 of this series. Three chapters in the present volume describe production of commercially important cell components. The most important of these are polysaccharides which Sutherland and Lawson describe in Chapter 9. There has been a small but sustained interest in large-scale production of microbial polysaccharides for over a quarter of a century, dating from the use of microbiologically produced dextrans as blood-plasma extenders (Grönwall and Ingelman, 1944). The past five years have witnessed a resurgence of interest in large-scale manufacture of microbial polysaccharides, not least because of the uncertainty associated with sources of supply of plant and seaweed polysaccharides.

Although there has for many years been an interest in the industrial production of fatty acids using micro-organisms, this has never been shown to be commercially viable. Questions of relative cost are clearly paramount, and the account by Ratledge in Chapter 7 describes these considerations in full and also future possibilities for the microbiological process. Finally, similar considerations apply to production of polyols by microbiological processes; these are described by the Spencers in Chapter 10 (pg. 393). Internal accumulation, and secretion, of polyols by osmotolerant yeasts has been known for a number of years, but the physiological significance of the process has only recently become apparent following Brown's elucidation of their role as compatible solutes (Brown, 1978).

REFERENCES

Bartlett, K. and Mercer, E. I. (1974). *Phytochemistry* **13**, 1115.

Brown, W. K. (1976). *Die Branntwein Wirtschaft* **116**, 216.

Brown, A. D. (1978). *In* 'Advances in Microbial Physiology', (A. H. Rose and J. G. Morris, eds.), vol. 17, Academic Press, London and New York.

Buchanen, R. E. (1918). *Journal of Infectious Diseases* **23**, 109.

Buchner, H., Longard, K. and Riedlin, G. (1887). *Centralblatt fur Bakteriologie, Parasitenkunde, Infektions Krankheiten und Hygiene* **2**, 1.

Bu'Lock, J. D. (1961). *In* 'Advances in Applied Microbiology', (W. W. Umbreit, ed.), vol. 3, p. 293. Academic Press, London and New York.

Bu'Lock, J. D. (1965). 'The Biosynthesis of Natural Products. An Introduction to Secondary Metabolism'. McGraw-Hill, London.

Bu'Lock, J. D. (1967). *In* 'Essays in Biosynthesis and Microbial Development', pp. 71. John Wiley and Sons, New York.

Bu'Lock, J. D. (1975). *In* 'The Filamentous Fungi', (J. E. Smith and D. R. Berry, eds), vol. 1, pp. 33–58. Edward Arnold, London.

Bunker, H. J. (1957). *Laboratory Practice* **6**, 36.

Chapman, A. G. and Atkinson, D. E. (1977). *In* 'Advances in Microbial Physiology', (A. H. Rose and D. W. Tempest, eds), vol. 16, p. 253. Academic Press, London and New York.

Cohen, S. N. (1975). *Scientific American* **233**, 24.

Curtiss, R. (1976). *Annual Review of Microbiology* **30**, 507.

Dagley, S. and Nicholson, D. E. (1970). 'An Introduction to Metabolic Pathways', 343 pp. Blackwell Scientific Publications, Oxford.

Dawes, I. W. and Sutherland, I. W. (1976). 'Microbial Physiology', 185 pp. Blackwell Scientific Publications, Oxford.

Demain, A. L. (1968). *Lloydia* **31**, 395.

Drake, J. W. (1970). 'Molecular Basis of Mutation', 273 pp. Holden and Day, San Francisco.

Dulaney, E. L., Stapley, E. O. and Simpf, K. (1954). *Applied Microbiology* **2**, 371.

Erwin, J. A. ed. (1973). 'Lipids and Biomembranes of Eukaryotic Micro-organisms', 354 pp. Academic Press, New York.

Goodwin, T. W. (1963). 'Biosynthesis of Vitamins and Related Compounds', 366 pp. Academic Press, London.

Grönwall, A. and Ingelman, B. G. A. (1944). *Acta Physiologica Scandinavica* **7**, 97.

Haskins, R. H. (1972). *In* 'Advances in Applied Microbiology', (D. Perlman, ed.), vol. 15, p. 415. Academic Press, London and New York.

Hastings, J. J. H. (1971). *In* 'Advances in Applies Microbiology', (D. Perlman, ed.), vol. 14, pg. 1. Academic Press, London and New York.

Hossack, J. A., Belk, D. M. and Rose, A. H. (1977) *Archives of Microbiology* **114**, 137.

Hungate, R. E. (1974). *American Society for Microbiology News* **40**, 833.

Hunter, K. and Rose, A. H. (1971). *In* 'The Yeasts', (A. H. Rose and J. S. Harrison, eds), vol. 2, pp. 211–270. Academic Press, London and New York.

Jackson, E. A. (1976). *Process Biochemistry* **11**(5), 29.

Jacob, F. and Monod, J. (1961). *Journal of Molecular Biology* **3**, 318.

Kluyver, A. J. (1931). 'The Chemical Activities of Micro-organisms', 109 pp. London University Press, London.

Kluyver, A. J. and Van Niel, C. B. (1956). 'The Microbe's Contribution to Biology', 182 pp. Harvard University Press, Cambridge, Massachusetts.

Lane-Claypon, J. E. (1909). *Journal of Hygiene* 9, 239.

Lehninger, A. L. (1970). 'Biochemistry', 833 pp. Warth Publishers Inc. New York.

Long, S. K. and Patrick, R. (1963). *In* 'Advances in Applied Microbiology', (W. W. Umbreit, ed.), vol. 5, pp. 135. Academic Press. London and New York.

MacDonald, K. D. ed. (1976). 'Second International Symposium on the Genetics of Industrial Micro-organisms', 630 pp. Academic Press, London.

Mandelstam, J. and McQuillan, K. (1973). 'Biochemistry of Bacterial Growth', 2nd edition, 582 pp. Blackwell Scientific Publications, Oxford.

Nägeli, K. W. (1877). 'Das Mikroscop', 2nd ed. 461 pp. Leipzig.

Neuberg, C. and Reinfurth, E. (1918). *Biochemische Zeitschrift* 331, 436.

Perlman, D. (1973a). *Process Biochemistry* 2 (7), 18.

Perlman, D. (1973b). *American Society for Microbiology News* 39, 648.

Perlman, D. (1977). *American Society for Microbiology News* 43, 82.

Rose, A. H. (1961). 'Industrial Microbiology', 286 pp. Butterworths, London.

Rose, A. H. (1976). 'Chemical Microbiology', 3rd edition, 469 pp. Butterworths, London.

Rose, D. (1976). *Process Biochemistry* 11, (2), 10.

Simpson, A. C. (1977). *In* 'Economic Microbiology', (A. H. Rose, ed.), pp. 537–593. Academic Press, London.

Slater, A. (1917). *Journal of Hygiene* 16, 100.

Sols, A., Gancedo, C. and Gancedo, G. (1971). *In* 'The Yeasts', (A. H. Rose and J. S. Harrison, eds), vol. 2, pp. 271–307. Academic Press, London and New York.

Stanier, R. Y. and Ornston, L. N. (1973). *In* 'Advances in Microbial Physiology', (A. H. Rose and D. W. Tempest, eds), vol. 9, p. 89. Academic Press, London and New York.

Vaněk, Z., Hošťálek, Z. and Cudlin, J. eds. (1973). 'Genetics of Industrial Micro-organisms', vol. 1, 496 pp. vol. 2, 510 pp. Elsevier, Amsterdam.

Vincent, W. A. (1971). *Symposium of the Society for General Microbiology* 21, 47.

Wakil, S. J. ed. (1971). 'Lipid Metabolism', 613 pp. Academic Press, New York.

Weete, J. K. (1974). 'Fungal Lipid Biochemistry', 393 pp. Plenum Press, New York.

World Wide Survey of Fermentation Industries (1971). International Union of Pure and Applied Chemistry. Information Bulletin. Technical Report No. 3. International Union of Pure and Applied Chemistry, Oxford.

2. Acetone–Butyl Alcohol Fermentation

JOHN J. H. HASTINGS

Formerly of Commercial Solvents (Great Britain) Limited

I. HISTORY

The story of the first use of anaerobic bacteria for production of acetone and normal butyl alcohol has been told on a number of occasions (Schofield, 1974), and always begins with an account of Weizmann's researches. In fact, Fernbach had already reported this behaviour in the literature in 1910, and it was almost by chance that Weizmann made use of his discovery to further his main line of study.

Chaim Weizmann was a Jewish chemist destined to become an international political figure. He was born in a village near Minsk and, after studying in Germany and Switzerland, he came to England at the age of 30 and eventually obtained a post in the University of Manchester under W. H. Perkin. For his personal research, he chose to study the preparation of synthetic rubber, and it was not long before he came to the conclusion that such a process could not succeed without a cheap and plentiful supply of normal butyl alcohol which, at that time, was not available in commercial quantities. He was fascinated by Fernbach's report and, although he was no microbiologist, he set about training himself in bacteriological skills as the first step on his journey. The isolation of anaerobes in pure culture was no easy matter with the techniques available at that time, and there is some doubt whether he in fact isolated his own culture or obtained it from an existing source. Be that as it may, he found himself with an organism that would ferment nearly 4% starch in a liquid medium, and required only simple inorganic salts such as ammonium sulphate and soluble phosphates as nutrients. Approximately 30% of the starch was converted to a mixture of organic solvents. Of this mixture about 60% was normal butyl alcohol, the main residual component being acetone in an amount of roughly 30% of the total.

So far so good. He could not foresee all of the problems of translating his laboratory work to large-scale industrial practice, but such calculations as he was able to make suggested that butyl alcohol could be obtained in this way at a cost that would not be prohibitive for his purpose. But, long before his efforts had made any substantial headway, the First World War (1914–1918) began, and his work was brought to a standstill. All scientists in Great Britain were required to undertake work that would make some direct contribution to the war effort.

At that time, the synthetic chemical industry in Britain could only be described as second rate. Other industry and commerce, and indeed the government itself, with a remarkable lack of foresight, had found it easier and more convenient to purchase a number of major organic chemicals from abroad, particularly from the more energetic marketing countries, including Germany, who had begun to conquer the World's export markets on a scale never before known. In certain cases, Germany had been our only reliable supplier and, when war started, her submarine campaign made importation from any outside source increasingly difficult. One special example of this was acetone, a solvent particularly important in the manufacture of cordite used in small-arms ammunition and as a propellant for heavy artillery shells. Acetone of a very high standard of purity was required. Not only was it in short supply World wide, but the material obtained from the destructive distillation of wood was far from meeting the required standards.

Weizmann immediately realized the possibility of his fermentation process as a source of pure acetone. He was energetic in bringing the matter to the attention of the proper authorities, and acetone, which he had previously considered to be a by-product of the process, became the major objective. An official team was gathered together under his leadership, and he was provided with all of the facilities that could be made available.

Progress was slow. Large-scale production was only a concept with very little knowledge of the engineering problems involved. Translation from the laboratory and pilot scale in what was really the birth of closed deep fermentation under sterile conditions presented many difficulties, and brought about many failures. But enthusiasm at least led to a recognition of the major problems to be overcome, and some degree of production was achieved, though well below the optimistic target figures that had been set. Of the quality of the acetone there was no doubt. Its fractionation from the mixed solvent yield was readily accomplished

Then for Weizmann and his team came a very bitter blow. The fermentation used starch from cereals as its raw material, and the time came when the government could not afford to release any more such material for this purpose. Rationing of foodstuffs was essential, and the fermentation could not be allowed to bring Britain nearer to starvation. In one last despairing move, children throughout England were asked to collect horse-chestnuts as a source of starch. They were used effectively on the laboratory scale but, on the large scale, their foaming characteristics made control of the fermentation virtually impossible, and the attempt was abandoned.

There was only one answer if the fermentation process was still to make a contribution to the war effort, and that was to transfer it to the North American continent, where cereals, particularly maize, were in plentiful supply. The United States of America, though sympathetic, was not yet a combatant, but a home for the process was found in Canada, and a number of young Canadian scientists joined the team, to form its backbone in years to come. Good progress was made, and there was soon no doubt that the process could be firmly established, particularly with the engineering facilities that enabled the design and construction of plant far superior to that which was available in Britain under the stress of war. Then the United States was drawn into the conflict, and the process so interested its scientific advisers that the whole project was transferred to the mid west, to the heart of the corn belt in Indiana. The Allied War Board initiated two plants in the neighbourhood of Terre Haute in Indiana, which was one of the industrial fermentation centres for the production of ethyl alcohol from corn.

Weizmann continued to inspire the project and to act as principal adviser. The plants were quickly completed and brought into use, but operations had hardly begun when the war came to an end with the signing of the armistice in November 1918. When peace was established a year later, the British government was most anxious to honour Weizmann for his work, but he refused all personal honours and rewards. Financially, the patent granted to him by the U.S. government placed him far beyond personal want when the process was transferred to commercial ownership, but his real desire was far more than this. He made it very clear to David Lloyd George, then Prime Minister of Great Britain, that his one wish was to see a home for the Jews established in Palestine and, when the Balfour Declaration made this possible, Weizmann became the leader of the whole Zionist organization. As a result, when the State of Israel was established, he became its first President.

II. MICROBIOLOGY

The organism used by Weizmann was named by him *Clostridium acetobutylicum*. Many variants have since been described in the literature, particularly in the patent applications that have been filed to cover various isolates. The organisms are anaerobic spore-forming bacteria, having a typical cigar shape with a greatest diameter about 1 μm and a length of 4 μm. In a liquid medium under the microscope, they appear as single cells, chains, and boat-shaped clusters. The spores are subterminal, ovoid, approximately 1 μm x 1.5 μm, and are formed in substantial numbers when the medium begins to become exhausted.

Weizmann's organism will ferment starch up to a concentration of 3.8% (w/v). Of this, approximately 30% is converted to a mixture of solvents, and the remainder appears as gas in the form of hydrogen and carbon dioxide. The ratio of these gases varies slightly during the course of fermentation but, overall, it is approximately 40% hydrogen and 60% carbon dioxide. As stated earlier, the solvent mixture consists mainly of normal butyl alcohol (about 60%) together with approximately 30% (v/v) acetone and 5 to 10% (v/v) of an intermediate boiling fraction consisting mainly of ethyl alcohol, isopropyl alcohol and mesityl oxide. The acetone–butyl alcohol ratio shows some variation with different strains of the organism, and commercial operation has made use of this by changing from one strain to another according to the state of the market for each solvent.

Isolation of such organisms is not difficult with modern anaerobic techniques. Spores of the organisms are widely distributed in soil. An aqueous soil suspension is heated sufficiently to kill vegetative organisms and plated out in a series of dilutions on a suitable solid medium, for example molasses-agar with mineral salts. After 60 hours anaerobic incubation at 30–32°C, the colonies are usually 2–3 mm in diameter, almost hemispherical, smooth and creamy-white in colour. At a later stage, the colonies darken slightly and grow in the form of a truncated cone, 2–3 mm high, with a concave top.

Cultures may be stored as spores on sterile sand or soil, and remain viable for several years. The sterilized soil is moistened with a spore suspension and air-dried under aseptic conditions. For large-scale fermentation, the culture is grown through as many as five stages of increasing volume. In the first stage, a potato-containing medium is employed in long test tubes. The heavy soil inoculum falls to the bottom of the tube and, as soon as germination and gassing have begun, the potato solids rise to form a loose plug in the upper part of the tube, thus helping to maintain anaerobic conditions. After inoculation, and before incubation, the tubes are heat shocked by placing in a water bath at 80°C for 2–3 minutes. This induces rapid germination of the spores.

In the second and subsequent culture stages, the normal molasses medium may be employed as used in the main fermentation. The inorganic nutrients added to the medium depend on the type of molasses in use but, in general, consist of ammonium sulphate, calcium carbonate and calcium superphosphate. The calcium carbonate is added as an agent to neutralize the acid radical after utilization of the ammonia. When a vigorously fermenting potato-tube culture is added to a culture flask of molasses medium, fermentation continues almost without a break and anaerobic conditions are soon established as the gases produced sparge the air from the flask. The flasks are kept still during incubation and are not placed on a shaking machine.

Soon after 1930, efforts were begun to find organisms that would ferment sugar. There were two reasons for this search. At the time, in nearly all the cane sugar-producing areas and particularly in the West Indies and Cuba, there was a glut of blackstrap molasses as a by-product from sugar processing. This was available at a price strongly competitive with starch. It was also thought possible that the more fluid sugar medium might permit fermentation of a higher concentration of carbohydrate, with a number of technical advantages that will be discussed later in this chapter.

Fortunately, the original operating centres in the middle west of America had maintained a large collection of cultures of different starch-fermenting bacteria that had been isolated over the years. A systematic survey of these quickly revealed that most if not all of them would ferment molasses sugars to a degree, and some of them would completely ferment as much as 6% sugar in the medium. A selected culture was rapidly brought into use, and many others were introduced later.

Attempts to ferment higher concentrations of sugar were unsuccessful, although a wide variety of conditions was employed. Extended studies, later confirmed by a number of other workers (Ryden, 1958), showed that all of these organisms had a maximum tolerance for normal butyl alcohol in the fermenting medium of approximately 1.2% (v/v), and this was the controlling factor in the concentration of sugar that could be used.

A molasses-containing medium with mineral salts and an initial pH value of 5.6–5.8 is an ideal medium for many aerobic organisms. Under anaerobic and aseptic conditions, contamination with aerobic bacteria is not a significant problem. Very few casual anaerobes can compete with the heavy inoculum and vigorous growth of the culture organism so that, in modern plant, the fermentations are reasonably free from bacterial contamination of all types. This is in contrast to strongly aerated antibiotic fermentations, where an outburst from minimal contamination is always a hazard.

The major problem with the acetone–butyl organisms is their susceptibility to bacteriophage, the effect of which can be extremely rapid and disastrous on the large scale. It was not until phage contamination had brought early American production to a standstill that the problem was recognized and efforts made to overcome it. Following extensive studies, a method of immunization was devised that is now classic procedure, involving serial transfer of the organism in the presence of phage. An important phenomenon which came to light was the specificity of phage against different strains. In practice it became usual to maintain a number of strains immunized against their own previous phage infection, so that a change could be made to a different immunized culture if trouble occurred. Ryden (1958) has discussed this issue at length.

III. DEVELOPMENT OF A MANUFACTURING PROCESS

When the British government established a team under Chaim Weizmann to develop the manufacture of acetone by this process, there was literally not a single fermentation vessel in Britain suitable for the purpose. The use of closed pressure vessels and operation under aseptic conditions had never been contemplated by the brewing or the potable-spirit industries. It is true that those pioneers who were experimenting with pure yeast cultures had developed metal culture vessels capable of being steam sterilized, but these were small tanks of up to ten hectolitres capacity, sufficient in size to prepare an inoculum for the large open fermentation vessels that were in common use, many of them constructed of wood or of slate. British beers at the time were made by top-fermentation yeasts, the major part of the yeast crop rising as a cream to the surface when the fermentation has passed its peak. General practice, which still widely continues, was to skim the yeast cream and use it as an inoculum for subsequent brews, and in a brewery operating to good standards of cleanliness vigorous growth of yeast restricted bacterial contamination to a minimum. Even when pure yeast cultures were introduced, they were not used for every fermentation but merely to provide an uncontaminated strain for a brewery which would then continue the standard practice of yeast-cream transfer as long as fermentations remained healthy.

Attempts made to use existing alcohol fermenters fitted with lids were a failure. They were not pressurized vessels, and steaming at atmospheric pressure was quite inadequate for sterilization. Before clostridia could produce enough gas to establish anaerobic conditions, aerobic infecting organisms played havoc in the culture and, even when fermentation managed to get under way, acid-producing anaerobic infectants frequently prevented formation of solvents.

With starch-fermenting organisms, the maximum concentration of acetone in a successful fermentation was approximately 0.4% (v/v). In order to obtain the output of acetone required, vessels of at least 500 hectolitres capacity were therefore needed. Fortunately it was found that mild steel did not inhibit the fermentation. To obtain complete sterilization of vessel and medium, a temperature of 121°C was required, equivalent to a steam pressure of 1.05 kg cm^{-2}, and pressure vessels were designed with an adequate margin of safety to meet this requirement. Following accepted engineering practice, fermentation vessels were constructed as vertical cylinders with hemispherical tops and hemispherical or conical bottoms, and such designs have continued in use. Now, although it was possible to prepare the medium and sterilize it in the vessel, it was not possible to inoculate say 2,000 hectolitres of medium with a culture flask from the laboratory. Even if successful, the lag phase would have been far too long for commercial operation. It was therefore necessary to elaborate a multistage system of inoculation having, for example, three stages in the laboratory and two stages in the plant itself, the last two stages being carried out in small steel pressure vessels of similar design to the fermenter itself. An acceptable scale of transfers could well be as follows: potato tube (15 ml) → liquid medium in flask (300 ml) → flask (2 litres) → steel vessel (2 hectolitres) → vessel (30 hectolitres) → fermenter (2,000 hectolitres).

Such a multistage system immediately raised problems of aseptic transfer. While there was no serious difficulty with the laboratory stages, which could be carried out in a bacteriological transfer room, transfer between vessels required piping connections and valves. To produce a sterile system before inoculation with the pure culture, all of these connections had to be simultaneously sterilized with the vessel. This immediately brought to light problems that were not perfectly solved even when the antibiotic industry came into the field of deep fermentation 30 years later.

When a system such as this comes into regular use, insoluble deposits from the medium begin to form in any crevice or low-lying point, and may build up to a thickness which will insulate the internal surface of the equipment so that it fails to reach the temperature required for sterilization. It was found that almost every type of valve available commercially was suspect in this respect. Gasketed joints were a particular danger and, when pumps were introduced for transfer of sterile medium, these were most unsatisfactory, particularly at glands. Considerable redesign of such equipment was necessary before regular reliable performance could be obtained, and a great deal of engineering research has continued through the years to design still better pumps, valves and other items that are required as connections. It was also found that any low-lying points in piping systems, where pockets of medium or condensate could remain stagnant, were a further hazard. Great care had to be taken in the lay-out of piping to ensure complete drainage. Gasketed joints were replaced wherever possible by welded connections, and special welding techniques were introduced to ensure a flush finish to interior surfaces. In sterilizing piping systems, it was found necessary to instal steam connections at all high points in the assembly, with drainage connections controlled by steam traps at every low point. When a whole battery of fermentation vessels is operated, with an appropriate number of seed vessels linked to inoculate any fermenter at choice, the piping system can become very complicated.

Achievements in this respect in the acetone fermentation were at least such that the antibiotic industry could commence deep fermentation under aseptic conditions with a great deal of confidence, even though it had new problems of its own to face. Nearly all antibiotic-producing organisms are strongly aerobic, requiring large volumes of air to be bubbled through the medium, and this air must be sterilized. In addition, many of the culture organisms produce a heavy mycelial growth, so that vigorous mechanical agitation is necessary to obtain complete fermentation of the medium. The acetone industry had some experience of mechanical agitation, as this was normally used in the seed vessels of the system, and it was able to advise on many of the pitfalls, but further extensive studies of high-power agitation had to be made before the penicillin fermentation, for example, reached its present efficiency. Large acetone fermentations were never mechanically agitated, simply because the enormous volume of gas produced by the fermentation was an effective agitant in itself, the fermentation literally appearing to boil during the whole of its active phase. Indeed, the major problem was found to be the control of foaming under such conditions. When uncontrolled and out of hand, the whole fermentation could be swept out through the vent pipe as a mass of gas and entrained liquid. Studies of antifoam agents and control techniques were of great help when other types of deep fermentation came into operation.

The gas produced in the acetone fermentation was in fact a most useful adjunct, not only for operational purposes but as an additional source of revenue in a variety of ways. It was not without its dangers. As stated earlier, the gas consists of approximately 60% carbon dioxide and 40% hydrogen, by volume. This mixture is explosive in air, and serious explosions have occurred in the industry on this account. When an attempt was made to start up the industry at Kings Lynn in Norfolk, England shortly after the First World War, the whole plant blew up shortly after commencing full-scale production, and there appears to be little doubt that a gas explosion was the cause.

When a fermenter has been sterilized under full steam pressure for an appropriate length of time, it must be cooled to operating temperature. Immediately the steam has reached condensation point a vacuum will occur unless the steam is replaced by an equivalent volume of air or gas. While the vessel is designed to withstand a pressure substantially in excess of that used for sterilization, it is not designed to withstand a vacuum, as this would be prohibitively expensive in a vessel of such size. It would be possible to replace the steam by sterile air, and the acetone fermentation would commence with a sufficiently heavy inoculum, eventually producing enough gas to sweep the air out of the vessel. But, during this period, the fermenter would be at the risk of explosion. Safe practice is therefore to use inert gas for initial cooling when starting up a plant, after which a large surplus of fermenter gas is available to maintain a closed anaerobic system. The gas is maintained under pressure and circulated through scrubbing towers or filters to ensure its sterility. It is then used both in cooling fermenters and for liquid transfer from one vessel to another by employing top pressure in the first vessel and a lower pressure in the second.

In the early days of the process, it was soon established that it was uneconomic to sterilize the medium in the fermenter itself. Not only did this necessitate providing the fermenter with an expensive cooling jacket or coils, but it also meant that the vessel was out of use for its prime function of fermentation for a substantial period of time within each cycle. Eventually separate pressure vessels were provided as 'cookers' to sterilize the medium, which was then pumped through coolers to the appropriate fermenter in the series. A battery of cookers, each of smaller volume than the fermenters, could be used to obtain a continuous flow of cooled sterile medium to the plant, the solid nutrients being added to each batch of medium when charging the cooker. With a starch-containing medium, the grade of calcium carbonate was not important but, with a medium containing molasses, the particle size and density were significant. Ground limestone rapidly settled out, and precipitated chalk was much more expensive. Fortunately, when the process was successfully restarted in England in 1935 (Hastings, 1971), very suitable supplies of sedimented chalk were available from chalk beds in the south of England, at a considerably lower cost than precipitated material. Attempts at continuous flash sterilization of the medium were unsatisfactory when it contained suspended solids. When a change was made to aqueous ammonia as a nutrient, a completely soluble medium could be prepared, paving the way for successful flash sterilization.

When growth starts in the fermenter after inoculation, the pH value of the medium falls steadily during the logarithmic phase of growth from an initial value of around 5.8 down 5.1–5.0, due to formation of acids which are the precursors of the final solvents. The pH value then rises to over 6.0 as the solvents are formed. Certain acid-producing anaerobic contaminants can develop rapidly at pH 5.0, in which case they continue to consume the carbohydrate and prevent the pH value from rising into the solvent-forming phase.

Studies in the British plant led to the ingenious solution both of nutrient addition and control of pH value. By supplying the total nitrogen requirement in the form of dilute aqueous ammonia during the early part of fermentation, it was possible to eliminate the combined use of ammonium sulphate and calcium carbonate. If ammonia was added, either continuously or in step additions, at a logarithmically increasing rate corresponding to the curve of logarithmic growth of the organism, the pH value could be prevented from falling to a trough from which it might never recover. This change led to much more reliable fermentations at a measurably lower cost, and the practice continued until the British plant was closed down in the early 1950s owing to competition from the petrochemical industry.

It must not be overlooked that fermentation is only half the story of the process. Recovery of the solvents by distillation is in fact one of the major costs of the overall operation. Up to the beginning of the Second World War, this was undertaken largely as a batch process. The fermented medium, containing approximately 2% of total solvents, was first continuously stripped through a perforated plate column to take off a concentrated mixture of all the solvents present. This concentrate was then distilled from batch stills through a fractionating column to yield three fractions, namely crude acetone, crude ethyl fraction and crude normal butyl alcohol. There was very little difficulty in obtaining sharp separation as the fractions have distinct boiling ranges sufficiently wide apart from one another. The acetone and normal butyl alcohol were then batch redistilled to give refined dry solvents of very high quality, remarkably free from other components. Their place in the British market was assured the moment they became available. The ethyl fraction, relatively small in volume, was shown to be an excellent solvent for the purposes for which methylated spirit is normally used. The United Kingdom excise authorities permitted its use in this field when suitably denatured.

During the Second World War, when government demands for acetone again rose to a very high level, the British fermentation plant was considerably expanded, far beyond the capacity of existing batch-distillation plant. Successful efforts were made to use multi-column continuous distillation units as employed for industrial alcohol, and both refined acetone and butyl alcohol were produced in this way. Continuous distillation therefore became the method of choice, though batch distillation was retained for the treatment of 'heads' and 'tails' withdrawn from the different fractionating columns. By fractionating in this way, it was possible to isolate small amounts of amyl alcohols and other high-boiling constituents that had not previously been identified from this fermentation.

IV. ECONOMIC ASPECTS

Economically, the acetone–butyl alcohol process has lived through a series of ups and downs. During the First World War, the acetone produced had to bear the whole cost of production, and butyl alcohol, which represented 60% of the total yield, was a useless by-product for which no significant industrial use had been developed.

Almost immediately after the war, butyl alcohol and butyl esters came to the front as solvents and thinners for the new cellulose lacquers that had been introduced for car painting and similar uses. The profitability of the process increased enormously, and was greatly helped by the introduction of sugar-fermenting organisms. The first result of fermenting 6% sugar in the medium instead of 3.8% starch was a 60% increase in plant capacity with no capital expenditure. Added to this was a decrease of the same order in the cost of sterilization and of process labour, per unit of product, and a similar reduction in steam costs at the first and most expensive stage of distillation, i.e. the stripping of dilute solvents from the fermented medium. These effects were very significant indeed, and led to a period of major prosperity for the industry. But even these advantages were not sufficient when competition from synthetic solvents became real and showed even lower production costs.

The fermentation industry did not give up without a struggle. Production efficiency based on raw materials was higher than ever before, and operating costs had been cut to a practical minimum. It was now time to make sure that every marketable by-product of the fermentation should make its contribution. The gases which made up 65–70% by weight of the fermented carbohydrate were the first objective. The hydrogen found a market in the manufacture of synthetic methanol and for hydrogenation of edible oils used in margarine manufacture. Carbon dioxide was used for the preparation of the compressed gas and for dry ice. The mixed gases as collected from the closed fermentation system were first passed through activated carbon adsorption towers, thus recovering a small but useful additional quantity of solvents. The resulting outlet gases were remarkably free from other gaseous impurities, and separation by scrubbing with water under pressure was a relatively simple operation.

Following the utilization of the fermenter gases came the discovery that the residual solids from the fermentation contained appreciable amounts of riboflavin (vitamin B_2) together with smaller quantities of other growth factors. Calculations showed that the market value of these components justified total evaporation of the spent medium after primary distillation, to give a protein-rich animal feed supplement that could be blended in small quantities with bulk animal feeds to provide a desirable vitamin level. In the late 1930s, the first plant designed for this purpose was erected in Peoria, Illinois in the United States of America, and continued to make a valuable financial contribution long after other commercial sources of riboflavin became available.

As long as competition from synthetic solvents was limited to acetone and butyl alcohol made from ethyl alcohol produced by fermentation, the acetone–butyl alcohol fermentation could have continued to compete. The petrochemical industry, using petroleum as the cheapest carbon source on the market, gave the death blow to the fermentation, at a time when there was a rapid and irrevocable change in the availability of molasses.

Over the last 60 years, the world cane-sugar industry has gone through cycles of over- and under-production. Up to the beginning of the Second World War, it always had a problem to dispose of blackstrap molasses, which accumulated in large quantities with each sugar harvest, and often reached the limits of available storage capacity, so that the fermentation industry was able to buy this material at rock bottom prices. The sugar producers were even prepared to convert their sugar surplus to so-called 'high test' molasses in order to take advantage of this outlet.

After the war, the practice developed in the farming industry, particularly in the United States of America, of using molasses as a substantial component of cattle feed. Their competitive buying in the open market soon led to the price of molasses rising to levels never before known, making it a much more expensive material for fermentation. It was at this time that the petrochemical industry moved in for the kill. Not that the fermentation industry is entirely dead, even in the 1970s. In those sugar-producing countries that have no extensive petrochemical industry, where the transport and distribution of molasses over a wide area raises problems, and where the internal economy of the country does not permit substantial importation of sophisticated chemicals, the fermentation industry still operates, for example in certain eastern European countries, using beet molasses, an excellent raw material, and in Egypt, using cane molasses close to its production source.

Using cheap petroleum, enormous plants have been constructed in a number of major industrial countries for synthesis of a wide range of chemicals, solvents and plastics, but many people have forecast that, when World resources of petroleum have been shown not to be limitless, a new era of fermentation will begin, using raw materials that are replaceable year by year. Few people could have foreseen that, in the 1970s and almost overnight, the era of cheap petroleum would be abruptly terminated. There are new calculations to be made, and it would be a brave man who would dare to be dogmatic on what the future holds.

REFERENCES

Hastings, J. J. H. (1971). *In* "Advances in Applied Microbiology", vol. 14, pg. 1.
McCutchan, W. N. and Hickey, R. J. (1954). *In* "Industrial Fermentations", (L. F. Underkofler and R. J. Hickey, eds.), vol. 1, pg. 1.
Ross, D. (1961). *In* "Progress in Industrial Microbiology", vol. 3, 71.
Ryden, R. (1958). *In* "Biochemical Engineering", (R. Steel, ed.), 125. Heywood, London.
Schofield, M. (1974). *Chemistry in Britain* 10, 432.
Weisgal, M. and Carmichael, J. (1962). "Chaim Weizmann", 364 pp. Weidenfeld and Nicolson, London.
Weizmann, C. (1915). British Patent, 4845.

3. Organic Acids

L. M. MIALL

Pfizer Central Research, Sandwich, Kent CT13 9NJ, England

I. INTRODUCTION

The compounds discussed in this chapter are either primary metabolites of micro-organisms or not more than one or two metabolic reactions removed from primary metabolites. Apart from lactic acid, they fall into two classes, compounds belonging to or closely related to the tricarboxylic acid cycle, and compounds obtained directly by oxidation of glucose. The latter naturally must be made from glucose; the former can be made from a variety of starting materials, not necessarily even confined to carbohydrates. All of these compounds are either regularly manufactured on a large scale, or have at one time been manufactured, or their manufacture has been actively considered, either for their direct use or for use as intermediates.

Standard fermentation equipment, principally agitated and aerated fermenters, can be used for the manufacture of all compounds referrred to except lactic acid, similar to that used for the manufacture of antibiotics (see Volume 3), amino acids (This vol. p. 210) or vitamins (This vol. p. 303). But being acids, it is advisable that the equipment used be of high-quality stainless steel. The items of plant required for removal of the mycelium or cells, rotary filters or centrifuges, are again essentially similar for all fermentation products. For processing the filtered broths to obtain the acids or their salts in pure form, again fairly standard equipment is necessary, namely holding tanks, reaction vessels, filters, evaporators, granulators, basket centrifuges, driers. With the exception of citric acid, which is made on a different scale from the other acids, and of lactic acid, one general purpose plant could be used to make on a campaign basis almost any of the compounds listed. Given the know-how, therefore, many companies in the fermentation field could fairly readily make these compounds, and just as readily stop manufacturing and use their plant for other purposes. For this reason, figures for manufacturing capacity are largely meaningless.

The techniques used for manufacturing these acids are again very similar. For those acids on or closely related to the tricarboxylic acid cycle, they involve a nearly complete blocking of the cycle (a complete block would almost certainly be lethal) either by obtaining a mutant deficient in an appropriate enzyme, or by deprivation of a coenzyme, use of an enzyme poison or some similar biochemical technique (see pg. 22).

The requirements of all of the organisms for chemical elements are also very similar. They will all need carbon, nitrogen (usually as ammonium ion or urea), potassium, phosphorus, sulphur, magnesium and trace amounts of iron, zinc, copper and manganese. If one of these trace metals is not listed as a requirement, it almost certainly indicates that it is present as an impurity in other medium constituents. Sodium, calcium and chlorine are other elements that may be required, with minute amounts of cobalt and molybdenum. The exact concentrations of all these elements for optimum production will almost certainly vary with the strain of organism used, and will need to be determined for it. For this reason, medium details have not regularly been listed; original papers should be consulted if such details are required. For commercial manufacture it is necessary to use as cheap a process as possible, which will almost inevitably involve use of crude raw materials. How this factor can be accommodated within the often stringent requirements for elements mentioned above is where the essential difficulties and the trade secrets of these manufacturing processes lie.

II. CITRIC ACID

$$
\begin{array}{c}
CH_2 . COOH \\
| \\
HO . C . COOH \\
| \\
CH_2 . COOH
\end{array}
$$

A. Historical Introduction

The history of citric acid production from aspergilli may be said to have begun with the publication of the paper by Currie (1917). Prior to this, it had been believed that citric acid was produced by penicillia and that aspergilli made oxalic acid, though a patent was taken out by Zahorski in 1913 covering a method for obtaining citric acid from sugar solutions with *Sterigmatocystis nigra*. Zahorski said this differed from *Aspergillus niger*, but it was later regarded as the same organism. Thom and Currie had shown in 1916 that a number of strains of black aspergilli made citric acid. Currie then carried this far further. Like most workers at that time, he grew his moulds in surface culture, and he was the first to demonstrate a number of now well known aspects of this fermentation. He worked out a medium with the composition (g/l): sucrose, 125–150; ammonium nitrate, 2.0–2.5; potassium dihydrogen phosphate, 0.75–1.0; magnesium sulphate heptahydrate, 0.2–0.25; with initial acidification to pH 3.4 to 3.5 with hydrochloric acid. This medium is the basis for that used ever since for citric acid fermentations based on pure sugar. He was the first to show the importance of using pure reagents for this fermentation, something that many subsequent workers neglected to do and which renders much of their work valueless. He at times used doubly distilled water and recrystallized his reagents, the sugar sometimes as often as five times in order to remove iron and thus to prove the necessity for its addition. He showed that it was necessary to add ferrous sulphate at 0.01 g/l for optimum yields and that larger quantities were deleterious. He showed that the highest yield of citric acid occurred when development of mycelium was restricted and not when it was stimulated; he showed that *A. niger* would form oxalic acid from citric acid. He worked also in shallow pans and he prepared several pounds of calcium citrate.

Currie subsequently joined Chas. Pfizer & Co. Inc. in Brooklyn, New York, and thus was subsequently partly responsible for the development of their citric acid process, which was first operated on a commercial scale in 1923.

Fernbach, Yuill and Rowntree & Co. Ltd. took out patents in 1927 which essentially were based on Currie's work, but which had as a novel feature prior acidification of the medium to pH 1.8, thereby it is said rendering sterilization unnecessary. Whether or not this is so, this initial acidification is necessary for high yields of citric acid. Yields up to 65% by weight on the sugar used (100–200 g/l) are claimed. This work formed the basis of the first process used by John & E. Sturge Ltd.

Raistrick examined the metabolic products of the black aspergilli as part of his monumental work on the biochemistry of moulds, and showed that these organisms produced large amounts of non-volatile acids (Birkinshaw *et al.*, 1931). Raistrick developed a process for citric acid manufacture but this work was never published, though it is referred to in a review (Clutterbuck, 1936) which mentions that, on a semi-large scale, overall yields of 87% of citric acid were obtained. It is not clearly stated, but is apparent from the text, that this is a percentage of the theoretical yield.

It is by no means always clear in publications what is meant by yield. This can be a percentage of the stoicheiometric yield, although it is more often a weight yield; but this is sometimes based on the total sugar input and sometimes on the sugar input less the residual sugar. With glucose as the substrate, it is often not clear whether the yield is based on anhydrous or hydrated glucose, nor is it always clear whether citric acid yield is based on the anhydrous or hydrated acid.

Doelger and Prescott (1934) carried out a thorough examination of the citric acid fermentation. They stress the importance of accurately standardized conditions for reproducible results, but surprisingly in view of previous work they did not purify the reagents used. They carried out a number of fermentations in aluminium pans and showed that, to avoid attack on the metal, it was necessary to use high-purity aluminium.

A factory to make citric acid was built in 1928 in Prague, and it is believed that it was at this plant that the treatment of molasses with ferrocyanide was introduced, though it is also described in a French patent (Mezzadroli, 1938). Subsequently, the Czech process was acquired by the German firm of Joh. A. Benckiser, which was operating a process using molasses treated with ferrocyanide when the British Intelligence Objectives Sub-Committee visited their Ladenberg factory in 1945. Another apparently independent development of this process is referred to by Gerhardt *et al.* (1946) who used ferrocyanide to treat beet molasses before fermenting with *A. niger* strain ATCC 1015 in surface culture. They state—'In the wine industry potassium ferrocyanide and ferricyanide have been used to remove iron from the product. Ferrocyanide was used for several years by Mr. William Eisenman of the Heyden Chemical Corporation for the processing of molasses in citric acid fermentations'. Gerhardt *et al.* (1946) acknowledge that they were given details of this procedure. They obtained about 45–50% yields of citric acid with much oxalic acid.

B. Trace Metal Requirements

Steinberg, in a series of papers between 1935 and 1939, examined the effect of a number of metals on growth of *A. niger*. He appreciated the necessity to treat nutrients extensively in order to attain a very high degree of purity, and extracted them with freshly distilled 95% (v/v) ethanol for six hours or longer in order to remove zinc and molybdenum. He also heated with basic magnesium carbonate at 100°C for one hour to remove trace metals (Steinberg, 1935). He showed that the addition of iron at 0.2 mg/l, zinc at 0.14 mg/l, copper at 0.04 mg/l and manganese at 0.02 mg/l were necessary and that, with ammonium nitrate as the nitrogen source, molybdenum at 0.01–0.02 mg/l was needed for growth and maximum acid production. He showed (Steinberg, 1936, 1937) that shortage of molybdenum had an effect when nitrate was the source of nitrogen but not when ammonium salts were used. Thus, with ammonium chloride or urea, lack of molybdenum made no difference to growth but, with sodium nitrate as nitrogen source, growth was only one tenth in the absence of molybdenum compared with that with molybdenum at 0.02 mg/l. Steinberg (1938) claimed that gallium at 0.01–0.02 mg/l was essential for maximum growth, a finding that has never been confirmed by other workers; D. Bertrand (1954) showed that at most only 0.0005 mg of gallium/l was necessary, and other workers have found it to be unnecessary. G. Bertrand (1941) also worked on this problem for many years. Using 4% succinic acid as the carbon source, and a medium with (g/l) ammonium chloride, 2.5; dipotassium hydrogen phosphate, 0.35; magnesium sulphate, 0.25; together with 0.2 p.p.m. iron, 0.18 p.p.m. zinc, 0.04 p.p.m. copper, 0.02 p.p.m. manganese and 0.02 p.p.m. molybdenum, he showed the need for vanadium at 0.001 p.p.m. for optimum growth. He purified his nutrients by treatment with cupferron (ammonium *N*-nitrosophenylhydroxylamine). The need for vanadium has also never been confirmed and, later, D. Bertrand (1966) stated that the concentration required was as little as 0.01 μg/litre. It was also claimed that zinc can partly, but not completely, be replaced by cadmium (Bertrand and de Wolf, 1955).

Information on the metal requirements of *A. niger* was used by Mulder (1948) as a method for determining the copper, magnesium and molybdenum requirements of soils and plant material. He compared the colour and abundance of spores produced on media containing the test materials with media containing known standards. This work was extended by Nicholas and Fielding (1951) of the Long Ashton Research Station in the University of Bristol in England, who developed extremely sensitive procedures for assay of magnesium, copper, zinc, manganese and molybdenum. The techniques required for freeing the test solutions from metals are very exacting, usually involving coprecipitation and extracting with solutions of metal-chelating agents. The work described above was not concerned with citric acid production, but the trace-metal requirements of the mould and the methods used for removing metals from solution are of interest to anyone working on the citric acid fermentation.

Of more direct interest is the work of Tomlinson and others in Vancouver (Tomlinson *et al.*, 1950; Kitos *et al.*, 1953) who pointed out that elements necessary for growth of *A. niger* must also be necessary for citric acid production. They accordingly studied the influence of zinc, iron, copper and manganese on citric acid yields, using Wisconsin strain 72/4 in surface culture, and paying due regard to the purity of the sugar and the other reagents used. They showed that, under these conditions, the optimum concentration of zinc is 1.0 p.p.m., of iron 0.1 p.p.m., of copper 0.05 p.p.m., and that 0.05 p.p.m. of manganese markedly lowers the citric acid yield, although about 0.2 parts per American billion of manganese are necessary. They pointed out the importance of working with chemically cleaned glassware. They showed that the optimum concentrations of these four metals were interdependent, and that the sensitivity to metals decreases with decreasing temperature. These papers, which were neglected by many subsequent workers, highlight the detailed care which must be taken to get meaningful results when studying citric acid production by fermentation. Chesters and Rolinson (1951), using a different strain of *A. niger*, obtained similar findings. Adiga *et al.* (1961) studied the adverse effect of cobalt, nickel and zinc on growth of *A. niger*, on glucose utilization and citric acid production, and on the antagonism of iron and magnesium to these toxic effects. They showed that both iron and magnesium overcame the growth inhibition and that magnesium, added in higher than normal concentrations, restored acid production, except with cobalt toxicity.

C. Biosynthetic Route

The mechanism of production of citric acid from sugar is now understood, at least in broad outlines. It is interesting to see how near the truth were some of the early speculations—and they could only have been wild speculations. Euler (1909) thought that sugar broke down to pyruvate and this to acetaldehyde, three molecules of which reacted to give citric acid. Raistrick and Clark (1919) were close to the mark in suggesting that citric acid is formed by condensation of oxaloacetic acid and acetic acid, but they thought that these arose from hexoses via α-γ-dioxo-adipic acid.

With the elucidation of the Embden-Meyerhof-Parnas (E.M.P.) pathway for breakdown of carbohydrate to pyruvate (Fig. 1), and the Krebs tricarboxylic cycle (Fig. 2), the likely route to citric acid in moulds was obvious. Jagannathan and Singh (1953) showed the presence in *A. niger* of all of the enzymes of the E.M.P. scheme. Shu *et al.* (1954) concluded that about 80% of the glucose is metabolized by this route, and McDonough and Martin (1958) showed that enzymes of the pentose phosphate cycle are also present in *A. niger* under conditions in which citric acid is made. Gluconic acid and 5-oxogluconic acid can be made under certain conditions by certain strains of *A. niger*, and mycelium from the mould grown on gluconic acid has been shown to contain ribose 5-phosphate, glycerol 3-phosphate and citric and malic acids (Bertrand and de Wolf, 1957).

But Leopold (1959) found that, during the early stages of a citric acid fermentation, more sugar is consumed than would correspond to the citric acid formed. I have several times suspected the converse of this—more citric acid formed late in the fermentation than can be accounted for by the sugar utilized—but this has never been clear-cut enough to be regarded as proved. Leopold (1959) suggested the intermediate formation of oligosaccharides, and Wells *et al.* (1936) once showed that addition of alcohol to a citric acid fermentation caused formation of a precipitate which, after several further precipitations, was shown to be a non-reducing carbohydrate giving glucose on acid hydrolysis.

One suggestion (Foster and Carson, 1950; Foster *et al.*, 1950) was that hexoses broke down *via* the E.M.P. pathway to pyruvate, which was decarboxylated to acetate or converted into some other compound with two carbon atoms, and that three of these reacted together to give citrate. This was possible in view of the experimental results with labelled acetate and carbon dioxide, but impossible in view of the high yields obtained by other workers, as the maximum molar yield by this scheme would be 67%, or 81.8% as a weight-yield of citric acid monohydrate from sucrose, and yields substantially higher than this have been obtained. Johnson and his coworkers studied the mechanism using *A. niger* 72/4 and conditions calculated

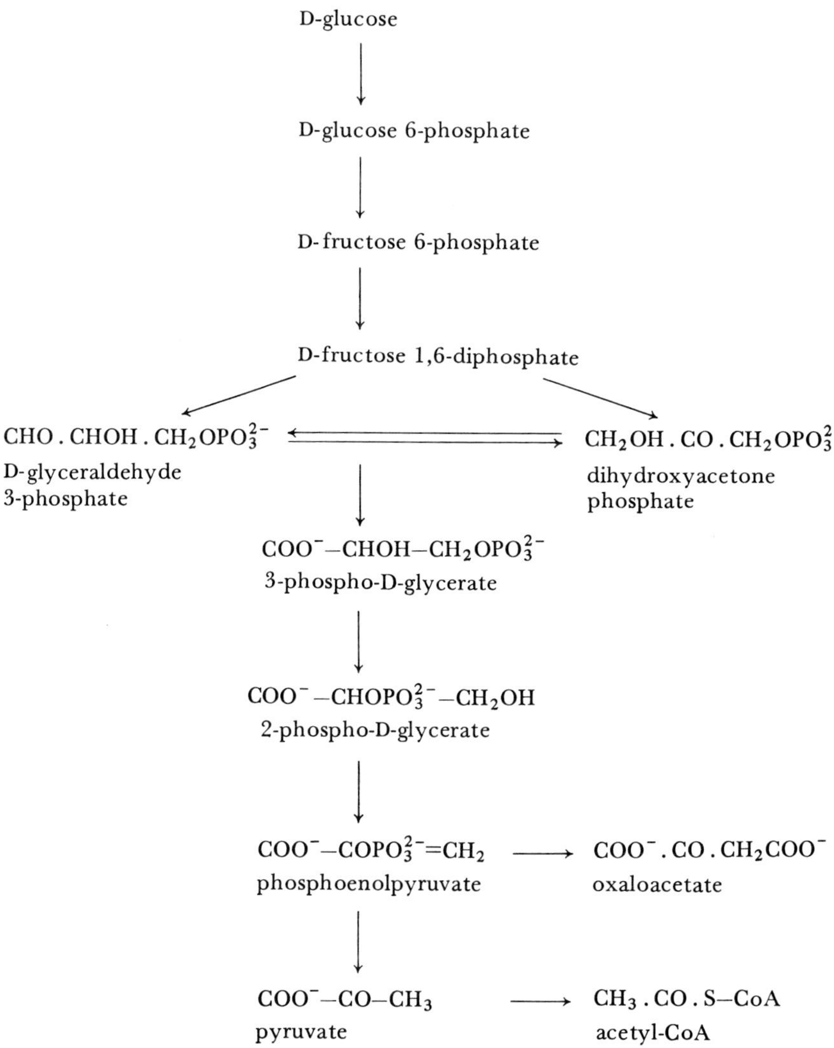

Fig. 1. Reactions of the Embden–Meyerhof–Parnas (EMP) pathway.

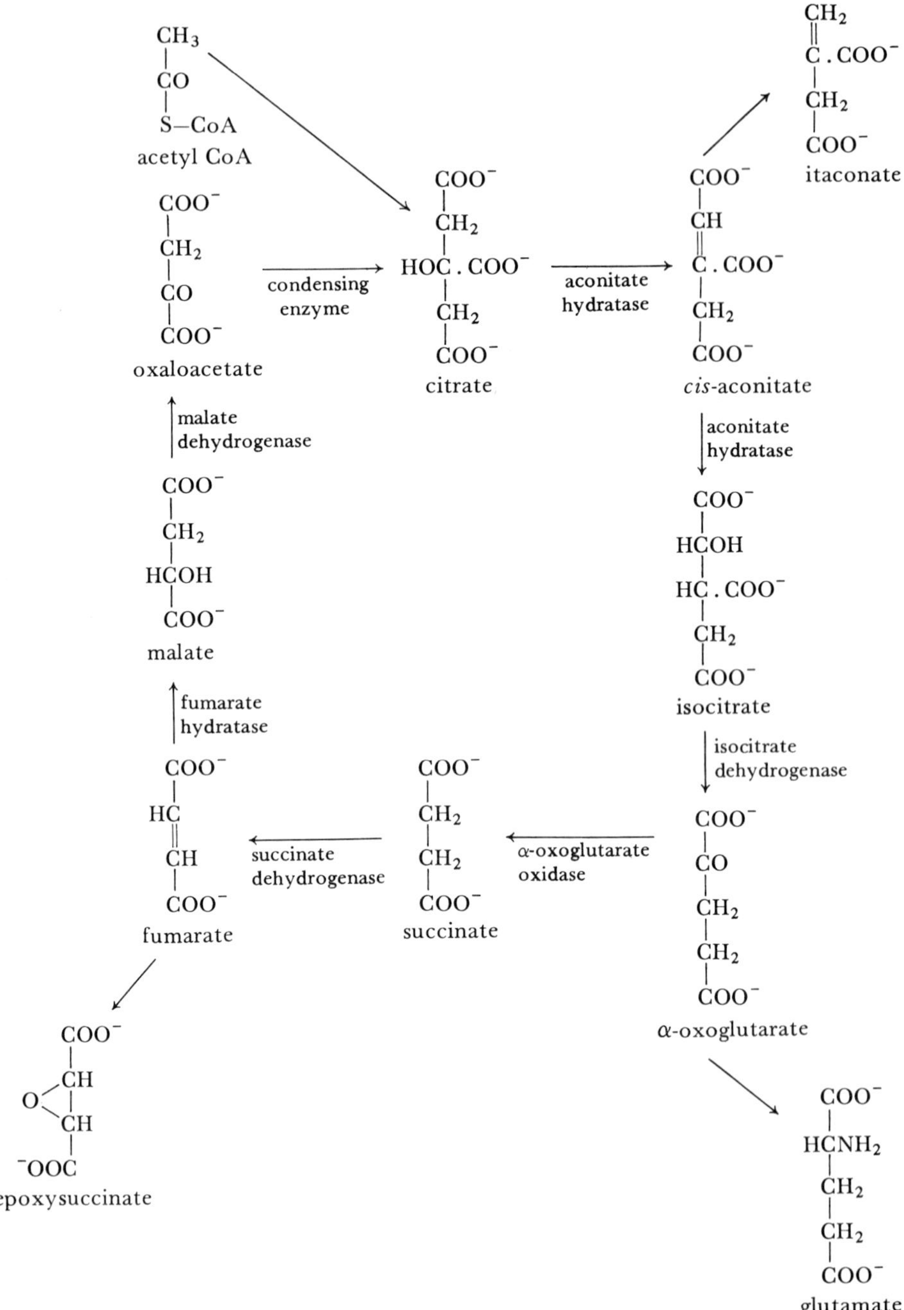

Fig. 2. Intermediates of the tricarboxylic acid cycle and related compounds.

to give high yields of citric acid. Labelled acetate was added to washed mycelium in a replacement medium, a maximum molar yield of 71.3% was obtained, and maximum radio-activity was in one of the terminal carbon atoms (Bomstein and Johnson, 1952). Later (Cleland and Johnson, 1954), using labelled glucose, it was shown that citric acid formation proceeds by a symmetrical split to two three-carbon compounds, carboxylation of one to a four-carbon compound, and decarboxylation of the other to a two-carbon compound, followed by condensation of these to citric acid. By fermentation of glucose in the presence of labelled carbon dioxide, it was shown that almost theoretical incorporation of carbon dioxide occurs, confirming earlier work by Martin *et al.* (1950) and by Lewis and Weinhouse (1951), who also showed that metabolism of carboxyl-labelled acetate by *A. niger* led to formation of citric acid with approximately one and a half times as much label in the primary carboxyl groups as in the tertiary.

It was at one time assumed that a six-carbon compound other than citric acid must first be made from the reaction between a two-carbon and a four-carbon compound. This was necessary in order to account for the asymmetrical distribution of radio-activity in citric acid made from radio-actively labelled compounds. But Ogston (1948) on theoretical grounds showed that citrate might be expected to react asymmetrically in enzymic reactions, and this was later confirmed for *A. niger* (Carson *et al.*, 1951). Woronick and Johnson (1960) and Johnson and Bloom (1962) studied carbon dioxide fixation in *A. niger*, and showed that there were two systems that can do this, one involving phosphoenolpyruvate and ADP, and giving oxaloacetate and ATP, and the other giving oxaloacetate directly from pyruvate and requiring ATP. Both systems require magnesium and potassium ions, and the second system definitely requires biotin.

The presence of all of the enzymes of the tricarboxylic acid cycle has been shown in *A. niger*, and the cofactors and required conditions for their operation studied (Martin, 1954; Ramakrishnan, 1954; Ramakrishnan and Martin, 1954a, b, 1955; Ramakrishnan *et al.*, 1955). Thus it has been shown that, during citric acid accumulation in media containing ferrocyanide-treated beet molasses, the specific activity of the condensing enzyme increases, but aconitate hydratase and isocitrate dehydrogenase activities are lost. In that aconitate hydratase needs iron for its activity, this is a logical picture. Neilson

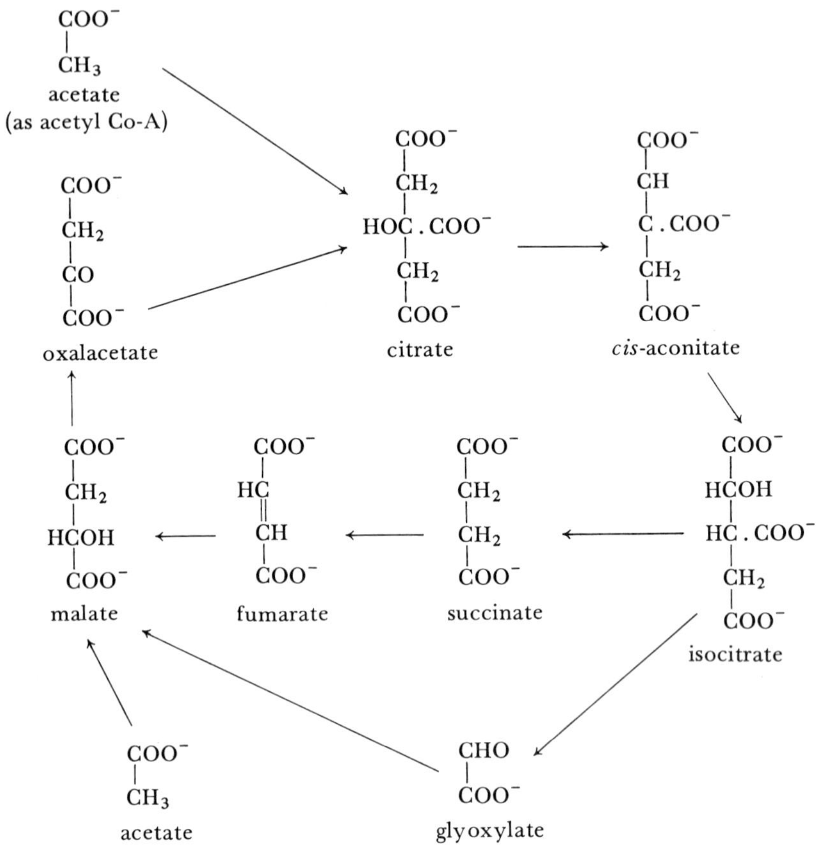

Fig. 3. Reactions of the glyoxylate cycle.

(1956) found that, when *A. niger* was grown on a medium that did not accumulate citric acid, two enzymes were found in the mycelium that acted on *cis*-aconitate. One resembled the aconitate hydratase of animal tissues in forming both citrate and isocitrate. The other, given the name aconitic hydrase, formed only citrate. Neither enzyme was found in mycelia grown in a chemically defined citric acid-producing medium. Addition of manganese salts led to the production of both enzymes, while addition of copper ions inhibited enzyme production. It has been found (Ramakrishnan and Martin, 1954c; Ramakrishnan, 1958), even after purification of the enzyme, that the condensing enzyme is inhibited by magnesium ions, a surprising finding in view of the tolerance for quite high concentrations of magnesium ions in citric acid fermentations; this finding remains unexplained.

The tricarboxylic acid cycle accounts for the formation of citric acid from carbohydrate, but not from acetate. Olson (1954) showed the presence in *A. niger* of isocitrate lyase, which catalyses the breakdown of isocitrate to glyoxylate and succinate, thus confirming much earlier work by Challenger *et al.* (1927) who showed that glyoxylate and oxalate were formed when *A. niger* grew in the presence of calcium acetate. Kornberg and Collins (1958) showed that labelled malate was formed in *A. niger* on incubation of labelled acetate with isocitrate, thus effectively confirming the operation of the glyoxylate cycle in the mould when grown on acetate. But isocitric lyase is not found (Collins and Kornberg, 1960) when glucose is the substrate, showing that this system does not then operate.

D. Submerged Culture

Methods of production described so far have been by the surface-culture technique, using either small flasks for laboratory-scale work, or vast numbers of metal trays for commercial production. Wells and Ward (1939) had gone so far as to state that all reliable evidence indicated the impossibility of using submerged-culture techniques for citric acid production, and had expressed the view that some vital derangement of the enzyme system was responsible. Even at the time this was published, the information was incorrect, as the fact that citric acid could be made by growing *A. niger* in deep culture had been shown by Amelung (1930) and by Perquin, working in Kluyver's laboratory. Amelung's work has been almost entirely neglected. He worked with a strain of *A. niger* var. *japonicus* and showed that, if a slow stream of air was passed through a 15 cm-deep solution containing per litre: 100 g sucrose, 5.0 g ammonium sulphate, 2.5 g potassium dihydrogen phosphate and 1.2 g magnesium sulphate, the mould grew beneath the surface and produced a considerable amount of citric acid, though not as much as in surface culture.

Kluyver and Perquin in 1933 invented the shaken-flask technique for emulating deep culture on a laboratory scale. Perquin's work on citric acid was published as a thesis in Dutch in 1938, just before the beginning of the 1939–1945 war, and it was consequently not widely read for a further six years, and even then the information only spread slowly. He showed that two strains of *A. niger* made citric acid in shaken-flask cultures, that limitation of phosphate was important, and that traces of material present in water might have a very adverse effect on acid production.

A major problem in making citric acid in deep culture is to restrict the growth of the mould, so that it does not get so thick as to be virtually unstirrable. Szucs (1946) patented a process in which *A. niger* is grown on a solution of sucrose and nutrient salts and the mycelium washed and transferred to a similar solution lacking phosphate. It was also necessary to aerate with oxygen or with oxygen-enriched air, a procedure later shown to limit growth. The adverse effect of high concentrations of phosphate has been recognized by a number of workers, and was shown by Martin and Steel (1955) to be connected with a greater production of gluconic and 5-oxogluconic acids. It also encourages oxalic acid production. Karow and Waksman (1947) claimed that, by using *Aspergillus wentii*, much better yields of citric acid could be obtained in deep culture than with *A. niger*, but again requiring oxygen or oxygen-enriched air.

The development of the citric acid fermentation owes much to the work of Johnson and his colleagues at the University of Wisconsin. They studied the effect of metals, including iron, zinc and manganese (Perlman *et al.*, 1946a). They showed the importance of using optimum concentrations of these ions, and that the optimum concentration varied with the strain of mould used. They also showed that treatment of commercial sugar solutions, including glucose, by passage through cation-exchange resins greatly improved yields. They showed that addition of Cuban high-test molasses to fermentations run on pure sugar was harmful even when only very small amounts were added (Perlman *et al.*, 1946b). Ferrocyanide treatment of the molasses partly alleviated this effect. It was in association with this work that *A. niger* strain 72/4 was developed. Previous workers at Wisconsin had used strain A.T.C.C. 1015, but in this work the mycelium was invariably found to be covered with spores, so the culture was plated out and one isolate found that gave the same results as those previously obtained with A.T.C.C. 1015.

This strain, 72/4, has been used for much of the subsequently published work on citric acid fermentations.

Shu and Johnson (1947, 1948a, b) were among the earliest workers to study production of citric acid by submerged fermentation. They first showed the importance of the composition of the inoculum, proving that enough manganese might be carried over adversely to affect citric acid yields, and then studied the interdependence of the various constituents of the medium. They used an alumina coprecipitation method to treat a pure-sugar medium and then added phosphate, zinc and iron. They obtained a 72% yield of anhydrous citric acid from sucrose in shaken-flasks in nine days. The process was scaled up (Buelow and Johnson, 1952) in 50-gallon glass-lined fermenters with glass-coated agitators and stainless-steel spargers, and yields of citric acid varied from 72 to 84% in about 200 hours. The importance of the aeration rate on the conversion rate was clearly shown.

E. Fermentations on Molasses

Miles Laboratories (1951, 1952a) patented the use of *A. niger* in deep culture and a medium in which the iron content was under one part per million, this concentration being obtained by passing a solution of pure sucrose or of invert molasses (high-test molasses) through a cation-exchange resin. The company had obtained a patent covering a similar treatment of carbohydrate solutions for use in surface fermentations in 1945, and here also had specifically referred to high-test molasses. The molasses solution is passed once or twice through columns of cation-exchange resin, the necessary sources of nitrogen, potassium, magnesium, zinc and other elements are then added, the solution sterilized and the fermentation carried out as usual. Ammonium carbonate was found to be the most satisfactory source of nitrogen, and it was claimed to suppress oxalic acid production. Later, for convenience in operation, the use of gaseous ammonia was recommended (Miles Laboratories, 1955). For submerged fermentation it is stated that the fermenters must be made from, or lined with, a material which is resistant to citric acid and will not contaminate the fermentation with iron. It is said that either glass-lined, rubber-lined or resin-lined equipment is satisfactory. The fermentations usually require 10–14 days to convert 120–150 g/l of sugar solution. The maximum yield quoted on the total sugar was 65.4% as citric acid monohydrate.

In another patent (Miles Laboratories, 1952b), the addition of morpholine at 100–1,000 p.p.m. is claimed to give improved citric acid yields, 80.4% being the highest value quoted, and later (Miles Laboratories, 1956a) it was claimed that copper, added in concentrations of up to 500 p.p.m., had the effect to some extent of counteracting the adverse effect of even as much as 10 p.p.m. of iron in the medium. Copper, at concentrations of between 5 and 75 p.p.m., has the added advantage of inhibiting growth of possible contaminating penicillia (Miles Laboratories, 1956b). The yields quoted in the various patents are sometimes contradicted in later patents. Thus in Miles Laboratories (1962) it was stated that, prior to the use of the patented process, yields of 65% to 75% were considered satisfactory, but that by addition of extra nitrogenous nutrients during the acid-producing stage the yield of citric acid is raised to 85% or greater. The addition of certain quarternary ammonium compounds is another way of obviating the effect of too much iron and getting increased yields; values of 92% are quoted (Miles Laboratories, 1969).

Lockwood and Batti (1965) of Miles Laboratories admit that the use of pure materials, either in pure form or purchased in crude form and subsequently refined, entails considerable expense and therefore that citric acid produced by such methods may not be priced competitively with that produced by other commercial processes. They find that less pure starting materials can be used if various potentially toxic organic compounds are added in sub-lethal quantities at the beginning of the fermentation and mould growth is thereby restricted. The compounds added are mostly phenols, substituted phenols or related aromatic compounds. Another patent covers the production of a carbohydrate solution by treating starch with liquefying amylase and amyloglucosidase and the conversion of this to citric acid with *A. niger* (Swarthout, 1966). By 1966 Miles Laboratories was reported to have switched to a glucose solution, produced from corn, as their basic raw material.

One member of the B.I.O.S. team, who inspected the German plants at the end of the 1939-1945 war, was G. A. Ledingham, at that time working with the National Research Council of Canada in

Ottawa. It can hardly be co-incidental that work on citric acid production was carried on for many years at Ottawa and led to one of the best of the published processes. This work was all done on Canadian beet molasses, treated with ferrocyanide, and was run in deep culture.

The first demonstration that citric acid could be obtained in deep culture from ferrocyanide-treated beet molasses was by Clement (1952). She obtained a 52% yield in shaken flasks and up to 64% in deep culture, but the results were very erratic. The best yields were obtained with strain, A-1-215, which was obtained from Germany by the B.I.O.S. team but, when this process was scaled up by Martin and Waters (1952), they used Wisconsin strain 72/4 (listed at Ottawa as A-1-233) and this strain was used for all subsequent work at Ottawa. They showed that yields of up to 72% could be obtained in 65 to 88 hours in an unagitated tower fermenter initially sparged with air and later with oxygen. This was a 3.5 litre submerged scale fermentation of ferrocyanide-treated beet molasses using a pellet-type inoculum.

This process was further scaled up to 40-litre units made of 15 cm-diameter Pyrex pipe (Steel *et al.*, 1955). No metal was allowed in direct contact with the fermenter. Even aeration disks made of 316 stainless steel gave poor results, and it was necessary to use sintered glass disks fitted into holes in a rubber-covered steel plate. Provided this was done, the fermentation scaled up satisfactorily, but it was shown that samples of molasses from different beet-sugar factories needed different treatment for optimum results and that samples from any one factory would vary considerably in behaviour from year to year. This variation in behaviour of molasses had also been shown by Bernhauer *et al.* (1949) and has been commented on by Kovats *et al.* (1957) and by Leopold and Fencl (1958), both sets of workers stating that it was not yet possible by simple analytical methods to determine how well a certain batch of molasses was suited to citric acid production.

The importance of a standard inoculum using mould particles with the correct morphology was shown by Steel *et al.* (1954) and later by Clark (1962a). The ideal is to have hard round pellets about 1.2 to 2.5 mm in diameter and with little filamentous growth. If growth is largely filamentous, citric acid production is poor. Conversely Takahashi *et al.* (1965a), using a glucose–salts medium in shaken flasks, state that the best acid production is achieved by filamentous growth of mould, which they achieved by addition of certain surface-active agents. It is difficult to reconcile these findings except by stressing that the medium, strain of organism, and fermentation conditions used by the Japanese group were all quite different from those of the Canadian workers.

The need to use oxygen for this fermentation obviously makes it an expensive process, but Clark and Lentz (1961) showed that the gas could be satisfactorily recirculated with no treatment other than scrubbing to remove carbon dioxide. Usually the mash was sparged with air for the first 24 hours while the mycelial pellets were developing, and with oxygen during the acid production stage. Clark (1962b) showed that, for optimum yields, and he claimed about 75% conversion in 140 hours, it was necessary to exert fairly strict control of the concentration of excess ferrocyanide, which needed values of about 20 p.p.m. at the start of the acid-producing stage of the fermentation, though values of up to 400 p.p.m. could be tolerated during the growth stage. But control of the ferrocyanide concentration was also necessary during growth of the inoculum, satisfactory pellets only developing in medium containing between 10 and 40 p.p.m. of ferrocyanide.

Martin (1955) also studied the effect of ferrocyanide, but came to no clear conclusion on whether this had a direct effect on the mould or an indirect effect through its action in removing trace metals. Clark *et al.* (1965), however, showed that treatment with ferrocyanide precipitated, at least in part, most of the metals present in beet molasses and most efficiently precipitated manganese, iron, zinc and copper, all known significantly to affect citric acid production. Horitsu and Clark (1966) made a more detailed study of the effect of ferrocyanide and showed that, at concentrations of over 30 p.p.m., this compound stimulated citric acid production in resting cells, but markedly inhibited the development of growing cells. They concluded that ferrocyanide benefits the fermentation indirectly by precipitating interfering metals or in some way making them biologically inactive. Clark *et al.* (1966) showed that addition of as little as 2 parts per billion of manganese ion to beet molasses treated with ferrocyanide caused a 10% decrease in citric acid yield and an undesirable change in morphology of the mould to a filamentous form. When one considers the virtual necessity of constructing production-scale fermenters of metal containing a small but appreciable proportion of manganese, and the not completely insignificant attack of citric acid on any metal, one reason for the difficulty in commercial development of this fermentation becomes apparent. Whether ferrocyanide has an additional effect is uncertain; Cejkova *et al.* (1966) claim that it has a direct positive effect on the fermentation.

Three patents have been taken out as a result of the Ottawa work. Martin (1956) claimed broad cover, Clark (1964) claimed a process with strict control of the ferrocyanide concentration, and Tveit (Svenska Sockerfabriks Aktiebolaget, 1964) claimed a process in which, after the pH value of the fermentation had dropped naturally to between 4.5 and 5.5, hydrochloric acid was added to bring the pH value to below 3.0, thereby restricting growth and stopping oxalic acid production. Tveit had worked at the Ottawa laboratories, and his process is a direct adaptation of that originally worked out by Martin and Waters (1952). A yield of from 90 to 100 g of citric acid from 100 g of sugar is claimed. It is not clear whether this is anhydrous or hydrated citric acid but, as the theoretical stoicheiometric conversion of sucrose to the hydrated acid is 122% and sugar has to be used both to provide carbon and energy for growth of the mould mycelium, this is a remarkably high conversion. It is, however, apparently necessary to run this process either using oxygen or air under pressure.

A procedure in which the pH value of a citric acid fermentation was deliberately lowered by adding various acids had, however, been previously claimed by Fried and Sandza (1959), of Stauffer Chemical Co. The examples given in their patent describe the use of both cane and beet molasses, and involve treatment with ferrocyanide, but not the use of oxygen or oxygen-enriched air. The yield on beet molasses was about 50%.

Another interesting process is that of Usines de Melle (1958). In this, the mould is first grown on a comparatively weak unbuffered solution of sugar or treated molasses until the pH value has dropped below 3.0, when a much stronger treated molasses solution is added. Growth of the mould is thereby controlled during the initial period both by restriction of nutrients and a rapid drop in pH value. During the last three days of the fermentation, a proportion of the broth can be run off and replaced with a treated molasses solution—a slight approach to a continuous fermentation.

Another type of continuous process, essentially using surface culture, is described by Zadrodzki and Krzysztofik (1953). Molasses is passed through an anion exchanger, acidified to pH 5.6, and then passed through a series of four half-filled tubes with intermediate anion exchangers to remove the citric acid formed. The tubes are inoculated with *A. niger*, and air passed in countercurrent flow. The retention time was four days in each tube, and the fermentation was operated for two months before any decrease in yield was observed.

Compared with the large number of publications describing the production of citric acid from beet molasses, relatively little has been published describing its production from blackstrap molasses. Clark (1962c) treated West Indian cane molasses with ferrocyanide and fermented it in deep culture. He showed the importance of accurate control of the ferrocyanide concentration, and obtained yields up to 80% in eight days. Of interest also is a paper by Yamada and Hidaka (1964) who found a strain of mould that gave 50–60% yields of citric acid from untreated blackstrap molasses. The addition of methanol increased the yield to a maximum of 83.5%.

This effect of methanol is of considerable interest. It was first reported by Sakaguchi and Baba (1942). A process for the manufacture of citric acid by fermentation worked out by Moyer (1953) depends on addition of lower alcohols to crude carbohydrate solutions before fermenting with *A. niger*. Moyer (1953) showed that addition of from 1–5% of methanol, ethanol, *n*-propanol or isopropanol or of methyl acetate to carbohydrate-containing media markedly increased yields. Addition of methanol reversed the adverse effects of iron, zinc and manganese on citric acid production, and made it possible to ferment otherwise untreated solutions of beet, cane or high-test molasses or gelatinized corn starch. In general, the cruder the material the more alcohol is required. This effect has been studied by other workers and the results can be summarized by stating that, in already purified solutions giving high yields of citric acid, addition of alcohols will be deleterious. But, with crude starting materials, alcohols added at the right concentration may markedly improve yields. This effect of alcohols is as yet unexplained, but both morphologically and biochemically it appears to be similar to the effects of trace-metal removal. In a process for making citric and isocitric acids from *n*-paraffins (Kimura and Nakanishi, 1973) with

Candida zeylanoides, the effect of added methanol is to increase citric acid and to decrease isocitric acid production. It acts, in fact, as an inhibitor of aconitate hydratase. But, if this is its action, it is difficult to explain why it increases the production of itaconic acid from *Aspergillus terreus* (Moyer, 1954) and even more difficult to explain its similar action on the production of epoxysuccinic acid by *Aspergillus fumigatus*, as this acid has been shown (Wilkoff and Martin, 1963) to be made by direct oxidation of fumaric acid.

F. Strain Improvement

Surprisingly little work has been done, or at least published, on improving citric acid-producing strains of *A. niger* by mutation. This is presumably in part due to the very high yields given by existing strains but, in that the fermentation is one of the most difficult to run, it is surprising that more has not been published on strain improvement. One Miles Laboratories patent (1951) mentions the use of a mutant strain of *A. niger*. James *et al.* (1956a) worked out a technique for testing cultures by growing them on paper disks soaked in a medium containing molasses, salts and bromocresol green. The size of the zone and acidity were assessed in relation to the size of the colony. Yields in deep culture were later found to be in good agreement with those predicted by this screening technique. The same authors (1956b) produced mutants of strain 72/4 by X-ray and ultraviolet irradiation, certain of which under some conditions gave better yields than the parent. Much the same was claimed by another Australian team, Trumpy and Millis (1963), who obtained a mutant, 72/44, which they claimed was more tolerant to trace metals than its parent, though it gave maximum yields of citric acid at low concentrations of trace metals. But they did not get such good yields with 72/4 as others have done, so one feels that their strain may have degenerated and they were simply re-isolating the original culture. Recently, weight yields of citric acid of 110 and 118% (the latter value seems impossibly high) have been reported from solutions containing 160 g sucrose per litre with added salts, in surface culture using *A. niger* mutants (Hannan *et al.*, 1973). These were obtained by gamma irradiation of a strain for which a maximum yield of 29% was reported.

G. Sporulation

It is essential in industrial fermentations to occupy the largest and most expensive fermenters for the shortest possible time; one way to achieve this is to grow a very active inoculum in smaller vessels. Workers at the University of Strathclyde in Glasgow, Scotland (Galbraith and Smith, 1969; Anderson and Smith, 1971, 1972) have studied sporulation of *A. niger* in submerged culture. They first devised a suitable medium based on 10 g glucose per litre with the usual macronutrients and small concentrations of salts of calcium, copper, iron, zinc and manganese. They showed that the presence of ammonium nitrate was inhibitory to sporulation, but that sodium nitrate or monosodium glutamate was not. Sodium glutamate was added at 5 g/l and ammonium sulphate at 1.98 g/l. The recommended procedure is to grow newly germinated conidia first at 44°C to produce swelling and then at 30°C. This gives maximum fresh conidial formation with minimum vegetative growth, which are the desired characteristics for subsequent acid production.

There have been many other publications on production of citric acid by *A. niger*, but most of them are of relatively little interest. As Meyrath (1967) has pointed out, many papers are worthless because the authors have apparently not taken into account the very exacting trace-metal requirements. Thus a statement, for example, that sucrose is more efficiently used than glucose may purely reflect the differences in metal content of the two samples of carbohydrate, and is useless unless steps have been taken either to remove trace metals or to ensure that they have the same metal content. Furthermore, much work has been done on low-yielding strains of mould, and methods of improving the yield from 40% to 50% are of relatively little interest when strains giving much higher yields are readily available.

Other compounds made in high yields by *A. niger* are gluconic acid and oxalic acid. Gluconic acid is made by specially selected strains and rarely is much gluconate found in the broth from a citric acid fermentation. Oxalic acid can be made in relatively high yield by selected strains and medium adjustment, with particular attention to high concentrations of phosphate, and is very regularly reported as a contaminant in fermentations making citric acid, but the reasons for its production are still somewhat obscure. Cleland and Johnson (1956), using *A. niger* 72/4, found that oxalic acid is formed in media containing pure sugar as a result of increased internal pH value and showed that, under these conditions, glucose is converted to gluconic acid and thence to oxalic acid. On the other hand, at low pH values, oxalate is formed chiefly by the oxidative splitting of oxaloacetate; it is suggested that oxalic acid is formed when the production of oxaloacetate is greater than that of acetyl-CoA. It is not convincing from the evidence presented that the splitting is oxidative, and an enzyme responsible for hydrolysing oxaloacetate to oxalate and acetate has been found in *A. niger* by Hayaishi *et al.* (1956), it needs manganese but no other cofactor for maximum activity. Emiliani and Bekes (1964) showed that a good citric acid-producing strain of *A. niger* was a good source of oxalate decarboxylase, which is produced at pH 1.1. At this pH value, oxalic acid was not found in the medium, but as the pH value was raised to 2.5 the enzyme activity decreased and increasing amounts of oxalic acid were found. It is commonly assumed that *A. niger* does not produce oxalic acid at low pH values, but it is suggested that it may well be destroyed as rapidly as it is formed.

In the past, oxalic acid was manufactured as a by-product of citric acid production. After removing mycelium, the next step was to add sufficient calcium salt to precipitate the oxalic acid present in the broth. With the use of improved strains of mould that make more citric acid and less oxalic acid, it is now seldom economically justifiable to recover oxalate. Direct manufacture of oxalates by fermentation is not competitive with purely chemical manufacturing processes.

H. Organisms Other Than Aspergilli

For many years it had become an article of faith in the industry that citric acid was best made by *A. niger* or organisms closely related to it, and possibly by *A. wentii*. This faith survived despite the elucidation of the tricarboxylic acid cycle as a major route for the metabolism of carbohydrates by almost all micro-organisms, and the likelihood that blocking the cycle either by mutation or chemically at the aconitate hydratase or isocitrate dehydrogenase step might enable citric acid to be accumulated. A number of strains of penicillia have also been shown to make citric acid (Kinoshita *et al.*, 1961), but penicillia suffer from the disadvantage that, in general, they produce acid much more slowly than aspergilli. Various yeasts can also be used to make citric acid, and it is claimed (Pfizer, 1970) that the fermentation time is shorter than with *A. niger*. In the examples given, a strain of *Candida guilliermondii* is grown on a medium containing, per litre, 150 g of glucose monohydrate or an equivalent amount of sugar as blackstrap molasses, and excretes 15–17 g of citric acid per litre. By the addition of sodium fluoro-acetate or certain other halogenated organic compounds, the citric acid concentration is increased to about 20 g/l. By the addition of *n*-hexadecylcitric acid or trans-aconitic acid, at about 0.2 g/l, the citric acid concentration is increased to about 110 g/l (Pfizer, 1972).

I. Hydrocarbon Fermentations

But of greater interest is the production of citric acid by fermen-tation of hydrocarbons. Until the publication of the work on microbiological production of protein by Champagnat *et al.* of British Petroleum (1963), nobody had thought seriously about the use of paraffins as substrates for microbiological processes. It had been long realized that there were micro-organisms that would grow on paraffins, but this information was used only by those who had to contend with microbiological contamination of stored petroleum products and ignored by the fermentation industry. With the publi-cation of the work from British Petroleum, there was, however, a realization that micro-organisms might make substances other than cells from normal paraffins, and much work was carried out to confirm this, particularly in Japan. Two publications (Takahashi *et al.*, 1965b, and Iguchi *et al.*, 1965) on the formation of L-glutamic acid in good yield by bacterial growth on normal paraffins were particularly important in the context being considered. If glutamic acid, only one enzymic step removed from α-oxoglutaric acid, a tricarboxylic acid-cycle component, could be made from paraffins, it should be possible similarly to make other compounds on or closely related to the cycle. Kyowa (1970a) published a patent claiming that one *Aspergillus* sp. and a number of strains of *Penicillium* made citric acid in relatively poor yield from paraffin solutions.

More interesting and more useful patents are those that cover the production of citric acid from hydrocarbons by bacteria or yeasts. Thus Kyowa Hakko Kogyo (1970b) patented the use of *Arthrobacter paraffineus*, and Takeda Chemical Industries (1971a) strains of corynebacteria. The latter claims conversion of paraffin to citric acid in about 100% yields (w/w). But most of the patents covering conversion of normal paraffins to citric acid refer to the use of various strains of *Candida*. The earliest patents all claim broadly the production of citric acid by submerged fermentation from C_9 to C_{20} normal paraffin mixtures using *C. lipolytica* and other *Candida* strains, with the addition of a nitrogen source, either inorganic or organic, and the usual nutrient salts. It is necessary to hold the pH value between 4.0 and 7.5 by addition of alkali. One patent (Takeda Chemical Industries, 1970) mentions a 114% yield of anhydrous citric acid. From later patents, it is obvious that under such conditions these organisms make in addition substantial amounts of isocitric acid. Thus a strain of *C. lipolytica* is mentioned which gives 49 g of citric acid per litre and 52 g of isocitric acid per litre from an *n*-paraffin mixture, and an unidentified *Candida* species, which gives isocitric acid but no citric acid (Takeda Chemical Industries, 1971b).

Methods of partially separating citric acid from isocitric acid depend on the lower solubility of the tricalcium salt of citric acid and, after precipitation and removal of this compound, the filtrate can be treated with cells of various yeasts that convert isocitrate to give equilibria containing much more citrate (Takeda Chemical Industries, 1971b, 1973). In essence this is an enzymic conversion involving aconitate hydratase in that aeration and hence growth of the organisms are unnecessary, and indeed the cells can be immobilized by addition of chloroform.

Various procedures have been devised for isolating and testing mutant strains of micro-organisms that will give improved yields of citric acid with less isocitric acid and other contaminating acids. In one patented procedure (Takeda Chemical Industries, 1972) the mutants are primarily tested for growth both on *n*-paraffins and on citrate and those selected for further study that grow on the former but not on the latter compound. An example is given of a mutant that gives a 138% yield of citric acid from a C_{13} to C_{15} *n*-paraffin mixture with 17% of isocitric acid, while the parent strain gave 61.2% yield of citric acid and 65% isocitric acid.

An alternative method of obtaining mutants low in aconitate hydratase activity is to select ones that are sensitive to monofluoroacetate which, after conversion to fluorocitrate, is a competitive inhibitor of aconitate hydratase (Akiyama *et al.*, 1973). The assumption was made, and was borne out by results, that strains with little aconitate hydratase activity would be particularly sensitive to fluoroacetate and would be particularly good producers of citric acid. Mutant strains of *C. lipolytica* were selected by virtue of their sensitivity to fluoroacetate at a concentration of about 0.01%. One gave a 145% (w/w) yield of citric acid from *n*-paraffins with negligible isocitric acid production. Yet another method of making citric acid from hydrocarbons with yeasts without production of isocitric acid (Benckiser, 1974) involves addition of an aconitate hydratase inhibitor, such as sodium fluoroacetate, and an uncoupling agent, such as 2,4-dinitrophenol. An uncoupling agent is a compound that permits an oxidation to proceed in the absence of phosphorylation that normally is coupled to production of ATP from ADP. With mutant strains of *C. oleophila*, which are blocked at the aconitate hydratase step, increased citric acid yields are claimed if an uncoupling agent is added 24 hours after inoculation. Another procedure for increasing citric acid production, which operates with yeast growing on hydrocarbons, is the addition of methanol (Kimura and Nakanishi, 1973). The effective concentrations of methanol and ethanol are however much lower in hydrocarbon fermentations, and higher alcohols such as lauryl, stearyl and oleyl are also effective, so the mechanism may not necessarily be the same as the action in carbohydrate fermentations. The papers and patents referred to represent only a few of the many that have appeared and are still appearing on methods of making citric and isocitric acids by batch fermentation from hydrocarbons.

Pfizer (1974) has patented a continuous process for this fermentation, using in the example given *C. lipolytica* ATCC 20228 growing on a normal paraffin mixture at a pH value of 3.5 for a total period of 304 hours. In this process, which operates in a single vessel, once the initial batch stage has been completed hydrocarbon and an aqueous salts solution are continually added and fermented broth continuously withdrawn. There is no recycling of cells or of hydrocarbon. Mitsui Sugar Co. (1974), as part of a more general claim covering the use of an organism they called *C. oleophila*, mentions a continuous fermentation running for 20 days in three fermenters connected in series at a weight yield of 148%.

Gledhill *et al.* (1973), using *C. lipolytica* ATCC 8661 growing on hexadecane or a C_{12}-C_{13} mixture at 28°C and with the pH value controlled at about 5.0 with ammonia, found that the rate of product formation declined markedly after about 100 hours running on an aqueous salts medium to which hydrocarbon was fed intermittently to maintain the concentration between 1 and 10 g/litre. The decline could not be attributed either to depletion of oxygen or of nutrients, and they therefore attributed it to product toxicity. Accordingly they periodically removed a portion of the broth, which might be as much as half of it, separated the cells and residual hydrocarbon by centrifuging, and returned these to the fermenter, making up to volume with additional aqueous medium. By these procedures, citrate production was maintained for 340 hours, but the production rate declined considerably during this semicontinuous operation.

Substrate costs are obviously one of the main considerations in any fermentation process. When the British Petroleum workers first studied production of protein from petroleum, straight-chain paraffins were an embarrassment to the petroleum producers and were very cheap. Because of their effect in raising the solidification temperature, it was necessary partially to remove them from aviation spirit, and at that time there were few other uses. Since then the increasing concern about contaminating the environment has made it undesirable to have more than the necessary minimum of smoky discharge from aircraft exhausts. For this purpose, the normal paraffin component of the fuel must be as high as possible. A delicate balance has therefore to be struck. Furthermore, uses for normal paraffins for manufacture of biodegradable detergents and for other purposes have greatly increased. Political difficulties in Libya, where the crude petroleum has the highest normal paraffin content, have further added to the problems. Straight-chain paraffins have ceased to be an undesirable by-product for which a use must be found, and become a material in short supply which must be specially made.

J. Manufacturers

Most commercial companies are highly reticent about their production capacity; quoted figures are apt to be both contradictory and highly misleading. Guesses by journalists, which may have little inspiration behind them, get repeated by others and gradually acquire an air of authority which is largely undeserved. It has been thought wiser to quote that the total World citric acid-manufacturing capacity is believed to be in the order of 200 million kilograms a year, with nearly half of that in Europe, and not to attempt further to divide this figure.

Pfizer Inc. is a leading manufacturer in the United States with plants in Brooklyn, New York and Groton, Connecticut, and in Europe with a plant at Ringaskiddy near Cork in Eire. Pfizer also has a fermentation plant in Argentina, and plants making citric acid from imported calcium citrate in Australia and Canada. Recently a new plant has come into operation at Southport, North Carolina, U.S.A. Miles Chemical Company is the only other manufacturer in the United States, with a plant at Elkhart, Indiana. Miles is a comparative newcomer to the industry, starting in 1958. It is believed to operate a submerged fermentation process based on glucose (probably a starch hydrolysate) which is pretreated by passing through a cation-exchange resin column. Previously Miles used high-test molasses from Cuba and the Dominican Republic. High-test molasses is not genuine molasses, but a concentrated sugar cane solution, half of which has been inverted. It is therefore a mixture of sucrose, glucose and fructose which does not readily crystallize. Miles has a subsidiary plant in Israel and joint ventures in Mexico (Quimica Mexama S.A.) and Colombia (Sucromiles). A few years ago there was another plant in the United States at Fieldsboro in New Jersey. It was originally built by the Bzura Company to make citric acid by submerged fermentation of blackstrap molasses. It was rumoured that the process involved addition of methanol. The Bzura company went bankrupt, and the plant was taken over by Stepan Chemical Co. but, after about two years of intermittent operations, the plant was again closed down. It was said that the fermentation step worked well; the problems were in clarification and recovery.

The only manufacturer in England is John and E. Sturge Ltd.

Sturge has made citric acid by fermentation for a long time, first by surface fermentation using a pure sugar medium, then using beet molasses. More recently, Sturge announced that it had introduced a new submerged fermentation process in 225,000-litre vessels that could run on either beet or cane molasses, but it has never been stated whether or not this has completely replaced the surface process. Sturge has since been taken over by the German company, C. H. Boehringer.

In Western Europe the biggest manufacturer now and for many years past is Citrique Belge, with a plant at Tirlement. The plant takes its molasses from the neighbouring S.A. Raffinerie Tirlemontoise, the major Belgian sugar company, but is independent of it. The company's process used to be by surface fermentation; nothing has been published to indicate that this has changed. Joh. A. Benckiser GmbH is another fairly large producer. It shut down its old plant at Ladenburg in Germany in 1971 and started up a submerged fermentation plant which is said to use cane molasses.

The French company Melle-Bezons is smaller; it is believed to operate the patented process already described (Usines de Melle, 1958). At one time it was stated that Melle-Bezons was to build a joint plant with Noury and van der Lande, a subsidiary of Akzo, who operate a surface fermentation plant at Devanter in Holland. But no more has been heard of this agreement, and it has now been announced that the Dutch plant will be run down over the next few years, as the process is obsolete and has become uneconomic.

Other European manufacturers include C. H. Boehringer in Germany, Biacor SpA in Italy, in which Sturge had a 40% share, Noury Rumianca in Italy, Ebro in Spain, Lessafre in France, Fursan in Turkey and A. G. Jungbunzlauer Spiritus-und-Chemisch Fabrik at Pernhofen in Austria. The Austrian plant closed down in 1967, but was modernized and re-opened in 1970. Jungbunzlauer Spiritus-und-Chemische Fabrik (1974a, b) has taken out patents on a submerged process using sugar solutions purified by passing through a cation-exchange resin and treated initially with sodium ferrocyanide and with more added later in the fermentation to stop excessive growth of mycelium. Essential trace metals are provided by attack by acid on the stainless steel of the fermenters!

There are two plants in Czechoslovakia, including the original pre-war one, two in Poland, one in Jugoslavia, one in Bulgaria and one being built in Romania. Russian production is only about 15 million kilograms in all, produced in three medium-sized plants and a number of very small ones. Strangely, none of the big Japanese companies which are active in fermentation technology makes citric acid on a large scale, though several of them, as already indicated, have done considerable work on processes starting with normal paraffins. Much of the citric acid in Japan is made by small companies from imported citrate and from sweet potato starch wastes.

K. Uses

A newcomer to the citric acid industry is Liquichimica Biosintesi. This company is at the time of writing erecting a plant at Reggio Calabria in Southern Italy to produce 50 million kilograms a year of sodium citrate from normal paraffins using a process obtained from Japan. This is primarily aimed for use in detergents. This is a relatively new use for citrates, and it remains to be seen whether the high hopes of vast sales will be justified. It arises from the opposition to the use of phosphates as detergent builders because this may lead to eutrophication, that is excessive growth of algae in rivers and lakes. This is no longer such a contentious issue as it was a few years ago, and it applies particularly where sewage is discharged into inland waters, such as the Great Lakes of North America. Elsewhere, the higher cost of citrates will be a disadvantage; even with the recent large increase in the cost of phosphates, it is difficult to believe that citrates can be competitive. There are, however, certain special detergent applications for citrates.

Citric acid is sold both anhydrous and as the monohydrate. It is the most widely used acidulant in the food industry, especially in soft drinks, sweets, jellies and jams. It is used in effervescent powders and has other uses in the pharmaceutical industry. Other industrial uses include boiler cleaning and metal cleaning generally. Sodium citrate is used in processed cheese manufacture and as a blood anticoagulant. Esters of citric acid are used as plasticizers.

III. ITACONIC ACID

$$CH_2$$
$$\|$$
$$C\text{--}COOH$$
$$|$$
$$CH_2\text{--}COOH$$

A. Surface Culture

The production of itaconic acid by moulds was first discovered by Kinoshita (1931) who named the organism *Aspergillus itaconicus*. In 1939, in the course of Raistrick's long study of mould metabolites, it was shown that a strain of *A. terreus* also produced itaconic acid when grown in surface culture on a Czapek-Dox medium (Calam *et al.*, 1939). *Aspergillus itaconicus* and *A. terreus* are unrelated; the former is a green mould belonging to the *A. glaucus* group that will only grow well on media that generate a high osmotic pressure; the latter is a well known brown mould, which, as its name implies, is a common soil organism frequently found on all types of vegetable material. But itaconic acid production is a microbiological curiosity. It is by no means characteristic of *A. terreus*; Calam *et al.* (1939) examined cultures of five other strains of the mould and failed to find itaconic acid. The organism reputed to be Kinoshita's *A. itaconicus*, after about 14 years' storage and transfer in culture collections, produced only traces of itaconic acid (Moyer and Coghill, 1945). One other *Aspergillus* species has been reported to make itaconic acid (Yuill, 1948). This had yellow- to ochraceous-coloured conidia, and was stated not to correspond with any published description. When grown on media containing 20 to 25% (w/v) sucrose, it surprisingly made itaconic acid or kojic acid depending on the temperature. At a higher temperature, itaconic acid was made in about 20% yield. Nothing has been published about it since. The only other organism reported to make itaconic acid is *Ustilago zeae* (Haskins *et al.*, 1955). Small quantities of the acid were identified as a metabolic product when screening for ustilagic acid production from cerelose.

The industrial processes for itaconic acid stem largely from the work at the Northern Regional Research Laboratory of the U.S.

Dept. of Agriculture at Peoria, Illinois which followed up the discovery of its production by *A. terreus*. Although this work was carried out about 30 years ago, it remains as the only thorough and detailed study of the itaconic fermentation to be published, and forms the foundation for most of the subsequent work with *A. terreus*. Moyer and Coghill (1945) screened 30 cultures of *A. terreus* in the Northern Regional Research Laboratory culture collection and found only one, namely N.R.R.L. 265, which gave promising yields. Lockwood and Reeves (1945) screened 308 strains isolated from soil samples. Eleven of these made itaconic acid in yields greater than 45% of the theoretical, which was accepted as one molecule of itaconic acid from one molecule of glucose consumed, a 72% weight yield. From these two screens, two cultures (N.R.R.L. 265 and 1960) were more thoroughly tested, and one (N.R.R.L. 1960) chosen for pilot-scale operations. This strain has been used for much of the subsequent published work on itaconic acid production.

The original work at Peoria used the surface-culture process scaled up in shallow aluminium trays measuring roughly 0.6 m x 0.9 m x 50 mm. The Northern Regional Research Laboratory had as its aim to find uses for maize and maize-derived products, so naturally glucose was the carbohydrate used. The medium developed (Lockwood and Ward, 1945) contained, per litre:

glucose monohydrate	165 g
ammonium nitrate	2.5 g
magnesium sulphate heptahydrate	4.4 g
sodium chloride	0.4 g
zinc sulphate heptahydrate	4.4 mg
corn-steep liquor (50% solids)	4.0 ml
nitric acid	1.6 ml

This gave an initial pH value of about 2.0. Despite the fact that distilled water was used, iron was not added. This and the necessary potassium and phosphate must have been present in the corn-steep liquor. The need for iron at about 5 mg/litre in the medium had been demonstrated by Lockwood and Reeves (1945). They also showed that, whereas at pH 2.4 too much iron (50 mg/litre) lowered the itaconic acid yield to a very low level, at pH 2.0 this did not happen. Lockwood and Ward (1945) quoted a best yield of 39% (w/w) on the glucose consumed or 33.4% on the total glucose (on an anhydrous basis).

B. Submerged Culture

Lockwood and Nelson (1946) converted the surface process to shake-flask operation, still using strain N.R.R.L. 1960. In this case, an initial pH value of 1.8 was optimum, and the concentrations of magnesium, iron and zinc required were lower. This is what one would expect as, in general, while excessive mycelial growth has a slightly adverse effect in surface culture, it can have a very adverse effect in submerged culture, as it will increase the viscosity of the medium, and restrict aeration, which in turn may lead to more mycelial growth. Thus, a greater deficiency in at least one medium component is almost certainly required for satisfactory product yields in submerged culture. The optimum medium developed contained, per litre: 74 g glucose, 2.5 g ammonium nitrate, 0.75 g magnesium sulphate heptahydrate, 0.15 g ferric tartrate and 4 ml corn-steep liquor. Surprisingly zinc was not required (it must have been present in the corn-steep liquor). The pH value was adjusted with nitric acid. The maximum yields achieved were 47.3% on the glucose consumed and 33.3% on the glucose supplied in a nine-day fermentation. This process was then scaled up in 20-litre fermenters by Nelson *et al.* (1952) and into 1,350 and 2,700-litre stainless steel fermenters by Pfeifer *et al.* (1952). On the largest scale, the inoculum was prepared in a 270-litre seed tank. The medium for the inoculum and for the production-scale fermenters was identical; its composition, per litre, was:

glucose monohydrate	66 g
ammonium sulphate	2.7 g
magnesium sulphate	0.8 g
corn-steep liquor	1.8 g

All other required nutrients were presumably supplied in the corn-steep liquor. The inoculum tank was seeded with spores of *A. terreus* N.R.R.L. 1960 which had been grown on malt agar, transferred to 750 ml medium in a small flask, and aerated for 48 hours at 34°C. After a further 48 hours in the inoculum tank, the main fermenters were seeded. Sterilization was either batchwise or continuous, and the fermentations were run best at 35°C under a pressure of 0.7–1.0 Kg per cm^2 (10–15 p.s.i.) with a power input to the agitator varying from 0.04–0.40 watts per litre (2.0 horse power per 1,000 U.S.

gallons). Soya-bean oil or an ethanol solution of octadecanol was used to control foaming. The best quoted weight yield from the anhydrous glucose supplied was as much as 64%, but increasing the glucose concentration to 10% lowered the yield to 45%. The time taken in the fermentation is not clearly stated, but one graph shows a 78-hour run starting with about 6.2% glucose. Elsewhere, a 72-hour total cycle time is mentioned including cleaning, sterilizing by steaming, filling and emptying.

In several runs, the air supply was interrupted for periods of from 15 to 60 minutes. This always completely stopped itaconic acid production, but this could be restarted, at a lower rate, after adding more nutrients and allowing the mould to proliferate. The fermentation could be prolonged by removing two-thirds of the medium when it had fermented completely and replacing with fresh medium.

In the initial stages of this work, the medium was sterilized at about pH 3.5 and then lowered to pH 1.9–2.1 with sulphuric acid before inoculation. Later it was found that the fermentation could be run quite satisfactorily if started at a pH value of 5.0, and that the rate of acid production was increased. The reason why this could be done under these conditions is nowhere stated. Possibly it was connected with the medium composition, though the trace-metal content of it is not given. Several attempts were made to replace the glucose used with less pure and cheaper sources of carbohydrate, including cane and beet molasses and hydrol, the crude glucose solution which has the same relationship to glucose as molasses does to sucrose. Significant amounts of itaconic acid were not made.

On the other hand, Pfizer had patented in 1948 a process for the production of itaconic acid by submerged culture from either sucrose or molasses using a strain of *A. terreus*. The weight yield quoted was 27.7% from pure sucrose in two weeks, and 24.5% from molasses in seven days. Few details are given, but an improved process was described later in very much greater detail (Nubel and Ratajak, 1964). This patent claims production of itaconic acid by submerged aerobic fermentation from a solution in which 10 to 30% of the carbohydrate is provided by beet molasses using either *A. terreus* or *A. itaconicus*. The production stage of the fermentation is carried out at 39–42°C, and the acid made is partially neutralized with lime. The beet molasses is mixed either with glucose, fructose or high-test cane molasses, giving final sugar concentrations of 100, 150 or 180 g/l. In one example, cane molasses is used to prepare the medium in the final fermenter, and beet molasses is used for the inoculum. Most of the nutrients required are present in the molasses, but zinc sulphate is added at 1 g/l, magnesium sulphate heptahydrate at 3 g/l, and copper sulphate pentahydrate at 0.01 g/l. The fermentation runs for three days on a 15% sugar solution, giving a quoted yield of about 47% on a weight basis.

Various other patents have been taken out covering modifications to the basic Peoria submerged process for itaconic acid production. One (Royal Norwegian Council for Scientific and Industrial Research, 1958) claims that addition of solid adsorbents, such as anhydrous calcium sulphate, kaolin, bleaching earth, Fuller's earth or anhydrous silicic acid, to a medium containing 10% pure sucrose improves both the yield and the rate of fermentation using *A. terreus* N.R.R.L. 1960. Average yields quoted are 55% by weight at a rate of 0.39–0.43 g/l/h in the presence of the adsorbents, compared with yields of 10–40% at rates up to 0.3 g/l/h in their absence. No suggestion is made of what harmful substances are being adsorbed.

Miles Laboratories (1963) patented a process which is obviously derived from their experience in citric-acid manufacture in that, in the quoted examples, they used decationized high-test molasses in 10% to 25% solution as the carbohydrate source. They claim that the essence of the process is that the medium shall contain (per litre) 350–3,500 mg of an alkaline earth metal and from 0.5 to 200 mg of copper ions and a similar concentration of zinc. This is a development of the medium of Lockwood and Ward (1945) already quoted, without the need to start at pH 2.0, with small concentrations of copper included, and with the discovery that magnesium can largely be replaced by calcium. Using *A. terreus* N.R.R.L. 1960 at 32–35°C, weight yields up to a maximum of 58.2% were obtained from a medium containing 126.4 g sugar per litre in about 7.5 days. Based on the sugar consumed, a 98.9% theoretical yield of itaconic acid is quoted. This is based on the assumption that one molecule of hexose gives one molecule of itaconic acid. It is impossibly high, because of the necessity to make biomass to effect the conversion, unless some possible recycling of the sixth carbon atom is assumed.

Batti (1964), also of Miles Laboratories, claimed that the pH value at which the fermentation is held is highly important for making itaconic acid uncontaminated with other products such as itatartaric and succinic acids. Above pH 5.0, acidic byproducts are not made, but below pH 3.0, with the strain of *A. terreus* used, about 15% of the acidity was in the form of unwanted products. Batti claimed that the pH value had to be held between 3.0 and 5.0 by addition of alkali for greater purity of the itaconic acid formed.

The fact that *A. terreus* mutants make not only itaconic acid, but also itatartaric acid, was shown by Stodola *et al.* (1945):

$$
\begin{array}{ccc}
CH_2 & & CH_2OH \\
\| & & | \\
C-COOH & \longrightarrow & HO-C-COOH \\
| & & | \\
CH_2-COOH & & CH_2-COOH
\end{array}
$$

From strain N.R.R.L. 265, an ultraviolet-induced mutant was isolated which when grown in a medium containing 220 g glucose per litre in surface culture gave about 5.8% of the total acids in the form of an equilibrium mixture of itatartaric acid and its lactone. Arpai (1959) showed that a cell-free extract of such a mutant could oxidize added itaconic acid to itatartaric acid. The enzyme itaconic oxidase has a relatively sharp peak of activity at pH 4 and was much less active at pH 2 or pH 7. It will be noted that Arpai's findings and that of Batti are apparently contradictory. They can possibly be explained by strain differences.

In another patent, von Fries (1966) claims the production of itaconic acid by fermentation with *A. terreus* using a crude carbo-hydrate-containing medium either pretreated by passing through a cation-exchange resin or by treatment with 5–50 mg/l of an alkali-metal ferrocyanide. The excess of ferrocyanide is subsequently decreased to less than 5 mg/l by addition of zinc sulphate. He claims a 60% conversion of technically hydrolysed starch (160 g/l) in six days and 61.6 to 66.2% yields from beet molasses.

Moyer (1954) showed that, as with citric acid, increased yields of itaconic acid could be obtained by addition of 1–4% of the lower alcohols to a medium containing glucose or sucrose with corn-steep liquor and salts. This procedure gives increased yields without the deliberate elimination of trace elements, and the process does not need to run at exceptionally low pH values; it was in fact run at pH 4.0–4.1 initially.

Galanti in 1959 showed that various phenols had an effect on production of itaconic acid, again using *A. terreus* N.R.R.L. 1960.

Addition of *m*- and *o*-nitrophenols and more definitely *p*-nitrophenol and 2,4,6-trinitrophenol at concentrations of 10^{-5} and 10^{-6} M increased itaconic acid production. Other strains that had lost the ability to make itaconic acid made it after addition of these phenols, which act as uncoupling agents in oxidative phosphorylation. Iwata Chemical Industry Co. (1973) claims much the same effect by addition of various anionic, cationic or amphoteric detergents such as propylene glycol monostearate to a fermentation using *A. terreus* N.R.R.L. 1960 and high-test molasses.

It will be seen, therefore, that many of the techniques used to improve citric acid production by *A. niger* have also been used at one time or another to improve itaconic acid production by *A. terreus*. These depend essentially either on removal of trace metals from solution, or on mitigating their effects. The latter can be effected by low pH values, by the presence of much larger quantites of dibasic metals, such as magnesium or calcium, by addition of lower alcohols, or by adsorption on to inert materials.

In opposition to this, it has been claimed that, with *A. itaconicus*, untreated molasses can be used as the substrate without any restrictions on the presence of heavy metals, range of pH value, or sugar concentration (Kinoshita and Tanaka, 1961). It is further claimed that itaconic acid could be made in favourable yicld by submerged fermentation using any strain of *A. itaconicus* isolated from natural sources. In that Thom and Raper (1945) considered that *A. itaconicus* was identical morphologically with *A. varians*, and that it differs only in its ability to produce itaconic acid, this statement is meaningless; furthermore, as already stated, the culture tested by Moyer and Coghill (1945) had apparently lost almost all of its itaconic acid-producing ability on storage. Kinoshita and Tanaka (1961) quote yields of up to 65.5% from a medium containing cane molasses with the equivalent of 150 g glucose per litre in seven days in shaken flasks.

But Kobayashi, who with Nakamura (1966) has made a long and thorough study of the itaconic acid fermentation, used *A. terreus*. He states that *A. itaconicus* is notable for its ability to decompose itaconic acid, and that *A. terreus* K26 is superior in every way. Kobayashi developed a semicontinuous process in which the mould is grown on a carbohydrate, corn-steep liquor and salts medium to which is added 35 to 50 g of itaconic acid per litre, which restricts mould growth. After a period, the mycelium is removed by filtration, part of the itaconic acid formed is 'extracted' (apparently by cooling, crystallizing and filtering), and the mother liquor and mycelium returned to the fermenter to which additional sugar and salts are added. By this procedure, running at pH 2.1–2.5, he claims that he could operate for a month on end with good productivity. Later he ran the process on a medium containing glucose obtained from hydrolysis of wood, and in Waldorf-type fermenters. He claims a four-fold increase in productivity over a batch process (Kobayashi, 1967; Kobayashi and Nakamura, 1966).

A more conventional continuous fermentation process is mentioned by Courtaulds (1967). In this, a solution containing 60 g glucose per litre with added ammonium sulphate, magnesium sulphate and corn-steep liquor was inoculated with spores of *A. terreus* and continuously fermented in two vessels in series, feeding in a solution containing 40 g glucose per litre and salts into the first vessel and transferring continuously to the second vessel. This patent primarily covers the equipment used, and few details are given of the process.

C. Biosynthesis

The best known work on biosynthesis of itaconate by *A. terreus* was carried out by Bentley and Thiessen (1957). They fed labelled glucose, succinate and acetate to the mould, and showed that the labelled carbon was incorporated into itaconate as would be expected if the pathway was via decarboxylation of cis-aconitate made by the tricarboxylic acid cycle. They found that there was marked sensitivity to fluoride and to iodoacetate, which was evidence for operation of the E.M.P. pathway, and showed the presence of α-oxoglutarate, succinate and malate, further evidence for operation of the tricarboxylic cycle, but very little citrate was detected and no pyruvate or oxaloacetate. Cell-free extracts of cis-aconitate decarboxylase were prepared from *A. terreus* mycelium which rapidly converted cis-aconitate to itaconate, but formation of carbon dioxide from isocitrate was about half that from cis-aconitate and two to three times that from citrate. This was regarded as evidence for the presence of aconitate hydratase. Small amounts of itaconate were formed from citric acid at a concentration of 20–30 g/l, but 100–150 g citric acid per litre were inhibitory to mould growth. It was suggested that permeability considerations were a factor.

Addition of fluoroacetate or fluorocitrate, both inhibitors of aconitate hydratase, caused no increase in citrate production. Addition of ferrocyanide, which it is said inhibits isocitrate dehydrogenase, increased itaconic acid synthesis. These results differed from those of Larsen and Eimhjellen (1955) who, with the same strain of *A. terreus* namely N.R.R.L. 1960, showed that fluoroacetate stimulated formation of itaconic acid both from glucose and from citric acid, raising the yield from the latter from 18 to 63%, where they failed to demonstrate conversion of cis-aconitate to itaconate. They could not find conversion of acetate, pyruvate or oxaloacetate to itaconate; in fact, these acids inhibited itaconic acid production. Jenssen *et al.* (1956) showed that itaconic acid was formed from citric and cis-aconitic acids under acid conditions, but not under neutral conditions.

Corzo and Tatum (1953), however, confirmed the incorporation of labelled acetate into the itaconate molecule, and showed the same with labelled citrate. Naim and Mansour (1966), working in surface culture with a different strain of *A. terreus*, obtained a 54% (w/w) yield from sucrose and 35.5% from citric acid.

Opposition to the view that itaconate is formed by a one-step side pathway from the tricarboxylic cycle was expressed by Lal and Bhargava (1962) who studied the effect of fluoride on *A. terreus* and its reversal by pyruvate. They showed that fluoride inhibits a step in the conversion of pyruvate more strongly than it inhibits enolase. They argued that all known enzymes on the suggested pathway from pyruvate to itaconate are less sensitive to fluoride than is enolase, and that some other pathway must operate. Shimi and Nour El Dein (1962) assumed that the route involves a condensation of acetate and succinate to give 1,2,3-propanetricarboxylic acid, which is dehydrogenated to aconitate and decarboxylated to itaconate. Nowakowska-Waszczuk (1973) supports these views on the grounds that she found that mitochondria of *A. terreus* N.R.R.L. 1960 and mutants therefrom could not utilize any tricarboxylic acid-cycle intermediates, and that with mycelium malate, fumarate and pyruvate were converted to itaconate, but citrate and succinate were not.

In that these findings, though confirmed by some, have been

disproved by others, there seems no reason to postulate an entirely novel biosynthetic pathway in place of the one shown to operate in the majority of cells, including many micro-organisms, and in particular aspergilli closely related in many ways to *A. terreus*. Surprisingly, what nobody has done is to attempt directly to demonstrate the presence or absence of the condensing enzyme in *A. terreus* mycelium making itaconic acid from carbohydrate, in the same way that Ramakrishnan and Martin (1954a) did with *A. niger* making citric acid. Unless this is done, and the enzyme shown to be definitely absent, there seems little reason not to accept the conclusions of Bentley and Thiessen (1957).

D. Manufacturers

There are relatively few manufacturers of itaconic acid. Pfizer is the only manufacturer in the U.S.A., and it also has a plant at Sandwich in England. Melle-Bezons has the only other plant in Western Europe. Itaconic acid is also made in Russia and by Iwata in Japan.

E. Uses

Over three-quarters of the itaconic acid made is used in styrene butadiene copolymers and for lattices and emulsions in general. The most important use for these is in carpet backing, with a significant proportion used for paper coating. Most of the remainder is used in acrylonitrile copolymers for synthetic fibre manufacture. A small amount of itaconic acid is incorporated and this gives improved properties including favourable dyeing characteristics. Other uses are all in the synthetic resin and related industries (Billington, 1969).

IV. FUMARIC ACID

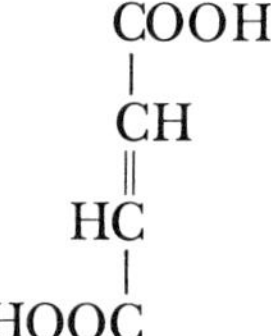

Most of the other acids in the tricarboxylic acid cycle have been made either on a laboratory or much larger scale by fermentation processes. Fumaric acid was at one time made on a commercial scale by Pfizer by fermentation with a strain of *Rhizopus*, but production was superseded by the economically more attractive synthesis by oxidation of benzene.

A. Production from Carbohydrates

Takahashi and his coworkers were the first to show that a number of *Rhizopus* species make fumaric acid (Takahashi and Sakaguchi, 1927), and the use of *R. nigricans* in submerged culture was patented later by Merck & Co. Inc. and by Chas. Pfizer & Co. Inc. (Waksman, 1943; Kane *et al.*, 1943). In these patents, the carbohydrate used was sucrose, glucose or starch and the fermentation was run in the presence of a neutralizing agent and various salts at 28–35°C. It is stated that, if the insolubility of calcium fumarate proves troublesome, alkalis giving more soluble salts can be used; magnesium carbonate is suggested. The development of this process highlighted for the first time many of the problems associated with submerged aerated mould fermentations; it was run at a neutral pH value and thus very liable to contamination. It was very liable also to foaming; the addition of components, in this case neutralizing agents, was necessary during the run, and the medium gave a very thick growth like porridge, and was difficult to aerate. Solution of these problems for the fumaric acid fermentation must have contributed appreciably to the ability shown both by Merck and Pfizer to develop the submerged penicillin process.

The fumaric acid fermentation was described by Foster and Waksman (1939a, b). Considerably later, this process was studied in detail at Peoria, Illinois, U.S.A. (Rhodes *et al.*, 1959, 1962) using two strains of *R. arrhizus*, N.R.R.L. 1526 and N.R.R.L. 2582. The latter strain had the advantage, not shown by many strains of *Rhizopus* used for fumaric acid production, of being able to synthesize invertase, and thus to use sucrose without prior hydrolysis. The fermentation was first studied in shaken flasks. The medium used for the production step contained varied amounts of glucose and calcium carbonate, together with (per litre): 1.0 g urea, 0.3 g potassium dihydrogen phosphate, 0.4 g magnesium sulphate heptahydrate, 0.044 g zinc sulphate heptahydrate, 0.01 g ferric tartrate, 0.5 ml corn-steep liquor and 0.15 g methanol. Fermentations were run for 4 to 7 days at 33–35°C depending on the initial glucose concentration; 160 g glucose per litre was the most concentrated medium that could be fermented in 7 days or less. The ratio of nitrogen to carbon provided was found to be an important factor on this scale, as the concentration of nitrogen was the growth-limiting factor, and the concentration of zinc (about 10 mg/l) was also critical. The fermentation was found to go equally efficiently with strain N.R.R.L. 2582 on high-test molasses. Addition of methanol caused an increase in yield of between 10 and 25%, depending on the strain and substrate used, being particularly effective with the molasses-based medium.

Methanol was omitted from the medium when the process was scaled up into 20-litre fermenters and commercial corn sugar used as the substrate. Yields were over 65% (w/w) from the sugar consumed at concentrations from 100 to 160 g/l. Neutralization of the acid formed was essential, and vigorous agitation was required in order to prevent the precipitated calcium fumarate giving too thick a culture The use of soda or potash as neutralizing agents was unsatisfactory, as production ceased once the concentration of soluble fumarates exceeded 40 g/l.

B. Biosynthesis

The maximum theoretical yield from glucose, on the assumption that one molecule of fumaric acid is formed from one molecule of glucose, is 63.3%. In the work at Peoria, yields up to 71% are recorded and it is stated that fumaric acid did not exceed about 80% of the total acid formed. With colleagues, I studied the fumaric acid fermentation in the period 1946 to 1951, and found the presence also of L-malic acid and small amounts of α-oxoglutaric acid, giving in total a 111.5% stoicheiometric yield from glucose. Even without taking into account the fact that an appreciable amount of glucose must be used to form biomass, the figures are absurd. The assumption therefore must be wrong. These results could, however, be explained if the glyoxylate bypass operated, when the reaction could be expressed as:

1.5 glucose → 3 pyruvate
3 pyruvate → 1 oxaloacetate + 2 acetate
1 oxaloacetate + 2 acetate → 2 malate (or fumarate).

But Romano *et al.* (1967) have shown that synthesis of isocitrate lyase is strongly repressed in *R. nigricans* with the high concentrations of glucose employed, thus rendering the operation of the glyoxylate route very doubtful. They present alternative evidence in favour of carbon dioxide fixation, without operation of the tricarboxylic-acid cycle. Thus fumaric acid made from radiolabelled glucose, but in the presence of unlabelled carbon dioxide, contained about 20% less label than if it were made completely from glucose. Overman and Romano (1969) showed that the main carbon dioxide-fixing system is an acetyl-CoA-dependent pyruvate carboxylase, which is made in high yield during rapid glucose utilization and precedes accumulation of fumarate. Essentially this confirmed much earlier work (Foster and Davis, 1948) in which a strain of *R. nigricans* was found that formed fumaric acid anaerobically from glucose in approximately 20% yield, and fumarate was shown to be formed by carbon dioxide fixation with pyruvate. It was also shown (Foster *et al.*, 1949) that fumaric acid could be made by *R. nigricans* from ethanol in a maximum yield of 72%.

C. Production from Hydrocarbons

Fumaric acid can also be made from *n*-paraffins. Yamada *et al.* (1970) isolated a species of *Candida* from soil, which they called *C. hydrocarbofumarica*, and which made fumaric acid in good yield from a C_{12}–C_{14} *n*-paraffin mixture. In subsequent studies (Furukawa *et al.*, 1970a) the yield was improved to 84% by weight from an 80 g mixture of paraffins per litre of an aqueous medium after seven days at 30°C. In many respects, the findings were the same as with the *Rhizopus* fermentation on carbohydrate in that it was necessary to operate with relatively low concentrations of product, that calcium carbonate was the best neutralizing agent, that potassium carbonate and sodium bicarbonate were unsatisfactory for this purpose and that yields were lower in the presence of more than trace quantities of iron, zinc and manganese.

D. Market

The market in U.S.A. for fumaric acid is believed to be in the region of 40–45 million kg and the European market 15 million kg, with uses in the plastics and food industries. The relative insolubility in water (1 part in 200 parts at 20°C) puts it at a disadvantage compared with other acids for many purposes. At the time of writing, it looks unlikely that there will be a future for fumaric acid production by fermentation, but raw material availability and costs are changing so rapidly that it is dangerous to be dogmatic. It is possible that, in the future, it may once more become an economic proposition to make fumaric acid by fermentation from an agricultural source rather than chemically from a fossil source, or it may be made from waste products primarily as a means of disposing of the wastes. In the past, for example, a method of making it from sulphite liquor has been studied, and 40% yields by weight from the total sugar obtained, with removal of 75% of the sugar content of the liquor (Romano, 1958).

V. L-MALIC ACID

Regulations promulgated by the European Economic Community, which allowed use only of L-malic acid, now permit the use of DL-malic acid as an acidulant in foods. In the U.K. and the U.S.A., use of synthetic DL-malic acid has been allowed for a longer time. The British market for the DL-acid is about 1–1.5 million Kg a year. In present circumstances, manufacture of L-malic acid by fermentation is not competitive with production by catalytic hydrolysis of maleic acid at high temperature and pressure. Even if L-malic is required, this could readily be made from fumaric acid enzymically, with or without isolation of the enzyme from a micro-organism producing it. What has been called a 'transcrystallization' process has been described using *Leuconostoc brevis*, which converts fumarate to L-malate in 99% yield in 24 hours. The greater solubility of calcium fumarate than calcium malate is said to force the reaction to completion. Similarly *Candida utilis, Pichia membranaefaciens* (Furukawa *et al.*, 1970b), *Pullularia pullulans* (Chugai Pharmaceuticals Co., 1968) and a species of *Brevibacterium* (Ozawa and Watanabe, 1968) have all been shown to convert fumarate to L-malate. Furukawa *et al.* (1970b) have shown that, by a double fermentation, first with *Candida hydrocarbofumarica* and then with *C. utilis*, a 72% weight yield of L-malic acid can be obtained in solution from *n*-paraffins. If fumaric acid is made using *R. oryzae*, the broth treated by addition of sodium arsenite to give a 0.01 M solution and the pH value raised to about 8.6, the fumaric acid is similarly almost completely converted to L-malic acid.

It has recently been shown (Tachibana and Murakami, 1973) that *Schizophyllum commune* can produce L-malate in high yield from ethanol. The sum of the organic acids produced, calculated as malate, and the carbon in the mycelium amount to over 100%. Little acid was formed in the absence of calcium carbonate, and here also it was concluded that L-malate is produced by a process involving fixation of carbon dioxide.

VI. OTHER TRICARBOXYLIC ACID-CYCLE ACIDS

Other tricarboxylic acid-cycle acids could readily be made by fermentation if required, but no commercial use has yet been found for α-oxoglutaric acid. Succinic acid can readily be made synthetically. When screening micro-organisms for glutamic acid production from *n*-paraffins, a number of strains of *Candida lipolytica* were found that made α-oxoglutaric acid, the best in 71% weight yield from 80 g of *n*-paraffins per litre with the usual salts and thiamin added. *Corynebacterium hydrocarboclastus* also makes α-oxoglutarate in good yields from *n*-paraffins, though in poor yield from glucose (Imada and Yamada, 1971). This is explained by the fact that it is necessary to control the concentration of thiamin, which is a cofactor of enzymes which catalyse not only the decarboxylation of α-oxoglutarate, but also that of pyruvate to acetyl-CoA. The latter reaction is not part of the metabolic pathway from paraffins but is of the pathway from glucose.

Similarly other strains of *Candida* have been shown to accumulate succinic acid when grown on *n*-paraffin (Sato *et al.*, 1972a, b). The best yield, 67%, was given by a strain of *C. brumptii*. Again yields were much better on paraffins than on glucose. Considerable amounts of α-oxoglutarate and L-malate were also produced, which is evidence in favour of a biosynthetic route via the TCA cycle, but this has not been proved and it is possible that succinate is made directly by diterminal oxidation of the paraffin.

α-Oxoglutarate can also be easily made from carbohydrates. A strain of *Pseudomonas fluorescens* (N.R.R.L.-B6) made this compound in over 50% theoretical yield from glucose in a chemically defined medium (Koepsell *et al.*, 1952). The route was shown to be via gluconate, 2-oxogluconate and pyruvate. The same process with the same micro-organism, which they now called *P. reptilivora*, was used to make α-oxoglutarate in shaken flasks from a solution containing, per litre, 90 g glucose, 2.4 g ammonium sulphate, 1.1 g dipotassium hydrogen phosphate, 0.5 g magnesium sulphate heptahydrate and 37.5 g calcium carbonate (Erickson *et al.*, 1965). Calcium α-oxoglutarate was recovered in 60.70% yield by acidification of the broth to pH 1, removal of cells, extraction into cyclohexanone, steam distillation of the extract, decolourization of the aqueous solution and precipitation of the calcium salt.

VII. EPOXYSUCCINIC ACID

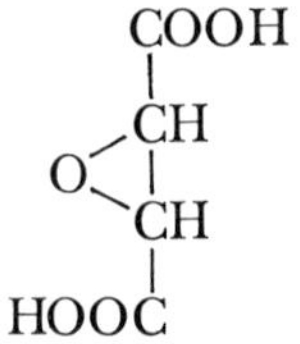

$$
\begin{array}{c}
\text{COOH} \\
|\\
\text{CH} \\
\text{O} \diagdown \quad | \\
\text{CH} \\
|\\
\text{HOOC}
\end{array}
$$

This acid (also called ethylene oxide dicarboxylic acid) has from time to time been studied for potential industrial use, including as an intermediate in the manufacture of tartaric acid. Its production by moulds was discovered by Sakaguchi *et al.* (1938) from organisms later identified by Raper and Thom (1949) as species of *Paecilo-myces*. It was later isolated as a metabolic product of *Aspergillus fumigatus* (Birkinshaw *et al.*, 1945; Martin and Foster, 1955).

In 1955, T. H. Anderson, a colleague of the writer, obtained a 39% weight yield from sucrose using a mutant strain of *A. fumigatus* in submerged culture. Wilkoff and Martin (1963), from work with radiolabelled compounds, concluded that epoxysuccinic acid is made by direct oxidation of fumaric acid. Epoxysuccinic acid can be further converted to mesotartaric acid by various micro-organisms, including *Pseudomonas putida*, from which an enzyme that catalyses this reaction has been obtained (Allen and Jakoby, 1969). Fumarate hydratase also catalyses this hydration (Albright and Schroepfer, 1970). Unfortunately mesotartaric acid, unlike the naturally occurring L(+)-tartaric acid, is of no commercial interest. It is significant that epoxysuccinic acid is another compound the yield of which is increased by addition of lower alcohols (Moyer, 1954).

VIII. KOJIC ACID

$$\text{HO}-\underset{\underset{\displaystyle\text{O}}{\|}}{\overset{\overset{\displaystyle\text{O}}{\|}}{\text{C}}}\cdots$$

One further acid should be mentioned here, even if at first sight it does not appear to fit happily in company with acids belonging to, or closely related to, the T.C.A. cycle. This is kojic acid, which has been made on a relatively small commercial scale for a number of years, and can be made in good yield by a number of aspergilli, particularly those of the *A. flavus-oryzae* group.

Work with the glucose radiolabelled in the C-1 position and again in the C-2 and C-3 positions (Arnstein and Bentley, 1953) indicated that 70–90% of the kojic acid made with *A. flavus-oryzae* must have come directly from glucose:

It appeared that there was also a subsidiary route involving breakdown of glucose to three-carbon compounds and synthesis of kojic
acid. Kojic acid is also readily made from ethanol, glycerol and
arabinose, but its biosynthetic route from these compounds is
unknown.

But from hexoses the techniques needed to make kojic acid in
good yield from *A. flavus* are very similar to those needed to make
citric or itaconic acids from other aspergilli, and there are obvious
relationships between the processes. Yuill's (1948) organism that
made either kojic or itaconic acids has already been mentioned, and
in my laboratory a strain of *A. flavus* was found that made kojic acid
in a medium containing pure sucrose and salts in surface culture,
although from molasses in submerged culture it made citric acid. This
finding may partially be connected with the fact that kojic acid
production has been shown to be markedly affected by agitation
(Camposano *et al.*, 1961). In shaken flasks, kojic acid was made in
30–40% yield from glucose, but in sparged aerated or rotary
fermenters only in 15% yield. Addition of fluoroacetamide largely
inhibits kojic acid production and the addition of lower alcohols
enhances it (Nonomura *et al.*, 1960; Kawade, 1972). The impression
is given that blocking the T.C.A. cycle in some way may activate the
direct conversion of hexose to kojic acid. The kojic acid fermentation has been most recently studied by Kitada *et al.* (1967, 1968)
who obtained a 54.2% yield with *A. oryzae* on a medium containing
100 g glucose per litre and salts in shake flasks, and who report that
yields up by one and a half times to twice as great were obtained in a
30-litre jar fermenter.

The chief problem in kojic acid manufacture is not the fermentation but the recovery step. Kojic acid gives a deep red colour with
as little as 0.1 p.p.m. of ferric iron and, therefore, to make product
of good colour, exacting purification techniques are called for.

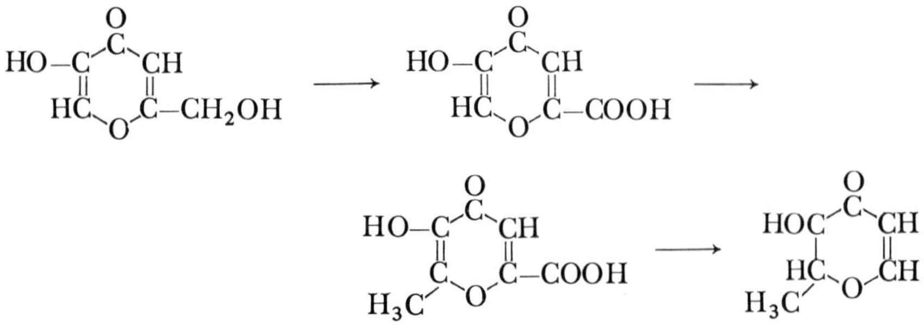

Because of the number of reactive groups on the molecule, kojic acid has a number of potential uses. One of these is exemplified in a paper by Ichimoto *et al.* (1965) showing the production of the flavouring agent maltol from kojic acid via comenic acid and 6-methylcomenic acid (see previous page).

IX. LACTIC ACID

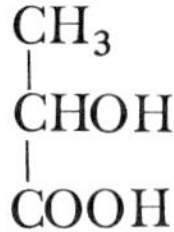

Lactic acid, which is sold as the optically inactive DL mixture of isomers, has for a long time been made by fermentation. It can also however be made at a competitive cost by purely chemical processes. Over the last ten to twenty years, manufacture by fermentation has been under heavy pressure from the chemical processes. There was at one time a suggestion that only the L-acid might be allowed for food use, which would have given a big advantage to manufacture by fermentation, but this is no longer likely to occur. It is probably because of the highly competitive position that details of the actual fermentation process or processes are not readily available.

$$C_6H_{12}O_6 \longrightarrow 2C_3H_6O_3$$

Lactic acid fermentation is anaerobic, but the organisms used are facultative anaerobes and strict exclusion of air is unnecessary. The fermentation is normally run at 45–50°C and at a slightly acid pH value, which considerably minimizes the normal problems of sterility. For these reasons, the fermentation is technically simple to operate compared with some of the processes already discussed. The organisms used are all species of *Lactobacillus*, most commonly strains of *L. delbruckii*. The organism will be chosen depending on the carbohydrate to be fermented and the product to be formed. Normally this, as already stated, will be the DL-acid, but certain micro-organisms make L-lactic acid. Starting materials may be glucose (starch hydrolysates), sucrose or lactose (whey) in various degrees of purity.

Details have been given (Inskeep *et al.*, 1952) of the process once used by American Maize Products for manufacture from corn sugar with *L. delbruckii*. The fermenters were of 25,000 or 113,500 litres capacity, and the acid was neutralized as it was formed with calcium carbonate, holding the pH value at 5.8–6.0. The temperature was held at 49°C, and the liquor was not sterilized. The medium contained, per litre, 150 g glucose, supplemented with 3.75 g of malt sprouts and 2.5 g diammonium hydrogen phosphate. Conversion took four to six days.

Another published process (Cordon *et al.*, 1950) used potatoes as the starting material and fungal amylase to convert the starch to sugar. The hydrolysis was very inefficient, giving only about 75% yield, and better enzymes are now available which would give far better conversion. But the fermentation step, using *L. delbruckii* and other lactobacilli in the presence of excess calcium carbonate, was very efficient giving a maximum yield of lactic acid of 91.2% based on original carbohydrate.

Another potential raw material which has been examined is sulphite-waste liquor (Leonard *et al.*, 1948). For the best results, the liquor was stripped with steam to remove sulphur dioxide and treated with alkali to precipitate lignins before fermentation. Malt sprouts were added to provide nitrogenous nutrients and accessory factors. The micro-organism used, *L. pentosus*, could ferment pentoses, which make up an appreciable proportion of the carbohydrate in sulphite liquor, though at a much slower rate than glucose. It operates at an optimum temperature of 30°C. After 40 to 48 hours, a maximum yield of 2.1% (w/v) was obtained from sulphite-waste liquor.

One of the first raw materials to be used for lactic-acid production was whey. It is necessary to use *L. bulgaricus*, which will grow on lactose, and not *L. delbruckii*, and to operate at about 43°C. Again the pH value is controlled by addition of calcium carbonate or a lime slurry. Production from whey has been carried out on a pilot-plant scale by continuous fermentation (Whittier and Rogers, 1931).

The exact pH value at which the fermentation is controlled is important from the point of view of fermentation rate. With enzymically-hydrolysed starch as substrate and *L. delbruckii* N.R.R.L. B-445, it was shown that with calcium carbonate as a neutralizing agent it was necessary to add accessory growth factors but, by controlling the pH value at 6.0, it was possible to get fast fermentations without adding accessory factors (Kempe *et al.*, 1950). With dextrose, the same organism gave a maximum production rate of 1.2–1.4 g/l/hour (Finn *et al.*, 1950).

Childs and Welsby (1966) of Bowmans Chemicals Ltd. describe a continuous process for lactic-acid manufacture using sucrose, glucose or a starch hydrolysate prepared by acid or enzymic hydrolysis. The fermentation is operated at about 50°C and the pH value controlled by ground limestone or precipitated chalk. Details are given of a process based on a hydrolysate of nine parts of maize starch and one of barley meal in 0.5 N sulphuric acid to give a medium with 180 g glucose per litre, which is neutralized with calcium carbonate, filtered and diluted to give 97 g total sugar per litre of medium. After addition of 0.5 g of ammonium sulphate and 0.004 g of sodium sulphite per litre, this is fermented with *L. delbruckii*. A mixture of strains is needed for most rapid conversion, the maximum production rate being 3.7 g/l/hour. With sodium carbonate as a neutralizing agent, the optimum pH value was 5.5 to 5.8, but with ground limestone 4.9 to 5.2. This was operated for as long as 64 days in two-litre glass vessels with a flow rate of 1.5 volumes per day. Despite the publication of this process, there has never been any other indication by Bowmans (later taken over by Croda) that a continuous process is used on a manufacturing scale.

The process used by Bowmans was described in several publications when their new plant was put into operation at Widnes in Great Britain in 1959 (e.g. Machell, 1959). The plant had a capacity of 2.4 million kg per year and operated with an overall theoretical yield of 80–85%. It used starch as the raw material, which was hydrolysed with sulphuric acid and neutralized with limestone before fermentation. One would guess that hydrolysis would probably now be carried out by enzymes. Fermentation was in wooden vats for five days at 50°C, but one 70,000-litre stainless-steel fermenter was installed. The fermented broth was boiled with lime to make calcium lactate, which was decomposed with sulphuric acid. The lactic acid solution was passed through columns of activated carbon and cation-exchange resins. Further purification was by extraction into isopropyl ether and re-extraction into water.

The problems in the manufacture of lactic acid are mostly in the recovery and not the fermentation step. The lactic acids can only be crystallized with great difficulty and in low yield, and in the purest forms are normally colourless syrups which readily absorb water.

They are usually sold as 65% aqueous solutions. Lactic acid is made in various grades with different degrees of purity. For the purer grades, it is necessary either to use a pure substrate for the fermentation step or to use an involved recovery procedure. Recovery methods include precipitation as the calcium salt and treatment of this with sulphuric acid, precipitation as the more easily purified zinc salt, and purification by solvent extraction or by distillation of the methyl ester.

The alternative purely chemical process is based on the conversion of lactonitrile, which is obtained as a by-product in the manufacture

$$CH_3 . CHOH . CN \longrightarrow CH_3CHOH . COOH$$

of acrylonitrile, or can be made from acetaldehyde and hydrogen cyanide. In another purely chemical process (Montgomery and Ronca, 1953), molasses is heated with lime at temperatures of 65–93°C and gives about half of the theoretical yield of lactic acid. This is only about half the yield obtained by fermentation, but the time is very much shorter. A process developed by Rhône-Poulenc for the manufacture of lactic and oxalic acids involves direct oxidation of propylene with nitrogen dioxide and oxygen to give γ-nitratopropionic acid which is then hydrolysed by water at a temperature of 100°C.

The market for lactic acid is relatively small and growing very slowly. The acid has uses in the food industry, including the manufacture of rye bread. Various esters are used as emulsifiers, and calcium lactate is used pharmaceutically as a source of calcium. There are also a number of miscellaneous uses, but the total capacity in Western Europe production is believed to be only about 10,000 million kg a year, about half of it at the Schiedam plant of the Dutch company Verenigde HVA. The only British company is Croda which is believed to make about two million kg a year. These companies use a fermentation process, but Rhône-Poulenc makes lactic and oxalic acids in France by the purely chemical process already described. This company has built a much larger plant, but so far only oxalic acid is being produced. Manufacture in the U.S.A. is on about the same scale as in Europe. The Monsanto Co. make it by a purely chemical process from lactonitrile; the Clinton Corn Processing Co. is believed to use a fermentation process.

X. GLUCONIC ACID

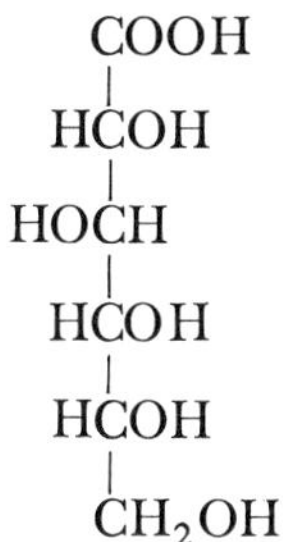

Production of gluconic acid from glucose is a simple conversion catalysed by glucose oxidase, an enzyme found in a very large number of micro-organisms. The conversion can also be carried out purely chemically. The problem in its microbiological production is essentially to use an organism and conditions in which conversion is rapid and complete, but stops at the gluconic acid stage and does not continue further. Most published work describes production with *Aspergillus niger*, but other organisms that can be used include *Penicillium chrysogenum* and other penicillia, various *Acetobacter* spp. (in particular *A. suboxydans*), various species of *Pseudomonas*, *Pullularia pullulans* and a species of *Moraxella*.

As with so many fermentations, much of the early development work on the subject was carried out by researchers working for the U.S. Department of Agriculture, in this case at the Bureau of Chemistry and Soils in Washington D.C.

At first, surface culture was used with *P. luteum purpurogenum* (May *et al.*, 1929). Then it was found that 80–87% yields of gluconic acid could be obtained in eight days submerged culture using media containing 200 g glucose per litre in the presence of calcium carbonate using *P. purpurogenum* (May *et al.*, 1934).

Later, the rotating-drum technique was used for the study of the submerged gluconic acid fermentation, now using a selected strain of *A. niger* (strain 67). In 1924, Bernhauer found a strain of *A. niger* that, if grown in the presence of calcium carbonate at low temperatures and with a restricted supply of nitrogenous nutrients, made gluconic acid in almost theoretical yields. Wells *et al.* (1937) also obtained almost quantitative conversion in submerged culture, and Moyer *et al.* (1937) showed that the use of pregerminated spores and fermenting under pressure gave very rapid conversions. Thus a medium containing glucose (150 g/l) and nutrient salts at a pressure of 2.11 kg cm^{-2} (30 p.s.i) was practically quantitatively converted to gluconic acid in 24 hours. The media developed were as shown in Table 1. The process was scaled up further into larger rotary drums 2.85 m long and 0.915 m internal diameter. In this vessel, 91 kg charges of glucose in 530 litres were converted in over 95% yield

Table 1

Media used for growing *Aspergillus niger* in the industrial production of gluconic acid. From Moyer *et al.* (1937)

Component	Concentration (g/l) used in medium for:		
	growth of spore-bearing mycelium	spore germination	gluconate production
Glucose	91.5	120	150
NH_4NO_3	0.45		
$(NH_4)_2HPO_4$		0.7	0.388
KH_2PO_4	0.072	0.3	0.188
$MgSO_4.7H_2O$	0.06	0.25	0.156
Peptone		0.15	
$CaCO_3$			100
Beer (ml/1)	60	67	

in under 24 hours. Optimum conversion was obtained with solutions containing 150 to 200 g glucose per litre at over pH 5.0, neutralizing with calcium carbonate (Gastrock *et al.*, 1938). It was also shown that the mycelium could be removed and re-used in 13 successive fermentations with no decrease in activity (Porges *et al.*, 1940). Conversion of high concentrations of glucose is normally inhibited by precipitation of calcium gluconate or by accumulation of free gluconic acid. Moyer *et al.* (1940) showed that, by adding boron compounds at levels of 500–2,500 mg of boron per litre and by using strains of *A. niger* that would tolerate such concentrations, it was possible to ferment media containing 200–350 g glucose per litre in the presence of excess calcium carbonate. Calcium borogluconate, unlike calcium gluconate, is a highly soluble compound, but its formation would complicate the recovery of gluconates and considerably add to the expense, so it is difficult to see a practical use for this procedure.

An early patent (Currie *et al.*, 1933) covers a conventional submerged culture process using *A. niger* or *P. luteum* in a medium containing glucose or carbohydrate that is readily converted to glucose. The description is not very clear, but it can be interpreted to indicate that a 90% yield of gluconic acid was obtained in 48 to 60 hours using 200 g glucose per litre and *A. niger* at 40°C in an agitated, aerated vessel. A neutralizing agent, such as calcium carbonate, was added to control the acid concentration. With such a published example, it is surprising that so much effort was spent in developing the mechanically much more difficult rotating drum process.

Gluconic acid was first made purely for conversion to calcium gluconate, which was used as a source of calcium in medicine and in veterinary practice, especially in the form of calcium borogluconate, which is used in the treatment of milk fever, a calcium deficiency in cows. Uses were later developed for the acid, either as such or in the form of the *delta*-lactone, for the sodium salt, largely as a sequestering agent, and for ferrous gluconate in iron therapy. To make sodium gluconate, a fermentation process was developed (Blom *et al.*, 1952) in which gluconic acid was continuously neutralized with soda as it was formed. This was run in a 1135-litre agitated tank 0.61 metre in diameter and 3.97 metres high with the following medium (per litre): glucose 220–350 g, urea 0.1 g, diammonium hydrogen phosphate 0.4 g, potassium dihydrogen phosphate 0.2 g, magnesium sulphate heptahydrate 0.16 g and corn-steep liquor 3.5 ml. The pH value was held at 6.0–6.5 and the temperature about 34°C. With mycelium used for the inoculum, there was virtually no lag period and the glucose was converted at a rate of 15 g/l/hour, a highly efficient process.

The ultimate development of the process using *A. niger* is described in a patent (Ziffer *at al.*, 1971). This shows that, by starting the run with only part of the total amount of glucose to be used and at least partially neutralizing the gluconic acid as it is formed, a very high glucose oxidase activity can be built up, such that more glucose can be added, preferably in increments, and remarkably high concentrations of gluconic acid and gluconates obtained in the broth. The best results are obtained if much of the gluconic acid is neutralized, but it is not necessary that all should be. Examples are given in which nearly quantitative conversions are achieved giving cultures containing the equivalent of over 600 g of sodium gluconate per litre, with up to about half of this in the form of the free acid. This can be achieved in about 60 hours running time. The process using *P. chrysogenum* has been studied again (Ambekar *et al.*, 1965) and it has been shown that waste mycelium from a penicillin fermentation can be used. A medium with 150 g glucose per litre can be used to obtain calcium gluconate in the presence of calcium carbonate in 90–95% yield in 72–96 hours.

There is little of interest in the very simple biosynthetic route from glucose to gluconic acid. What would be of interest, but has never been studied, is why with *A. niger* the normal glycolytic pathway and tricarboxylic cycle are blocked in the gluconate-producing strains.

The fact that certain *Acetobacter* spp. make gluconic acid from glucose has been known since the beginning of the Century. The use of *A. suboxydans* for this purpose was first patented by Verhave in 1930. Currie and Carter patented a process in 1933 based on the old method for making acetic acid in which a medium containing 200 g glucose per litre together with other nutrients and a neutralizing agent flowed through a tower packed with wood shavings or coke which had been inoculated with suitable strains of *Acetobacter*, while air was passed upwards through the packing. Bernhauer and Knoblock (1938) showed that, whereas in acid solution *A. suboxydans* converts glucose to gluconic acid, in the presence of calcium carbonate the conversion goes one step further to give calcium 5-oxogluconate. They stated that calcium or potassium gluconates were converted to the 2-oxogluconate, but it is now known that different strains of *A. suboxydans* can make either 2- or 5-oxogluconate or mixtures of the two.

The Dutch company Noury and van der Lande (1962) has patented a process for the crystallization of gluconic acid as a monohydrate at a temperature not above 30°C. It is interesting that the gluconic acid used in the example is described as made by oxidizing glucose (250 g/l) with *A. suboxydans*.

The use of *Pseudomonas ovalis* for conversion of glucose to gluconate was first developed at Peoria (Lockwood *et al.*, 1941). They examined 16 strains of *Pseudomonas*; most of them made

2-oxogluconate from glucose, but a strain of *Ps. ovalis* (N.R.R.L.-B-8) made only gluconic acid. Subsequently Tsao and Kempe (1960) showed that this conversion went in 99% yield, and that the rate varied directly with the efficiency of aeration. They proposed this system as an ideal one for the determination of oxygen-transfer rates in fermenters.

Humphrey and Reilly (1965) studied the kinetics of the same conversion, initially studying the rate of acid production. When they began the work, they were obviously unfortunately ignorant of earlier work and did not initially appreciate that this was a two-step conversion, with glucono-δ-lactone as an intermediate. They were in fact measuring the rate of hydrolysis of glucono-δ-lactone, which in their system was slower than the oxidation step. Because the lactone accumulated, they assumed lactonase was absent at pH 5.8, at which acidity they were operating. Jermyn (1960) had previously demonstrated that *Ps. fluorescens* produces a glucono-δ-lactonase. A fraction from disintegrated cells grown on glucose, which he purified one hundred-fold, accelerated the spontaneous hydrolysis of the lactone, which anyhow was felt to be too slow to participate in a sequence of enzymic reactions. The last word on this subject is from Bull and Kempe (1970) who showed that there was a critical level at which the oxidation rate is independent of the dissolved oxygen value, but can still be increased by increasing the rate of agitation. In these circumstances, the concentration of glucono-δ-lactone in the broth increases. They postulate that this is caused by an increased gas–liquid interfacial area.

The conversion of glucose to gluconate is highly efficient and goes rapidly, as already indicated, with a number of different organisms. The recovery of gluconates, most of which are very soluble or difficult to crystallize from impure solutions, is a bigger problem. To avoid difficult recovery procedures it is essential that the fermentations produce a high yield, with high concentrations of glucose and pure raw materials. On the other hand, one European manufacturer, Roquette Frères, is believed to make gluconate as a by-product in the manufacture of a polyol solution as described in a patent issued to them (1971). In this, a starch suspension is partially hydrolysed with amylase to give a solution with a dextrose equivalent of about 20–25. This is then fermented with *A. suboxydans*, after addition of the necessary salts, and the glucose converted to gluconic acid, leaving a solution with a dextrose equivalent of about 10. The term 'dextrose equivalent' is the value obtained by a total reducing-sugar determination; it will measure the reducing end-groups of maltose and higher saccharides as well as glucose. After clarification, the gluconic acid made is removed by passing through an anion-exchange resin, and the residual liquid hydrogenated to give a polyol solution for use in confectionery. Sodium gluconate is recovered from the resin with caustic soda solution, and isolated as such or converted to glucono-δ-lactone. A similar principle is involved in a patent by Murtaugh and Mahieu (1960). This involves the use of invert sugar as the raw material, and *A. niger* to convert the glucose to gluconic acid, which is recovered, giving fructose as a by-product.

European gluconate producers other than Roquette Frères are Pfizer in Eire, Benckiser in Germany and Noury and van der Lande, a subsidiary of Akzo, in Holland. Akzo has two plants making calcium gluconate, and Avebe, the Dutch co-operative potato-starch producers, has one plant making sodium gluconate. It has been reported that the two Dutch companies are to erect a joint plant to make three million kg of gluconates a year at Groningen. The older Dutch plants would then close down. In the United States, producers are Grain Processing, Pfizer, the Premier Malt Division of Pabst Brewery, and Stauffer. In Japan, gluconates are made by Fujisawa and Kyowa Hakko.

It is essential that fermentation processes for making gluconate from glucose are highly efficient, because there are several high-yielding purely chemical processes. These include electrolytic oxidation in the presence of bromine (Isbell *et al.*, 1932), and various palladium-catalysed air oxidation processes. In one of these (Buckley and Embree, 1957) air is passed through a solution containing 380–390 g of glucose per litre at 30°C with the pH value held at 10.0 with caustic soda, and it gives sodium gluconate in about 90% yield. In another (Acres and Budd, 1970) a continuous process in a column reactor is described using oxygen and a solution containing 180 g of glucose per litre at 45–50°C, neutralizing with sodium carbonate or bicarbonate, and giving a 100% yield.

Finally there is the possibility of using an enzymic procedure for making gluconates. Such a method is described by Baker (1953). An aqueous solution of glucose is stirred but not aerated, and an enzyme mixture which contains both glucose oxidase and catalase activity is added. A stoicheiometric excess of hydrogen peroxide is added slowly, and the catalase splits this to give oxygen atoms. Conversion goes in high yield with a high concentration of glucose, but the cost of hydrogen peroxide would make the process uncompetitive. In a more recent adaptation of this process (C.P.C. International, 1972), an electrodialysis step follows the enzymic conversion. Gluconate passes an anion-permeable membrane and goes to a further recovery operation, while the enzyme and unconverted glucose are returned for re-use. Glucose oxidase has been immobilized by binding it to glass, to nickel oxide or to nickel on silica alumina (Herring *et al.*, 1972). The last-mentioned catalyst is described as having superior physical characteristics compared with conventional immobilized enzymes with good activity and stability and at a cost reckoned at less than two U.S. dollars per kg. One can predict that a future process for making gluconates may well involve an immobilized glucose oxidase.

XI. OXOGLUCONIC ACIDS

Further bacterial oxidation of gluconate leads to 2- or 5-oxogluconate or, with *Acetobacter melanogenum* (Katznelson *et al.*, 1953) and *Pseudomonas albosesamae* (Wakisaka, 1964), to 2-5-dioxogluconate.

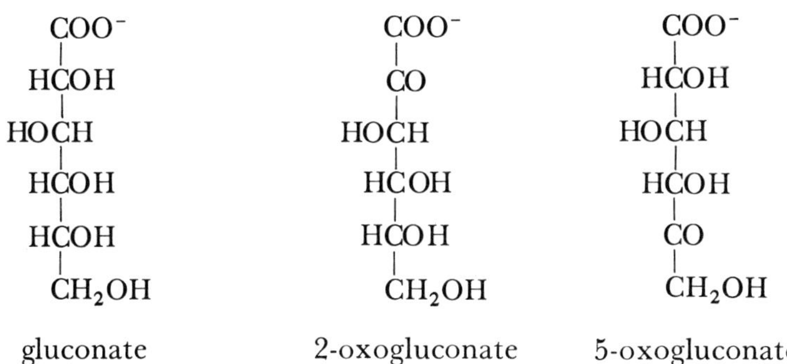

Whether 2- or 5-oxogluconate is made depends partly on the medium and pH value, but much more on the organism used. In unpublished work carried out many years ago, I examined 15 different strains of *A. suboxydans* grown on glucose in the presence of calcium carbonate and on calcium gluconate. One strain gave neither 2- nor 5-oxogluconate, seven gave 2-oxogluconate, but no

5-oxogluconate, six gave more 2-oxogluconate than 5-oxogluconate in varying proportions, one only (A.T.C.C. 621) gave about twice as much 5-oxogluconate as 2-oxogluconate. Later, a strain was obtained from Kluyver's collection, called by him *A. suboxydans* var. *amaltasiferum*, which gave about a 90% theoretical yield of 5-oxogluconate from glucose.

5-Oxogluconate has never been manufactured on a commercial scale, but is of interest as a possible starting material for the manufacture of L(+)-tartaric acid, which can be made from it either by air oxidation or by oxidation with nitric acid, both processes involving the use of a vanadium salt as a catalyst. Stubbs *et al.* (1940), working at Peoria, were the first to scale up this fermentation, using a strain of *A. suboxydans*, running in rotary aluminium fermenters and in an aluminium vat-type fermenter. They showed that it is necessary to carry out the conversion at a neutral pH value, using calcium carbonate to neutralize the acid formed. The glucose is first converted to calcium gluconate, and only when the glucose is exhausted is the gluconate further oxidized to give calcium 5-oxogluconate. Calcium gluconate is relatively insoluble and calcium 5-oxogluconate is very insoluble (0.2 g/100 ml), and solutions with more than 120 g of glucose per litre cannot be successfully converted because the broth gets too thick. This fermentation was later studied by Kheshgi *et al.* (1954) with a better-yielding strain of *A. suboxydans*, N.R.R.L. B-72, which gave up to 85% of the theoretical yield in 5–6 days. They attempted to solubilize calcium gluconate by the addition of borate, but this decreased calcium 5-oxogluconate production.

2-Oxogluconic acid is made on a commercial scale as an intermediate in the production of erythorbic acid. It is usually made by oxidation of glucose with strains of *Pseudomonas*. Unlike *A. suboxydans*, *Pseudomonas* spp. apparently convert glucose directly to 2-oxogluconate, which appears at the beginning of the fermentation, and not only when the glucose is exhausted. This process was scaled up by Pfeifer *et al.* (1958), running it in 225, 1,125 and 2,250-litre fermenters, on a medium containing (per litre): 110 g commercial glucose, 5 g corn-steep liquor, 2.0 g urea, 0.6 g potassium dihydrogen phosphate, 0.25 g magnesium sulphate heptahydrate and 27 g calcium carbonate. Three different strains of *Pseudomonas* were tested, *Ps. fluorescens* N.R.R.L. B-10 gave the best reults. The average yield of calcium 2-oxogluconate was 74% by weight. It was recovered by filtration, concentration and crystallization. At one time, it was believed that *Pseudomonas* spp. produced only 2-oxogluconate, but Stewart (1959) found that two out of 50 strains of *Pseudomonas* studied made the 5-oxo compound. Lockwood *et al.* (1941) found 16 strains of *Pseudomonas* which made 2-oxogluconate directly from glucose in over 80% yield.

Some people now regard *A. suboxydans* and other *Acetobacter* spp. with restricted oxidizing capacity as belonging to a different genus, namely *Acetomonas* included under the family Pseudo-monadaceae and not as *Acetobacter* in the family Acetobacteriacae. Others use the name *Gluconobacter.* One significant difference is that, in *Acetomonas* spp., the tricarboxylic acid cycle is not functional.

Misenheimer *et al.* (1965) studied the production of calcium 2-oxogluconate with *Serratia marcescens*, N.R.R.L. B-486. This process was run in 20-litre fermenters and gave a 95–100% yield in 16 hours from a medium containing 120 g glucose per litre with added salts, and with calcium carbonate as the neutralizing agent. With continuous feeding of glucose, a concentration of 180 g/litre could be as efficiently converted in 24 hours, and up to 240 g of glucose per litre gave 85 to 90% yields of calcium 2-oxogluconate in 32 to 40 hours. Ten strains of *S. marcescens* all gave over 80% yields. It is claimed that both yield and productivity are greater than when using *Pseudomonas* spp.

XII. ERYTHORBIC ACID

The normal method for manufacturing erythorbic acid (isoascorbic acid, D-araboascorbic acid), which is used as an antioxidant and stabilizer in foods, is to convert 2-oxogluconic acid to its methyl ester and then treat this with methanol and sodium hydroxide.

There is, however, a direct microbiological route for the preparation of erythorbic acid from glucose (Takahashi, 1969) using a strain of *Penicillium notatum.* A number of other species of *Penicillium* make it also. A strain was obtained following a mutation and screening programme which made erythorbic acid in a medium containing 80 g glucose per litre and salts in about 40% yield in five

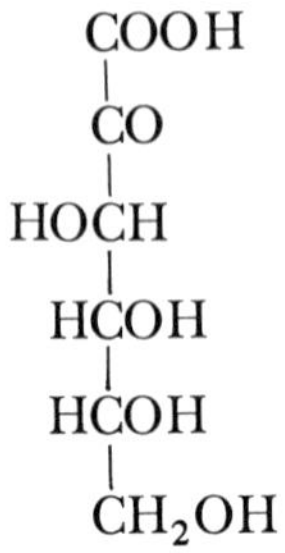

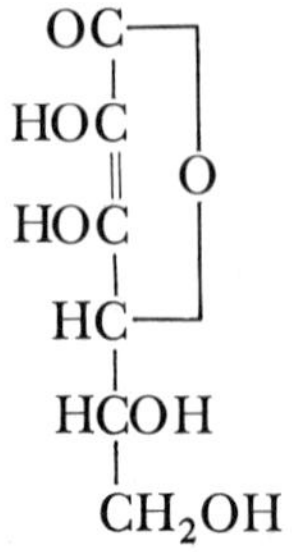

$$
\begin{array}{c}
\text{COOH} \\
| \\
\text{CO} \\
| \\
\text{HOCH} \\
| \\
\text{HCOH} \\
| \\
\text{HCOH} \\
| \\
\text{CH}_2\text{OH}
\end{array}
\qquad\qquad
\begin{array}{c}
\text{OC} \\
| \\
\text{HOC} \\
\| \quad\text{O} \\
\text{HOC} \\
| \\
\text{HC} \\
| \\
\text{HCOH} \\
| \\
\text{CH}_2\text{OH}
\end{array}
$$

2-oxogluconic acid   erythorbic acid

days, and in the same yield from 120 g glucose per litre in 12 days. The biosynthetic route is believed to be:

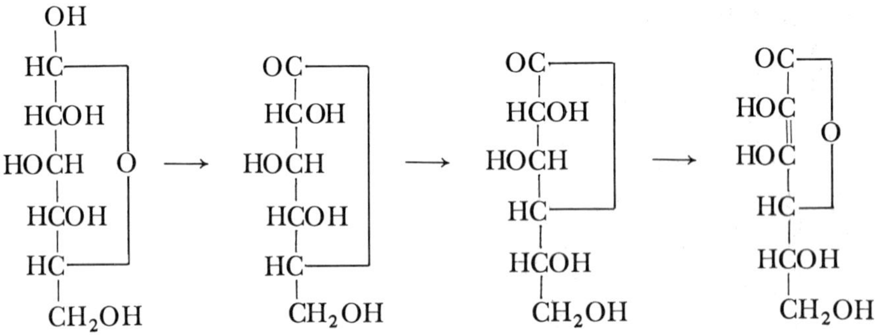

$$
\begin{array}{c}
\text{OH} \\
| \\
\text{HC} \\
| \\
\text{HCOH} \\
| \\
\text{HOCH} \quad \text{O} \\
| \\
\text{HCOH} \\
| \\
\text{HC} \\
| \\
\text{CH}_2\text{OH}
\end{array}
\longrightarrow
\begin{array}{c}
\text{OC} \\
| \\
\text{HCOH} \\
| \\
\text{HOCH} \\
| \\
\text{HCOH} \\
| \\
\text{HC} \\
| \\
\text{CH}_2\text{OH}
\end{array}
\longrightarrow
\begin{array}{c}
\text{OC} \\
| \\
\text{HCOH} \\
| \\
\text{HOCH} \\
| \\
\text{HC} \\
| \\
\text{HCOH} \\
| \\
\text{CH}_2\text{OH}
\end{array}
\longrightarrow
\begin{array}{c}
\text{OC} \\
| \\
\text{HOC} \\
\| \quad\text{O} \\
\text{HOC} \\
| \\
\text{HC} \\
| \\
\text{HCOH} \\
| \\
\text{CH}_2\text{OH}
\end{array}
$$

glucose   glucono-δ-   glucono-γ-   erythorbic
                       lactone   lactone   acid

Conversion of glucono-δ-lactone to glucono-γ-lactone is believed to be a non-enzymic step. The equilibrium between these compounds is well on the side of the δ-lactone and conversion is relatively slow, so this might explain the slow speed of this fermentation.

XIII. TARTARIC ACID

As already mentioned, the interest in the possible large-scale production of 5-oxogluconate lay in its conversion to tartaric acid. One method of doing this was by aeration of an acid or alkaline solution of 5-oxogluconate in the presence of certain catalysts, vanadium salts being by far the most effective (Pasternack and Brown, 1940).

A number of scattered references to the production of tartaric acid by fermentation with *A. suboxydans* or *Ps. fluorescens* have appeared in the literature. In 1943, Kamlet patented a procedure in which L(+)-tartaric acid was produced when a glucose solution was fermented aerobically with a strain of *A. suboxydans* (A.T.T.C. 621) and held at a pH value of 5 to 7 by addition of sodium carbonate. The rate of production was increased in the presence of a vanadium compound. No yield was quoted. Jackson *et al.* (1949) reported the production of tartaric acid by the action of *Ps. fluorescens* on potassium 5-oxogluconate. I was unable to repeat this using the same culture, and Jackson later stated that he could not obtain tartaric acid in shaken flasks and only in small quantities in fermenters operating under pressure. Lockwood and Nelson (1951) patented a similar process claiming a 22.4% theoretical yield after 17 days fermentation under a pressure of one kg cm^{-1} (15 p.s.i.) in a medium containing 92.5 g of potassium 5-oxogluconate per litre and urea, potassium phosphate, magnesium sulphate and corn-steep liquor. Khesgi *et al.* (1954) demonstrated the production of tartaric acid chromatographically from a 5-oxogluconate-producing strain of *A. suboxydans*. They also added vanadium pentoxide to their media.

Yamada *et al.* (1969) screened about 3,600 strains of fungi and 4,000 strains of bacteria and yeasts for tartaric acid production, using chromatography for qualitative analysis and a polarographic method for quantitative estimation. Six strains of bacteria, all isolated from Japanese persimmon, using a mannitol-based medium, were shown to produce L(+)-tartaric acid. Four of these were identified as *Gluconobacter suboxydans*. Subsequently, a number of other strains of *Acetobacter* and *Gluconobacter* were shown to produce tartaric acid. The highest yield quoted in this paper is 5.9 g of tartaric acid per litre from a medium containing (per litre): 100 g glucose, 20 g calcium carbonate and other salts, including 0.001 g ammonium vanadate, at pH 6.0 after six days in shaken cultures.

Kotera *et al.* (1972) reported higher yields of tartaric acid from mutant strains of *A. suboxydans* and quoted a concentration of 11.6 g/l. They found a new compound produced from 5-oxogluconate and intact cells in the presence of ammonium vanadate which they identified as 1,2-dihydroxyethyl hydrogen L(+)-tartrate and called 'pretaric acid'. They suggested that the course of the reaction is:

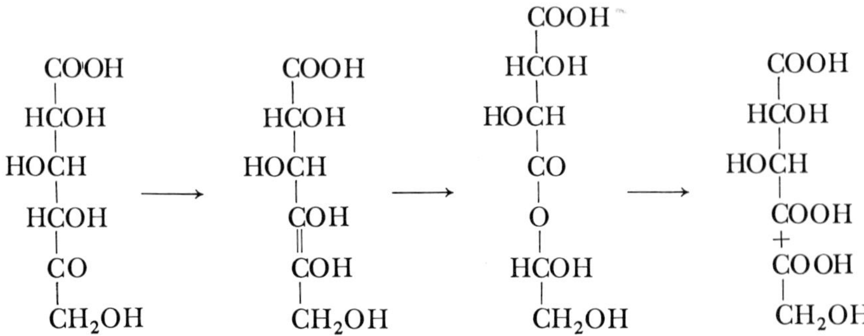

Tartaric acid was produced from pretaric acid by intact cells of *G. suboxydans*, but pretaric acid was found to be chemically unstable and, when allowed to stand in water solution at room temperature, it decomposed to tartaric and glycollic acids.

In a later paper (Kodama *et al.*, 1972), further mutants were tested and one of them, N-9338, gave a maximum yield of 14.6 g from 50 g of glucose, 35.0% of the theoretical yield, in six days in shaken flasks. The amount of 2-oxogluconate dropped sharply as the amount of tartaric acid increased with successive mutations, indicating that the later mutants were increasingly losing the 2-oxogluconate pathway. The mutants were stated to require *p*-aminobenzoic acid and pantothenic acid for their growth, and a patent claiming the production of L(+)-tartaric acid from *A. suboxydans* with a medium containing pantothenic acid has been taken out (Yamada *et al.*, 1971). But the fact that *A. suboxydans* requires pantothenic acid for growth has been known for 30 years (Underkofler *et al.*, 1943).

Yamada's team has also obtained mutants from known high 5-oxogluconate-producing strains of *A. suboxydans*, in particular mutants that tolerate increased acidity in general and increased concentrations of glycollic acid in particular. Glycollic acid, which, as stated above, is formed as a by-product, has been shown to inhibit bacterial growth.

REFERENCES

Acres, G. J. K. and Budd, A. E. R. (1970). British Patent 1,208,101

Adiga, P. R. Sivarama Sastry, K., Venkatasubramanyam, V. and Sarma, P. S. (1961). *Biochemical Journal* 81, 545.

Akiyama, S., Suzuki, T., Sumino, Y., Nakao, Y. and Fukuda, H. (1973). *Agricultural and Biological Chemistry* 37, 879.

Albright, F. and Schroepfer, G. J. (1970). *Biochemical and Biophysical Research Communications* 40, 661.

Allen, R. H. and Jakoby, W. B. (1969). *Journal of Biological Chemistry* 244, 2078.

Ambekar, G. R., Thadani, S. B. and Doctor, V. M. (1965). *Applied Microbiology* 13, 713.

Amelung, H. (1930). *Chemiker-Zeitung* 54, 118.

Anderson, J. G and Smith, J. E. (1971). *Journal of General Microbiology* 69, 185.

Anderson, J. G. and Smith, J. E. (1972). *Canadian Journal of Microbiology* 18, 289.

Arnstein, H. R. V. and Bentley, R. (1953). *Biochemical Journal* 54, 493, 508, 517.

Arpai, J. (1959). *Journal of Bacteriology* 78, 153.

Baker, D. L. (1953). United States Patent 2,651,592.

Batti, M. A. (1964). United States Patent 3,162,582.

Benckiser, Joh, A. GmbH. (1974). United States Patent 3,843, 465.

Bentley, R. and Thiessen, C. P. (1957). *Journal of Biological Chemistry* 226, 673, 689, 703.

Bernhauer, K. (1924). *Biochemische Zeitschrift* 153, 517.

Bernhauer, K. and Knobloch, H. (1938). *Naturwissenschaften* 26, 819.

Bernhauer, K., Rauch, J. and Gross, G. (1949). *Biochemische Zeitschrift* 319, 499.

Bertrand, D. (1954). *Comptes Rendus Hebdomadaires des Séances de l'Académie des Sciences* 239, 1704.

Bertrand, D. (1966). *Comptes Rendus Hebdomadaires des Séances de l'Académie des Sciences* 262, 2350.

Bertrand, D. and de Wolf, A. (1955). *Comptes Rendus Hebdomadaires des Séances de l'Académie des Sciences* 240, 1626.

Bertrand, D. and de Wolf, A. (1957). *Comptes Rendus Hebdomadaires des Séances de l'Académie des Sciences* 244, 2985.

Bertrand, G. (1941). *Comptes Rendus Hebdomadaires des Séances de l'Académie des Sciences* 213, 254.

Billington, R. H. (1969). *Chemical Processing (London)* 15 (9), 8.

Birkinshaw, J. H., Charles, J. H. V., Raistrick, H. and Stoyle, J. A. R. (1931). *Philosophical Transactions of the Royal Society of London, Series B* 220, 27.

Birkinshaw, J. H., Bracken, A. and Raistrick, H. (1945). *Biochemical Journal* 39, 70.

Blom, R. H., Pfeifer, V. F., Moyer, A. J., Traufler, D. H., Conway, H. F., Crocker, C. K., Farison, R. E. and Hannibal, D. V. (1952). *Industrial and Engineering Chemistry* 44, 435.

Bomstein, R. A. and Johnson, M. J. (1952). *Journal of Biological Chemistry* **198**, 143.

Buckley, J. S. and Embree, H. D. (1957). British Patent 786,288.

Buelow, G. H. and Johnson, M. J. (1952). *Industrial and Engineering Chemistry* **44**, 2945.

Bull, D. N. and Kempe, L. L. (1970). *Biotechnology and Bioengineering* **12**, 273.

Calam, C. T., Oxford, A. E. and Raistrick, H. (1939). *Biochemical Journal* **33**, 1488.

Camposano, A., Chain, E. B. and Gualandi, G. (1961). *Rendiconti Istituto Superiore di Sanita* **24**, 234.

Carson, S. F., Mosbach, E. H. and Phares, E. F. (1951). *Journal of Bacteriology* **62**, 235.

Cejkova, A., Rybarova, J. and Sestakova, M. (1966). *Folia Microbiologica* **11**, 223.

Challenger, F., Subramanian, V. and Walker, T. K. (1927) *Journal of the Chemical Society*, 200.

Champagnat, A., Vernet, C., Lainé, B. and Filosa, J. (1963). *Nature, London* **197**, 13.

Chesters, C. G. C. and Rolinson, G. N. (1951). *Journal of General Microbiology* **5**, 553.

Childs, C. G. and Welsby, B. (1966). *Process Biochemistry* **1**, 441.

Chugai Pharmaceutical Co. (1968). Japanese Patent 68/20,705.

Clark, D. S. (1962a). *Canadian Journal of Microbiology* **8**, 133.

Clark, D. S. (1962b). *Industrial and Engineering Chemistry, Product Research and Development* **1**, 59.

Clark, D. S. (1962c). *Biotechnology and Bioengineering* **4**, 17.

Clark, D. S. (1964). United States Patent 3,118,821.

Clark, D. S. and Lentz, C. P. (1961). *Canadian Journal of Microbiology* **7**, 447.

Clark, D. S., Ito, K. and Tymchuk, P. (1965). *Biotechnology and Bioengineering* **7**, 269.

Clark, D. S., Ito, K. and Horitsu, H. (1966). *Biotechnology and Bioengineering* **8**, 465.

Cleland, W. W. and Johnson, M. J. (1954). *Journal of Biological Chemistry* **208**, 679.

Cleland, W. W. and Johnson, M. J. (1956). *Journal of Biological Chemistry* **220**, 595.

Clement, M. T. (1952). *Canadian Journal of Technology* **30**, 82.

Clutterbuck, P. W. (1936). *Journal of the Society of Chemical Industry* **55** (11), 55T.

Collins, J. F. and Kornberg, H. L. (1960). *Biochemical Journal* **77**, 430.

Cordon, T. C., Treadway, R. H., Walsh, M. D. and Osborne, M. F. (1950). *Industrial and Engineering Chemistry* **42**, 1833.

Corzo, R. H. and Tatum, E. L. (1953). *Federation Proceedings. Federation of American Societies for Experimental Biology* **12**, 470.

Courtaulds Ltd. (1967). British Patent 1,085,901.

C. P. C. International, Inc. (1972). British Patent 1,276,245.

Currie, J. N. (1917). *Journal of Biological Chemistry* **31**, 15.

Currie, J. N. and Carter, R. H. (1933). United States Patent 1,896,811.

Currie, J. N., Kane, J. H. and Finlay, A. (1933). United States Patent 1,893,819.

Doelger, W. P. and Prescott, S. C. (1934). *Industrial and Engineering Chemistry* **26**, 1142.

Emiliani, E. and Bekes, P. (1964). *Archives of Biochemistry and Biophysics* **105**, 488.

Erickson, E. R., Berntsen, R. A., Eliason, M. A. and Peterson, M. E. (1965). *Journal of Agricultural and Food Chemistry* **13**, 452.

Euler, H. von (1909). "Grundlagen und Ergebnisse der Pflanzenchemie". Brunschweig, Band III.

Fernbach, A., Yuill, J. L. and Rowntree and Co., Ltd. (1927). British Patents 266,414, 266,415.

Finn, R. K., Halvorson, H. O. and Piret, E. L. (1950). *Industrial and Engineering Chemistry* **42**, 1857.

Foster, J. W. and Waksman, S. A. (1939a). *Journal of the American Chemical Society* **61**, 127.

Foster, J. W. and Waksman, S. A. (1939b). *Journal of Bacteriology* **37**, 599.

Foster, J. W. and Davis, J. B. (1948). *Journal of Bacteriology* **56**, 329.

Foster, J. W. and Carson, S. F. (1950). *Journal of the American Chemical Society* **72**, 1865.

Foster, J. W., Carson, S. F., Anthony, D. S., Davis, J. B., Jefferson, W. E. and Long, M. V. (1949). *Proceedings of the National Academy of Sciences of the United States of America* **35**, 663.

Foster, J. W., Carson, S. F. and Jefferson, W. E. (1950). *Federation Proceedings. Federation of American Societies for Experimental Biology* **9**, 172.

Fried, J. H. and Sandza, J. G. (1959). United States Patent 2,910,409.

Fries, H. von (1966). West German Patent 1,219,430.

Furukawa, T., Nakahara, T. and Yamada, K. (1970a). *Agricultural and Biological Chemistry* **34**, 1402.

Furukawa, T., Nakahara, T. and Yamada, K. (1970b). *Agricultural and Biological Chemistry* **34**, 1833.

Galanti, M. (1959). *Archives Internationales de Physiologie et de Biochimie* **67**, 112.

Galbraith, J. C. and Smith, J. E. (1969). *Journal of General Microbiology* **59**, 31.

Gastrock, E. A., Porges, N., Wells, P. A. and Moyer, A. J. (1938). *Industrial and Engineering Chemistry* **30**, 782.

Gerhardt, P., Dorrell, W. W. and Baldwin, I. L. (1946). *Journal of Bacteriology* **52**, 555.

Gledhill, W. E., Hill, I. D. and Hodson, P. H. (1973). *Biotechnology and Bioengineering* **15**, 963.

Hannan, M. A., Rabbi, F., Faizur Rahman, A. T. M. and Choudhury, N. (1973). *Journal of Fermentation Technology* **51**, 606.

Haskins, R. H., Thorn, J. A. and Boothroyd, B. (1955). *Canadian Journal of Microbiology* **1**, 749.

Hayaishi, O., Shimazono, H., Katagiri, M. and Saito, Y. (1956). *Journal of the American Chemical Society* **78**, 5126.

Herring, W. M., Laurence, R. L. and Kittrell, J. R. (1972). *Biotechnology and Bioengineering* **14**, 975.

Horitsu, H. and Clark, D. S. (1966). *Canadian Journal of Microbiology* **12**, 901.

Humphrey, A. E. and Reilly, P. J. (1965). *Biotechnology and Bioengineering* **7**, 229.

Ichimoto, I., Fujii, K. and Tatsumi, C. (1965). *Agricultural and Biological Chemistry* **29**, 325.

Iguchi, T., Takeda, I. and Seno, S. (1965). *Agricultural and Biological Chemistry* **29**, 589.

Imada, Y. and Yamada, K. (1971). *Agricultural and Biological Chemistry* **35**, 18.

Inskeep, G. C., Taylor, G. G. and Breitzke, W. C. (1952). *Industrial and Engineering Chemistry* **44**, 1955.

Isbell, H. S., Frush, H. L. and Bates, F. J. (1932). *Industrial and Engineering Chemistry* **24**, 375.

Iwata Chemical Industry Co. Ltd. (1973). Japanese Patent 4,852,990.

Jackson, R. W., Koepsell, H. J., Lockwood, L. B., Nelson, G. E. N. and Stodola, F. H. (1949). Abstract No. 7/12, First International Congress of Biochemistry, Cambridge.

Jagannathan, V. and Singh, K. (1953). *Enzymologia* **16**, 150.

James, L. V., Rubbo, S. D. and Gardner, J. F. (1956a). *Journal of General Microbiology* **14**, 223.

James, L. V., Rubbo, S. D. and Gardner, J. F. (1956b). *Journal of General Microbiology* **14**, 228.

Jenssen, E. B., Larsen, H. and Ormerod, J. G. (1956). *Acta Chemica Scandinavica* **10**, 1047.

Jermyn, M. A. (1960). *Biochimica et Biophysica Acta* **37**, 78.

Johnson, M. J. and Bloom, S. J. (1962). *Journal of Biological Chemistry* **237**, 2718.

Jungbunzlauer Spiritus-und-Chemische Fabrik. (1974a). British Patent 1,342,297.

Jungbunzlauer Spiritus-und-Chemische Fabrik. (1974b). British Patent 1,342,311.

Kamlet, J. (1943). United States Patent 2,314,831.

Kane, J. H., Finlay, A. and Amann, P. F. (1943). United States Patent 2,327,191.

Karow, E. O. and Waksman, S. A. (1947). *Industrial and Engineering Chemistry* **39**, 821.

Katznelson, H., Tanenbaum, S. W. and Tatum, E. L. (1953). *Journal of Biological Chemistry* **204**, 43.

Kawade, S. (1972). Japanese Patent 7,227,036.

Kempe, L. L., Halvorson, H. O. and Piret, E. L. (1950). *Industrial and Engineering Chemistry* **42**, 1852.

Kheshgi, S., Roberts, H. R. and Bucek, W. (1954). *Applied Microbiology* **2**, 183.

Kimura, K. and Nakanishi, T. (1973). British Patent 1,332,180.

Kinoshita, K. (1931). *Acta Phytochimica* **5**, 271.

Kinoshita, S. and Tanaka, K. (1961). British Patent 878,152.

Kinoshita, S., Tanaka, K. and Akita, S. (1961). British Patent 878,151.

Kitada, M., Ueyama, H., Suzuki, E. and Fukimbara, T. (1967). *Journal of Fermentation Technology* **45**, 1101.

Kitada, M., Ueyama, H., Suzuki, E. and Fukimbara, T. (1968). *Journal of Fermentation Technology* **46**, 437.

Kitos, P. A., Campbell, J. J. R. and Tomlinson, N. (1953). *Applied Microbiology* **1**, 156.

Kluyver, A. J. and Perquin, L. H. C. (1933). *Biochemische Zeitschrift* **266**, 68.

Kobayashi, T. (1967). *Process Biochemistry* **2** (9), 61.

Kobayashi, T. and Nakamura, I. (1966). *Journal of Fermentation Technology* **44**, 264.

Kodama, T., Kotera, U. and Yamada, K. (1972). *Agricultural and Biological Chemistry* **36**, 1299.

Koepsell, H. J., Stodola, F. H. and Sharpe, E. S. (1952). *Journal of the American Chemical Society* 74, 5142.

Kornberg, H. L. and Collins, J. F. (1958). *Biochemical Journal* 68, 3P.

Kotera, U., Kodama, T., Minoda, Y. and Yamada, K. (1972). *Agricultural and Biological Chemistry* 36, 1315.

Kovats, J., Lewicka, M. and Kaminski, S. (1957). *Przemysl Spozywczy* 11, 156.

Kyowa Hakko Kogyo K. K. (1970a). British Patent 1,187, 334.

Kyowa Hakko Kogyo K. K. (1970b). British Patent 1,187,610.

Lal, M. and Bhargava, P. M. (1962). *Biochimica et Biophysica Acta* 58, 628.

Larsen, H. and Eimhjellen, K. E. (1955). *Biochemical Journal* 60, 135.

Leonard, R. H., Peterson, W. H. and Johnson, M. J. (1948). *Industrial and Engineering Chemistry* 40, 57.

Leopold, J. (1959). *Nahrung* 3, 987.

Leopold, J. and Fencl, Z. (1958). *Chemische Technik* (Berlin) 10, 507.

Lewis, K. F. and Weinhouse, S. (1951). *Journal of the American Chemical Society* 73, 2500.

Lockwood, L. B. and Reeves, M. D. (1945). *Archives of Biochemistry* 6, 455.

Lockwood, L. B. and Ward, G. E. (1945). *Industrial and Engineering Chemistry* 37, 405.

Lockwood, L. B. and Nelson, G. E. N. (1946). *Archives of Biochemistry* 10, 365.

Lockwood, L. B. and Nelson, G. E. N. (1951). United States Patent 2,559,650.

Lockwood, L. B. and Batti, M. A. (1965). United States Patent 3,189,527.

Lockwood, L. B., Tabenkin, B. and Ward, G. E. (1941). *Journal of Bacteriology* 42, 51.

Machell, G. (1959). *Industrial Chemist* 35, 283.

Martin, S. M. (1954). *Canadian Journal of Microbiology* 1, 6.

Martin, S. M. (1955). *Canadian Journal of Microbiology* 1, 644.

Martin, S. M. (1956). United States Patent 2,739,923.

Martin, S. M. and Waters, W. R. (1952). *Industrial and Engineering Chemistry* 44, 2229.

Martin, S. M. and Steel, R. (1955). *Canadian Journal of Microbiology* 1, 470.

Martin, S. M., Wilson, P. W. and Burris, R. H. (1950). *Archives of Biochemistry* 26, 103.

Martin, W. R. and Foster, J. W. (1955). *Journal of Bacteriology* 70, 405.

May, O. E., Herrick, H. T., Moyer, A. J. and Hellbach, R. (1929). *Industrial and Engineering Chemistry* 21, 1198.

May, O. E., Herrick, H. T., Moyer, A. J. and Wells, P. A. (1934). *Industrial and Engineering Chemistry* 26, 575.

McDonough, M. W. and Martin, S. M. (1958). *Canadian Journal of Microbiology* 4, 329.

Meyrath, J. (1967). *Process Biochemistry* 2 (10), 25.

Mezzadroli, G. (1938). French Patent 833,631.

Miles Laboratories Inc. (1945). British Patent 572,383.

Miles Laboratories Inc. (1951). British Patent 653,808.

Miles Laboratories Inc. (1952a). British Patent 669,733.

Miles Laboratories Inc. (1952b). British Patent 672,128.

Miles Laboratories Inc. (1955). British Patent 738,940.

Miles Laboratories Inc. (1956a). British Patent 742,972.

Miles Laboratories Inc. (1956b). British Patent 743,395.

Miles Laboratories Inc. (1962). British Patent 908,024.
Miles Laboratories Inc. (1963). British Patent 943,665.
Miles Laboratories Inc. (1969). British Patent 1,145,520.
Misenheimer, T. J., Anderson, R. F., Lagoda, A. A. and Tyler, D. D. (1965). *Applied Microbiology* 13, 393.
Mitsui Sugar Co. Ltd. (1974). British Patent 1,354,192.
Montgomery, R. and Ronca, R. A. (1953). *Industrial and Engineering Chemistry* 45, 1136.
Moyer, A. J. (1953). *Applied Microbiology* 1, 1.
Moyer, A. J. (1954). United States Patent 2,674,561.
Moyer, A. J. and Coghill, R. D. (1945). *Archives of Biochemistry* 7, 167.
Moyer, A. J., Wells, P. A., Stubbs, J. J., Herrick, H. T. and May, O. E. (1937). *Industrial and Engineering Chemistry* 29, 777.
Moyer, A. J., Umberger, E. J. and Stubbs, J. J. (1940). *Industrial and Engineering Chemistry* 32, 1379.
Mulder, E. G. (1948). *Analytica Chimica Acta* 2, 793.
Murtaugh, J. J. and Mahieu, J. J. (1960). United States Patent 2,949,389.
Naim, M. S. and Mansour, I. S. (1966). *Journal of Microbiology. United Arab Republic* 1 (1), 91.
Neilson, N. E. (1956). *Journal of Bacteriology* 71, 356.
Nelson, G. E. N., Traufler, D. H., Kelley, S. E. and Lockwood, L. B. (1952). *Industrial and Engineering Chemistry* 44, 1166.
Nicholas, D. J. D. and Fielding, A. H. (1951). *Journal of the Horticultural Society* 26, 125.
Nonomura, S., Ichimoto, I. and Tatsumi, C. (1960). *Journal of the Agricultural Chemical Society of Japan* 34, 732.
Noury and van der Lande, N. V. (1962). British Patent 902,609.
Nowakowska-Waszczuk, A. (1973). *Journal of General Microbiology* 79, 19.
Nubel, R. C. and Ratajak, E. J. (1964). British Patent 950,570.
Ogston, A. G. (1948). *Nature, London* 162, 963.
Olson, J. A. (1954). *Nature, London* 174, 695.
Overman, S. A. and Romano, A. H. (1969). *Biochemical and Biophysical Research Communications* 37, 457.
Ozawa, T. and Watanabe, S. (1968). Japanese Patent 68/28,948.
Pasternack, R. and Brown, E. V. (1940). United States Patent 2,197,021.
Perlman, D., Dorrell, W. W. and Johnson, M. J. (1946a). *Archives of Biochemistry* 11, 131.
Perlman, D., Kita, D. A. and Peterson, W. H. (1946b). *Archives of Biochemistry* 11, 123.
Perquin, L. H. C. (1938). "Bijdrage tot de kennis der oxydative dissimilatie van *Aspergillus niger* van Tieghem". Meinema, Delft.
Pfeifer, V. F., Vojnovich, C. and Heger, E. N. (1952). *Industrial and Engineering Chemistry* 44, 2975.
Pfeifer, V. F., Vojnovich, C., Heger, E. N., Nelson, G. E. N. and Haynes, W. C. (1958). *Industrial and Engineering Chemistry* 50, 1009.
Pfizer Inc. (1948). British Patent 602,866.
Pfizer Inc. (1970). British Patent 1,182,983.
Pfizer Inc. (1972). British Patent 1,293,786.
Pfizer Inc. (1974). British Patent 1,369,295.
Porges, N., Clark, T. F. and Gastrock, E. A. (1940). *Industrial and Engineering Chemistry* 32, 107.

Raistrick, H. and Clark, A. B. (1919). *Biochemical Journal* 13, 329.

Ramakrishnan, C. V. (1954). *Enzymologia* 17, 169.

Ramakrishnan, C. V. (1958). *Nature, London* 182, 1601.

Ramakrishnan, C. V. and Martin, S. M. (1954a). *Chemistry and Industry* 160.

Ramakrishnan, C. V. and Martin, S. M. (1954b). *Canadian Journal of Biochemistry and Physiology* 32, 434.

Ramakrishnan, C. V. and Martin, S. M. (1954c). *Nature, London* 174, 230.

Ramakrishnan, C. V. and Martin, S. M. (1955). *Archives of Biochemistry and Biophysics* 55, 403.

Ramakrishnan, C. V., Steel, R. and Lentz, C. P. (1955). *Archives of Biochemistry and Biophysics* 55, 270.

Raper, K. B. and Thom, C. (1949). "A Manual of the Penicillia", pg. 690. Baillière, Tindall and Cox, London.

Rhodes, R. A., Moyer, A. J., Smith, M. L. and Kelley, S. E. (1959). *Applied Microbiology* 7, 74.

Rhodes, R. A., Lagoda, A. A., Misenheimer, T. J., Smith, M. L., Anderson, R. F. and Jackson, R. W. (1962). *Applied Microbiology* 10, 9.

Romano, A. H. (1958). *Tappi* 41, 687.

Romano, A. H., Bright, M. M. and Scott, W. E. (1967). *Journal of Bacteriology* 93, 600.

Roquette Frères (1971). French Patent 2,054,825.

Royal Norwegian Council for Scientific and Industrial Research (1958). British Patent 795,401.

Sakaguchi, K. and Baba, S. (1942). *Bulletin of the Agricultural Chemical Society of Japan* 18, 85.

Sakaguchi, K., Inoue, T. and Tada, Y. (1938). *Journal of the Agricultural Chemical Society of Japan* 14, 362.

Sato, M., Nakahara, T. and Yamada, K. (1972a). *Agricultural and Biological Chemistry* 36, 1969.

Sato, M., Nakahara, T. and Yamada, K. (1972b). *Agricultural and Biological Chemistry* 36, 2025.

Shimi, I. R. and Nour El Dein, M. S. (1962). *Archiv für Mikrobiologie* 44, 181.

Shu, P. and Johnson, M. J. (1947). *Journal of Bacteriology* 54, 161.

Shu, P. and Johnson, M. J. (1948a). *Journal of Bacteriology* 56, 577.

Shu, P. and Johnson, M. J. (1948b). *Industrial and Engineering Chemistry* 40, 1202.

Shu, P., Funk, A. and Neish, A. C. (1954). *Canadian Journal of Biochemistry and Physiology* 32, 68.

Steel, R., Martin, S. M. and Lentz, C. P. (1954). *Canadian Journal of Microbiology* 1, 150.

Steel, R., Lentz, C. P. and Martin, S. M. (1955). *Canadian Journal of Microbiology* 1, 299.

Steinberg, R. A. (1935). *Journal of Agricultural Research* 51, 413.

Steinberg, R. A. (1936). *Journal of Agricultural Research* 52, 439.

Steinberg, R. A. (1937). *Journal of Agricultural Research* 55, 891.

Steinberg, R. A. (1938). *Journal of Agricultural Research* 57, 569.

Stewart, D. J. (1959). *Nature, London* 183, 1133.

Stodola, F. H., Friedkin, M., Moyer, A. J. and Coghill, R. D. (1945). *Journal of Biological Chemistry* 161, 739.

Stubbs, J. J., Lockwood, L. B., Roe, E. T., Tabenkin, B. and Ward, G. E. (1940). *Industrial and Engineering Chemistry* 32, 1626.

Svenska Sockerfabriks Aktiebolaget (1964). British Patent 951,629.

Swarthout, E. J. (1966). United States Patent 3,285,831.

Szucs, J. (1946). British Patent 577,490.

Tachibana, S. and Murakami, T. (1973). *Journal of Fermentation Technology* 51, 858.

Takahashi, J., Hidaka, H. and Yamada, K. (1965a). *Agricultural and Biological Chemistry* 29, 331.

Takahashi, J., Kobayashi, K., Imada, Y. and Yamada, K. (1965b). *Applied Microbiology* 13, 1.

Takahashi, T. (1969). *Biotechnology and Bioengineering* 11, 1157.

Takahashi, T. and Sakaguchi, K. (1927). *Bulletin of the Agricultural Chemical Society of Japan* 3, 59.

Takeda Chemical Industries Ltd. (1970). British Patent 1,211,246.

Takeda Chemical Industries Ltd. (1971a). British Patent 1,244,824.

Takeda Chemical Industries Ltd. (1971b). British Patent 1,257,900.

Takeda Chemical Industries Ltd. (1972). British Patent 1,297,243.

Takeda Chemical Industries Ltd. (1973). British Patent 1,326,785.

Thom, C. and Currie, J. N. (1916). *Journal of Agricultural Research* 7, 1.

Thom, C. and Raper, K. B. (1945). "A Manual of the Aspergilli", pg. 142. The Williams and Wilkins Co. Baltimore.

Tomlinson, N. and Campbell, J. J. R. (1951). *Journal of Bacteriology* 61, 17.

Tomlinson, N., Campbell, J. J. R. and Trussell, P. C. (1950). *Journal of Bacteriology* 59, 217.

Trumpy, B. H. and Millis, N. F. (1963). *Journal of General Microbiology* 30, 381.

Tsao, G. T. and Kempe, L. L. (1960). *Journal of Biochemical and Microbiological Technology and Engineering* 2, 129.

Underkofler, L. A., Bantz, A. C. and Peterson, W. H. (1943). *Journal of Bacteriology* 45, 183.

Usines de Melle (1958). British Patent 797,390.

Verhave, T. H. (1930). German Patent 563,758.

Wakisaka, Y. (1964). *Agricultural and Biological Chemistry* 28, 369.

Waksman, S. A. (1943). United States Patent 2,326,986.

Wells, P. A. and Ward, G. E. (1939). *Industrial and Engineering Chemistry* 31, 172.

Wells, P. A., Moyer, A. J. and May, O. E. (1936). *Journal of the American Chemical Society* 58, 555.

Wells, P. A., Moyer, A. J., Stubbs, J. J., Herrick, H. T. and May, O. E. (1937). *Industrial and Engineering Chemistry* 29, 653.

Whittier, E. O. and Rogers, L. A. (1931). *Industrial and Engineering Chemistry* 23, 532.

Wilkoff, L. J. and Martin, W. R. (1963). *Journal of Biological Chemistry* 238, 843.

Woronick, C. L. and Johnson, M. J. (1960). *Journal of Biological Chemistry* 235, 9.

Yamada, K. and Hidaka, H. (1964). *Agricultural and Biological Chemistry* 28, 876.

Yamada, K., Minoda, Y., Kodama, T. and Kotera, U. (1969). *In* "Fermentation Advances", (D. Perlman, ed.), pg. 541. Academic Press, New York.

Yamada, K., Furukawa, T. and Nakahara, T. (1970). *Agricultural and Biological Chemistry* 34, 670.

Yamada, K., Minoda, Y., Kodama, T. and Kotera, U. (1971). United States Patent 3,585,109.

Yuill, J. L. (1948). *Nature, London* **161**, 397.

Zadrodski, S. and Krzysztofik, W. (1953). *Acta Microbiologica Polonica* **2**, 209.

Zahorski, B. (1913). United States Patent 1,066,358.

Ziffer, J., Gaffney, A. S., Rothenberg, S. and Cairney, T. J. (1971). British Patent 1,249,347.

4. Acetic Acid: Vinegar

R. N. GREENSHIELDS

Department of Biological Sciences, University of Aston in Birmingham, Birmingham, England

When Jesus had received the vinegar, he said, 'It is finished', and he bowed his head and gave up his spirit. John XIX. 29

I. INTRODUCTION

A. General

Arthur Harden (1911) has said that 'History finds man in the possession of alcoholic liquors' but since alcohol forms one of the main nutrients for the acetic-acid bacteria, the manufacture of vinegar therefore probably also ranks as one of mankind's oldest fermentation activities.

Fig. 1. The oldest method for producing 'vin aigre'. Acetobacters were probably introduced into the cup of wine on the legs of vinegar flies (*Drosophila* spp.).

Wine left open to the air rapidly 'mothers' (forms a skin) and sours, thus providing the oldest technique of providing 'vin aigre'. This term literally means 'sour wine' according to its derivation from the French (vin—wine, aigre—sour or sharp). Hence also, by analogy, the product derived from beer became known as 'ale—gar' (Fig. 1). Wine-making is an art dating back at least ten thousand years, thus vinegar is quite likely as old since it would arise readily by spoilage.

Certainly it is known and recorded that the Babylonians in 5,000 B.C. made vinegar as an end-product of a wine from the date palm (Huber, 1927).

Vinegar is described in the Bible, and the first quotation is one of four references to vinegar in each of the Old and New Testaments (The Holy Bible). Vinegar certainly had its recognized place amongst the products of the alchemist. The alchemical symbols ⵜ, ✕, + were used while the characteristics ⵜ, ✳, ✚ were used for distilled vinegar or acetic acid, although the alchemist's distilled vinegar is not the same as that called distilled vinegar today (Mitchell, 1926). According to Lemery (1720) there were three sorts of alchemical liquors known as Spirit, the Spirit of Animals, the Burning Spirit of Vegetables and the Acid Spirit. The first was typified by the Spirit of Hartshorn, the second by the Spirit of Wine, while the last 'the Spirit of Vinegar, Tartar and Vitriol, is an Acid Essential Salt dissolved and put in fusion by the fire, as I shall prove when I speak of Vinegar'.

B. Uses

The history of the use of vinegar is remarkable in that, like alcohol, it is a philosopher's stone used by man for almost all purposes. Many references occur dating back to antiquity on its multifarious uses as a condiment, as a food preservative, as a medicinal agent and an antibiotic, and even as a cleaning agent (The Jewish Encyclopedia, 1906; Francis Adams, 1849; Thompson, 1931; Ochs, 1950). The medicinal use of vinegar is certainly of considerable historical importance since, in its antagonistic action to many micro-organisms (antibiosis), vinegar was possibly the first known antibiotic. In Biblical times, records show it was included in wet dressings on wounds, while Hippocrates in 400 B.C. had also used it likewise on his patients, and the Assyrians successfully treated chronic middle-ear diseases when other methods failed. Medicinal vinegars, with formulations such as antiseptic vinegar, antiscorbutic vinegar, camphorated vinegar, colchicum vinegar, squill vinegar, black drop vinegar and syrup of vinegar, have been recorded for a variety of diseases including the plague, lameness, poison ivy, shingles, night sweats, burns, varicose veins, impetigo and ringworm (Dussance, 1871; Fetzer, 1930; Ochs, 1950; Jarvis, 1959).

Use of vinegar as a food condiment and preservative is also a matter of record. Credit must be given to the Babylonians for developing applications in this respect which are still popular today (Huber, 1927). Addition of herbs and spices to improve flavour and aroma is a characteristic example. Tarragon, ruta, absynth, lavender, mint, celery, pontulaca and saffron are recorded, and may have assisted in the preservative action when vinegar was used in pickling. In cooking, it was frequently used in highly seasoned foods particularly those of the Orient where legumes form a large part of the diet. Meat and fish were also preserved by pickling but only for the more festive occasions. Pickling with vinegar is now almost World-wide. French cooking is noted for the quality of its sauces many of which are flavoured with vinegar, and the use of oil in their food is also modified by its incorporation. More recently, the American and British people 'cut' the fat content of fried fish and chips, and enhance their flavour by sprinkling vinegar on them.

C. Malt Vinegar

An introduction to vinegar would not be complete without some consideration of the history of malt vinegar, since there are special considerations with respect to its development within a tax situation. Not unexpectedly, the manufacture of malt vinegar in England evolved from brewing (Mitchell, 1926) since it was the obvious way of disposing of sour beer or ale. Certainly, since Englishmen consumed large quantities of beer and ale, it is logical that the vinegar should be obtained from this source rather than from the more usual one of wine. The product became known as 'alegar'—a word which has the same relation to ale as vinegar had to wine. The date of the establishment of the first vinegar factory in England is uncertain, but there was a so-called vinegar yard in Castle Street in Southwark in 1641. The legality of this situation was considered by the Revenue Act of Parliament in 1673 during the reign of Charles II, when products obtained by souring beer or ale, or as a waste material, were termed 'vinegar beer' and were taxed with a duty of sixpence per barrel. 'The tax was applied to every barrel of vinegar-beer brewed by any common brewer in any common brew-house'. This duty was frequently evaded and, in 1696, William III imposed a penalty of 40 shillings for every barrel of vinegar concealed from the gaugers or sent out of the brewhouse without due notice to the Excise Officers. A further Act, passed in 1710, increased the duty on vinegar to 9d per barrel, a rate that remained through the following century. In 1796, during the reign of George III, vinegar makers were not allowed to have a distillery on the premises, and a licence was required stating whether vinegar was to be manufactured from malt or corn or from molasses or sugar. Such changes continued in the Tax laws concerning vinegar up until the last century, but vinegar is now without a direct tax and this is reflected in its price. It has remained a cheap basic food commodity whose price has remained remarkably stable over the last half century.

D. Comment

More unusual uses have also been recorded, not the least of which is the solvent power of the acetic acid. One may quote the well known anecdote of Cleopatra related by Pliny (Holland, 1964) when 'to gain a wager she would consume at a single meal the value of a million sisterces, she dissolved pearls in vinegar which she drank' . . . and in Hannibal's march over the Alps to Rome, Titus Livius reported that, to crack rocks blocking the road, Hannibal's soliders poured vinegar on heated limestone boulders which 'rendered them soft and crumbling. They then opened a way with iron instruments through the rock . . . so that not only the beasts of burden, but also the elephants, could be led down' (Spillan and Edmonds, 1895).

Thus, the beginnings of vinegar are indeed a reflection of the beginnings and history of mankind. Today, the preservation and flavouring of food depend to a large extent on the use of vinegar. The value of this liquid can best be summed up by Mahomet who has said 'If there is no vinegar in a house it is a sin: there is no blessing neither'! (Allgeier *et al.*, 1974).

II. DEFINITION

Whilst the oldest method of producing vinegar has been called the 'let alone process', that is wine allowed to sour naturally, this situation is no longer true, and vinegar is deliberately made by quite sophisticated fermentation methods. Thus it is essential to define its manufacture since much variation in the methods and materials is made. Vinegar can be defined as the product of a double fermentation. Firstly, there is an alcoholic fermentation of carbohydrate-containing materials by various strains of *Saccharomyces*. The alcohol so produced is then oxidized to acetic acid under aerobic conditions by a suitable culture of acetic-acid bacteria. Vinegar has been manufactured from an amazing diversity of raw materials, including fruits, berries, apples, pears, honey, molasses, rice, cereals, melons, bananas, pineapple wastes, coconuts, sugar cane, potatoes, beets, malt and related grains, and whey, from the upper portions of the plant (flowers and fruits) down through the stem (juices) to the underground parts (tubers and roots). This change occurred over the centuries and reflects what raw materials were available for alcohol production and thus vinegar production. Broadly, we can classify the materials into four groups, namely fruits and their juices, cereals, sugar syrups and lastly alcohol since this may also be derived from ethylene, a byproduct of the oil industry. Beers and wines, therefore, largely serve as the alcoholic base, although it is not unnatural for almost any material which contains alcohol to be used. Certainly, the alcoholic source for acetification reflects the main alcoholic beverage (and its raw materials) of a country or province.

Thus, it is found that wine vinegar is common in the Continent of Europe and around the Mediterranean, malt vinegar in the United Kingdom, rice vinegar in Japan and China, pineapple vinegar in Malaysia, and a variety of vinegars in Africa. The exception to this is spirit vinegar derived from pure alcohol produced by distillation of alcoholic fermentations. However, spirit vinegar is a relatively modern innovation since it is not so much used as a direct condiment but as an acetic means for pickling, sauce or food manufacture. Its use is now World-wide and merely depends on the availability of a distillery. It can be seen from this variation in vinegar type that some classification and definition could be conceived. In the United States, the Food and Drug Administration adopted definitions and standards for vinegar as early as 1936, although previous analyses of cider vinegar had been used as definite guide lines (Brooks, 1927).

Subsequent to the discovery in 1840 or 1850 that acetic acid could be synthesized by the destructive distillation of wood, attempts were made to imitate vinegar with 'pyroligneous acid' or 'wood vinegar'. As a result of the competition from this cheap imitation, the issue of definition came to law in the United Kingdom. Several court cases were fought, but a clear definition and standard were finally set by a High Court decision in the case of Kat. v Diment (1950) which confirmed Sir Lawrence Dunne's written decision in the original court case (Vinegar Brewers' Federation, 1952).

It is not unexpected therefore that British malt vinegar is defined as being derived from an alcoholic wash (or 'charging wort' as it is called), fermented from barley or malted barley, and converted by the enzymes of malt (Ministry of Agriculture, Fisheries and Food, 1971). This is reproduced below:

Food Standards Commitee Report on Vinegars

Appendix II. Paragraphs 106 and 107 of the Food Standards Committee's Report on Claims and Misleading Descriptions (published in 1966).

VINEGAR

106. Vinegar is a product of the alcoholic and acetous fermentation of a sugar-containing solution without any intermediate distillation, except in the case of spirit vinegar as defined in (c) below. It may be made from various materials and may be flavoured or unflavoured. The following are descriptions of unflavoured vinegars at present used:

(a) *Wine vinegar.* Vinegar made by the double fermentation process set out above, the alcoholic fermentation being from grapes.

(b) *Malt vinegar.* Vinegar derived without intermediate distillation, wholly from malted barley, with or without the addition of entire cereal grain, malted or otherwise, the starch of which has been converted to sugar (saccharified) by the diastase of malt.

(c) *Spirit vinegar.* Vinegar made by the acetous fermentation of a distilled alcoholic fluid itself produced by fermentation.

(d) *Fruit wine vinegar.* Vinegar made from a mixture of non-grape, or grape and non-grape fruit wines.

(e) *Cider vinegar.* Vinegar made from the alcoholic fermentation of the juice of apples, followed by the acetous fermentation.

(f) *Distilled vinegar.* Vinegar made by the distillation of malt vinegar.

107. We consider that there is a need for clear nomenclature in this field and we *recommend* that the names listed above should become obligatory. We do not consider that there should be a similar control on the names of flavoured vinegar. It should also be forbidden to use the word 'Vinegar' to describe solutions of acetic acid, whether flavoured or not, and we recommend accordingly.

(H.M.S.O. London. 1971.)

Currently, the United Nations Food and Agriculture Organization Joint World Health Organization Food Standards Programme is gathering information on the vinegar throughout the World, and is attempting to unify the various definitions and standards of the various countries. The Codex Alimentarius Commission at the Tenth Session in Rome 1–12th July, 1974 produced a relevant document ALINORM 74/32 June 1974 for Agenda Item 33 entitled 'Background Document on Vinegar, Eggs and Salt'. However, most of the information came from Europe on definitions, types, manufacture and processes, characteristics and composition, nature of regulations, production and trade. The Secretariat therefore was unable to make a recommendation as to whether standards should be elaborated for vinegar on a European, regional or World-wide basis. Nevertheless, much knowledge is given in this document which is pertinent here and, for the first time, hints at the possibility of proper universal standards.

A brief summary relates the following:

Accepting the term 'vin-aigre' to give the main characteristic property of the liquid vinegar and accepting the basic double fermentation manufacturing procedure, a general definition could read as follows:

> Vinegar is a liquid produced from a suitable raw material containing starch or sugar or starch and sugar by the process of double fermentation, alcoholic and acetous, and which contains a specified amount of acetic acid.

However, although this definition accords with the regulations of many countries, some countries, e.g. Austria, allow for the use of the name 'vinegar' for dilutions of chemically produced acetic acid (*saureessig*) and for mixtures of diluted acetic acid with vinegar (*essigverschnitte*). In other cases (e.g. New Zealand), diluted acetic acid for food use may be denominated 'imitation vinegar'. Otherwise, synthetic acetic acid in most countries, although allowed for human consumption, must be declared as such.

The following was found to be the broad classification of the types of vinegar encountered.

3.1. SYNTHETIC ACETIC ACID

As mentioned above, dilutions of acetic acid are not considered as vinegar in many countries. Pure acetic acid is obtained by dry distillation of wood and subsequent distillation of the intermediate wood vinegar with sulphuric acid, by oxidation of ethanol or from acetylene.

3.2. VINEGARS OBTAINED BY FERMENTATION OF SUITABLE RAW MATERIALS

3.2.1. *X-vinegar:* where X means the name of the raw material used. Commonly used raw materials are: cull apples, peels and cores—Cider vinegar; sugar syrup, molasses, refiners syrup—sugar vinegar; fruits and fruit waste (peels, cores etc)—Fruit vinegar. Other sources are honey, whey, palm juice and 'marc'. All of them are produced by the double fermentation process without intermediate distillation. A modification consists in the fact that some countries (e.g. France, Austria and Spain) allow as raw material substances such as beer, fruit wines etc. which have already undergone the alcoholic fermentation stage. The acetic acid content varies from not less than 3.8 of acetic acid (calculated as anhydride) in 100 ml (Belgium) to not less than 6% acetic acid (France). Many countries require the name of the raw material to be added to the generic denomination—vinegar (e.g. United Kingdom, Austria, Netherlands and Spain).

3.2.2. *Wine Vinegar*

Especially in wine-producing countries, the manufacture of wine vinegar is controlled by strict regulations. The product may be obtained only by acetous fermentation of wine ('natural wine' in Spain). Residues from wine production may not be used. France requires compliance with the E.E.C. Regulation No. 816/70.

3.2.3. *Malt Vinegar*

The United Kingdom has a proposal for the production of malt vinegar, stating that the raw material should be only malted barley, with or without the addition of cereal grain, the starch of which has been converted only by diastase of the malted barley. Manufacturing requirements are the same as under 3.2.1.

3.2.4. *Grain Vinegar*

Again the United Kingdom has proposed to apply the name grain vinegar to all vinegars produced from cereals in which the starch has not been converted into sugar by the diastase of malted barley. Other converting procedures are mineral or enzymatic hydrolysis. Manufacturing requirements are the same as under 3.2.1.

3.2.5. *Spirit Vinegar, Distilled Vinegar*

This product is obtained by the acetous fermentation of alcoholic distillates. However, the requirements for the origin of these solutions vary widely. The United Kingdom proposes only the use of distillates from a product of alcoholic fermentation. The distillation procedure serves the purpose of concentrating the vinegar stock to produce a vinegar with a higher acidity (10–13%) for the manufacture of other foodstuffs such as mayonnaises, pickles, sauces, salads etc. Only a small amount is used for direct consumption. Other countries do not define the origin of the ethanol so that industrial alcohol is used to a great extent (e.g. Austria, U.S.A. and France). In the U.S.A. distilled vinegar is produced in normal strength or as concentrated vinegar. The U.K. proposal requires the addition of 'distilled' to the denomination if the distillation takes place after the acetous fermentation process.

3.2.6. *Flavoured Vinegars*

Flavoured vinegars are prepared from various types of vinegar. Commonly added flavouring substances include tarragon, garlic, table salt and spices such as chilli, mint, thyme. Flavouring procedures are the following: addition of flavour extracts, direct extraction of flavour by vinegar, and addition of artificial flavours. The generic name of the flavouring substances has then to be added to the denomination of the vinegar, e.g. garlic malt vinegar (e.g. United Kingdom). The presence of salt is expressed by the term 'salted' (e.g. Belgium and the Netherlands).

3.2.7. *Blends of Vinegar*

Sometimes vinegars of various origins are blended. Most of the legislation require in such cases that the denomination contains the name of the components used (e.g. United Kingdom and Austria). In Austria where also the use of diluted acetic acid is allowed under the denomination vinegar, these blends are declared as 'vinegar blends' (Essigverschnitte). Other preparations are called special vinegars and consist of a mixture of spirit vinegars with other vinegars with or without the addition of flavouring extracts.

It is interesting to see that these types do not markedly differ from those of the Food Standards Committee, and nominally accept the U.K. definition for malt vinegar. However, the reservation of the Secretariat must be observed in that insufficient data World-wide are available to make suitable recommendations on a European, regional or World-wide basis at this stage, although they feel that there is a *prima facie* case for standards for vinegar.

III. MATERIAL REQUIREMENTS

From an examination of the definition of vinegar and consideration of the various types, it is obvious that the following material requirements of vinegar manufacture are worthy of special attention. These are firstly, the organisms, and secondly treatment of the raw materials and alcoholic fermentation with the acetification or acetation, particularly the process techniques on which it depends.

A. The Organisms

1. The Yeasts

As previously stated, the yeasts involved in vinegar manufacture are generally those used in alcoholic fermentation used to make the alcoholic beverage of the country concerned, particularly since excess or spoilt alcoholic beverages are often a convenient raw material for vinegar. However, most modern vinegar manufacturers deliberately make alcoholic washes for acetification (as is the case for charging wort in the U.K.) and thus the yeast is chosen deliberately. For wine vinegars, *Sacch. ellipsoideus* (= *Sacch. uvarum*), a wine yeast, is used and found to improve the flavour of the final product. Fermentations are carried out at 24–32°C. For malt vinegars, pressed yeasts (*Sacch. cerevisiae*) derived from ale breweries are preferred, but have to be renewed each time because the yeast dies at the end of the fermentation or at least is in such a poor and attenuated condition that it is unable to reproduce in fresh wort. Partly, this is due to infections which occur in the wort, both lactic and acetic, but also due to the extent to which the fermentation is taken. Frequently, *Sacch. diastaticus* is also added since fermentation of the residual carbohydrates by *Sacch. cerevisiae* may not be complete or fast enough. Fermentation times of only 48–72 h are preferred with temperatures in the range of 21–30°C. The 'boiling' fermentations of *Sacch. cerevisiae* are not always to be desired since there is a loss of alcohol with worts of high specific gravity. *Saccharomyces carlsbergensis* can then be used to advantage although the fermentations may be slower.

2. *The Bacteria*

The acetic-acid bacteria have a more important role to play in vinegar production than yeast but, like yeast, also have a considerable scientific history. They were first recognized and isolated in 1837 by F. T. Kützing who obtained the organism from naturally fermented vinegar, although Boerhaave (according to Lafar, through Asai (1968)) considered mother of vinegar to be vegetable in character. Many strains have subsequently been obtained from such sources and have been responsible for remarkable advances in biochemistry and microbiology. The mass of bacteria forming a film on wine and beer was studied by Persoon (1822), who proposed the name *Mycoderma* (viscous film), Desmazières (1826) distinguishing wine films as *M. vini* and beer films as *M. cerevisiae*. Kützings' name of *Ulvina aceti* was given because he thought they were a kind of alga, Thompson (1852) later changing the name to *M. aceti*. The first systematic studies on these bacteria were made by Pasteur (1868) who showed that they alone were responsible for acetification of alcohol although he did not identify them as bacteria. This task was left to Knieriem and Mayer (1873), and also repeated by Cohen (1872). Hansen

(1879) classified the bacteria into three kinds, namely *Bacterium aceti*, *B. Pasteurianum* and *B. kutzingianum*. Since then, virtually a pitched battle has raged over their classification, and it is not entirely solved to everyone's satisfaction to this day. A complete review of this situation has been clearly made by Asai (1968) but only a summary can be made here. *Mycoderma* (Persoon, 1822), *Ulvina* (Kutzing, 1837), *Umbina* (Naegeli, 1857), *Termobacterium* (Zeidler, 1896), *Acetobacterium* (Ludwig, 1898), *Acetomonas* (Orla-Jensen, 1909) and *Bacterium* (Beijerinck, 1900) have all been used as generic names for the acetic-acid bacteria. The generic designation *Acetobacter* was suggested in 1900 by Beijerinck, and *A. aceti* (Beijerinck) first appeared as a synonym for *B. aceti* (Hansen). In 1935, Asai proposed to differentiate the acetic-acid bacteria into two genera, namely *Acetobacter* and *Gluconobacter* gen. nov., the latter having only a limited ability to oxidize ethanol, accumulating gluconic acid rather than acetic. Most classifications until this time were based on taxonomic studies on bacteria isolated from alcoholic beverages or vinegar. Asai, however, included the oxidative bacteria living in various fruits, isolating 38 strains most of which exhibited quite different characteristics from the usual acetic-acid bacteria and lacked the ability to form films.

In 1954, Leifson observed the existence of peritrichous flagella on some commonly recognized species of *Acetobacter* and concluded that they be further subdivided into two genera, namely *Acetobacter* and *Acetomonas* gen. nov. The latter genus is identical with *Gluconobacter* not only with respect to flagellation but also to physiological and biochemical properties.

Also in 1953, a systemization based upon nutritional requirements was presented by Rainbow and Mitson and later Brown and Rainbow in 1956. One group with a predominantly lactate metabolism was termed 'lactaphilic', the other with a predominantly glucose metabolism was termed 'glycophilic'.

Several families have been proposed accommodating the genus *Acetobacter*, namely Oxydobacteriaceae, Nitrobacteriaceae and in a single genus in the family Acetobacteriaceae. Until recently, the Pseudomonadaceae was the only acceptable family for the genus *Acetobacter* (Bergey's Manual of Determinative Bacteria, 1957) but recent studies by Leifsen (1954), Asai, (1958) and Shimwell (1959), separately, have shown *A. aceti* as peritrichously flagellated so that it cannot be included in the family Pseudomonadaceae. Thus the proposal by Asai (1935), that the genus *Acetobacter* be placed in the family Acetobacteriaceae and the genus *Glucanobacter* in the Pseudomonadaceae as it appeared in Bergeys' Mannual (1939), has been revived.

A further problem and complication has been the variability of strains of acetic-acid bacteria first reported by Tošić and Walker (1946). This has caused considerable difficulty in arriving at sub-divisions in any classification; since the first proposal to divide the genus *Acetobacter* on the basis of biochemical properties by Frateur (1950a, b), four groups, namely *peroxydans, oxydans, mesoxydans* and *suboxydans*, considerable overlapping variations in the individual properties were found. Shimwell and Carr's (1964) work clearly showed that biochemical, physiological and morphological variants could be detected and probably were related by mutation. De Ley (1961) proposed abolishing specific names for all acetic-acid bacteria since there are no abrupt changes in the sequence of strains, and it was difficult to draw demarking lines in the gradation of properties defining the limits of species. Rainbow (1966) re-affirmed his proposed systemization of the bacteria into two groups (lactaphiles and glycophiles) based on their morphological, nutritional, bio-chemical and metabolic characters. Asai *et al.* (1964) confirmed Shimwell and Carr's and De Ley's findings as did Scopes (1962). Asai (1968) thus proposes the adoption of *A. aceti* Beijerinck as the type (and sole?) species in the genus *Acetobacter*, and *G. oxydans* (Henneberg) Asai as the type (and sole) species in the genus *Gluconobacter* because of their historical position in the literature. Other existing species should be considered varieties of the type species for each genus. This differentiation is identical with Leifson's (1954) and De Ley's (1961) classification except that *biotype* instead of *genus* was used.

3. *Other Organisms*

These are those organisms associated with spoilage although, if vinegar-charging wort is stored, they may contribute to a decrease in the carbohydrate content and an increase in desirable acidity. Here, lactic-acid bacteria are the most common although 'wild' yeasts, *Mycoderma aceti* and secondary fermentative yeasts may also be involved. Spoilage of the vinegar occurs due to 'vinegar eels', which are the nematode worms *Anguillula aceti* particularly in fruit or wine vinegars and in the quick generators. 'Wine flowers' or *Mycoderma* films occur freely and grow rapidly when charging wort or even vinegar is exposed to the air, often imparting musty odours and flavours. 'Mother of vinegar' or *Acetobacter xylinum*, which forms cellulosic films both colloidal and fibrous, occurs in the same way particularly in warm weather. 'Mothering' is perhaps the most common spoilage phenomenon of vinegar, and occurred frequently before the advent of sterile filtration and pasteurization. Some manufacturers add 1% (w/v) sodium chloride to combat mothering but also to enhance flavour. A mere decade ago, the public were quite used to decanting or filtering vinegar but now it is expected to remain clear for up to two years. Mites and vinegar flies (species of *Drosophila*) also occur in the factories or where vinegar is stored if the areas and equipment are not kept scrupulously clean.

B. Treatment of Raw Materials

Since fermentation of alcoholic washes derived from raw materials other than cereals are similar to those fermentations used to produce the alcohol associated with the beverage, and since low-grade or spoilt (partially acetic) samples are frequently used, a consideration of wine, cider and similar raw materials and their fermentations is of little significance. For malt vinegar, more can be detailed since the technique for producing the alcoholic wash or charging wort differs from that for ale or lager brewing although the basic processing equipment is largely the same. Surprisingly, little research or development has been done on this aspect of vinegar manufacture until recently.

Traditionally and in common with the brewing fraternity where, over the last 70 years, the main ingredient has been malted barley, in vinegar brewing, an all-malt mash has generally been used (Maule and Greenshields, 1970, 1971). A high diastatic malt is frequently chosen since the wort is not boiled and the enzymes remain active throughout most of the subsequent fermentation. Adjuncts are now used and up to 40% of the mash may be substituted with barley. Combinations with maize or wheat have also been used, but the definition for malt vinegar should strictly be adhered to in that the enzymolysis be accomplished by malt enzymes (fungal enzymes or acid should not be used). Normally, worts with specific gravities of 1.040 to 1.060 (10 to 15° Balling or 10 to 15%, w/v, sugar) are produced and fermented immediately without boiling (since hops are not required), thus allowing the diastase enzymes to survive and pass into the fermenting vessel where their action continues.

Until recently, therefore, the mashing process was quite conventional with milling and mashing into a tun with hot liquor. The basic equipment of a mill, grist case, spiral masher and mash-tun is therefore common to the brewing industry. Long mash stands and good sparging ensured maximum conversion and extraction. The protein breaks of traditional beer brewing due to hop boiling do not occur and indeed are not necessary since, in the later stages of the process, the high concentrations of alcohol and acid cause adequate precipitation and separation.

Modern mashing techniques have been introduced in some instances. Wet-milling and high-speed lauter-tun mashing as found in modern breweries have been used and these systems allow use of high proportions of barley or maize adjuncts.

A fully continuous system is also now in commercial operation (J. Yates, personal communication). This process consists of hammer milling various types of grain and sieving into fine and coarse streams. The fine stream passes further into the process. The course stream is further milled by a pin-mill and again sieved into a fine and coarse stream. The coarse stream is mostly husk and skin with some protein, and is sold as bran. The fine streams are combined and treated with alkali containing electrolytes. The protein fraction becomes partially soluble as a colloid, and can be separated from the starch granules by two high-shear separators and centrifuging. The separated starch with some protein is then gelatinized in a continuous scraped-wall heat exchanger and finely ground barley malt added to effect enzymic conversion under continuous controlled conditions. Two enzymic conversions are accomplished at two different pH values, the first in which viscosity is lowered and protease activity encouraged, the second involving diastatic conversion of the starch. This gives a highly converted extract ideally suited for good alcohol conversion by selected yeast strains. Separation of unconventional material is minimal and accomplished by continuous centrifuging. The protein fraction is used for other food-processing purposes, and helps to offset the cost of the process and price of the cereal. It is claimed that lower grade malts and barleys can be used in the process since the protein content is of little concern, thus giving even cheaper raw material feedstocks. Many of the handling problems are minimized by this process, and the accepted but wasteful brewing byproducts such as spent grains do not arise and are replaced by valuable raw materials for other uses.

C. Alcoholic Fermentation

The process for production of alcohol both in the wine and the malt vinegar industries has a history of 'laissez-faire'. This is due mainly to the fact that there is a marked seasonal demand for the vinegar, not unlike that in the cider and to some extent the wine industry. Conversion of sugar to alcohol should also be complete as possible and, since stored wash benefits by some infection, the wash can be stored for long periods before being acetified. Batch fermentations with little control over temperature have been common although more recently it has been shown that, in the fermentation itself, a low acidity due to infective organisms encourages attenuation and economic recovery of the alcohol. However, since 1960 a re-appraisal of the techniques used in the brewing industry re-awakened possibilities of improvements. High-speed batch and semicontinuous fermentation methods were studied but did not attain much success due to the problems of infection which are inherent in the production of charging wort. Apart from the fact that vinegar factories are heavily contaminated with the acetic-acid bacteria used in the acetification step, the fact that the wort is not boiled allows a further vector of infection into the fermenting vat. Lactic-acid bacteria (and coliform bacteria) from the grain survive the mashing process and can rapidly multiply to spoil any continuous-fermentation technique which does not take stringent precautions for maintenance of sterility. High-speed batch production has survived this difficulty providing that not more than three brews of fresh wort are added to fermenting wort in one fermenting vessel. Even then, care has to be taken to ensure that lactic acid does not accumulate (above 0.5–1.0%, w/v) to an extent where the final attenuation of the wort below a specific gravity of 1.000 is affected, the yeast being particularly sensitive at this point in the fermentation when the alcohol concentration is high and the rate of fermentation is slow, the last few degrees below a specific gravity of 1.000 to 0.994 being particularly crucial economically to the malt-vinegar process.

The brewing industry has adopted many of the new fermentation techniques, among them being the tower continuous-fermentation system (Royston, 1963; Shore and Royston, 1968) which has made a decisive impact. Its successful application in the brewing industry (Shore and Royston, 1968; Klopper *et al.*, 1965; Greenshields and Smith, 1971) attracted the attention of the vinegar-producing industry since the advantage of a continuous highly efficient fermentation with low residence times, i.e. reducing a 3–5 day fermentation to a few hours under a variety of conditions, fitted the requirements of vinegar manufacture. One instance of its application has been made and, apart from a few minor details, the system does not differ in principle and operation from its use in brewing. The design of tower fermenters and their associated equipment has been described elsewhere in detail for laboratory, pilot- and production-scale systems (Ault *et al.*, 1969; Greenshields and Smith, 1971). The fermenter consists of a vertical, cylindrical tower with a conical bottom. Above the tower section, the vessel opens into a large settling zone, which contains a yeast separator. Generally an overall aspect ratio of 10:1 with an aspect of 6:1 on the tubular section is used. The separator, which can be of various forms, provides a volume in the tower free from rising gas, such that the yeast may settle and return to the main body of the tower, and the relatively clear alcoholic wash or charging wort can be removed. A flocculent and sedimentary yeast is essential for producing an alcoholic fermentation in the tower because, otherwise, the yeast would be washed out and an insufficient concentration of yeast maintained in the tower. An average yeast concentration of 25% (w/w), expressed as centrifuged wet weight, is generally attained with values as high as 30 to 35% (w/w) at the bottom of the tower and as low as 5 to 10% (w/w) at the top.

Yeasts used for tower fermentation can be divided into three main groups based on their flocculent and sedimentary properties as assessed by laboratory tests (Greenshields *et al.*, 1971) and by their behaviour in the laboratory tower fermenter. These groups are: (1) non-flocculent, (2) flocculent—physically limited; these yeasts attain concentrations of 20 to 30 (w/w). At any particular specific gravity of the wort there is a critical volumetric efficiency. Volumetric efficiency is equivalent to space-velocity, i.e.:

$$\frac{\text{volume of wort}}{\text{days}} \times \frac{1}{\text{void volume of tower}}$$

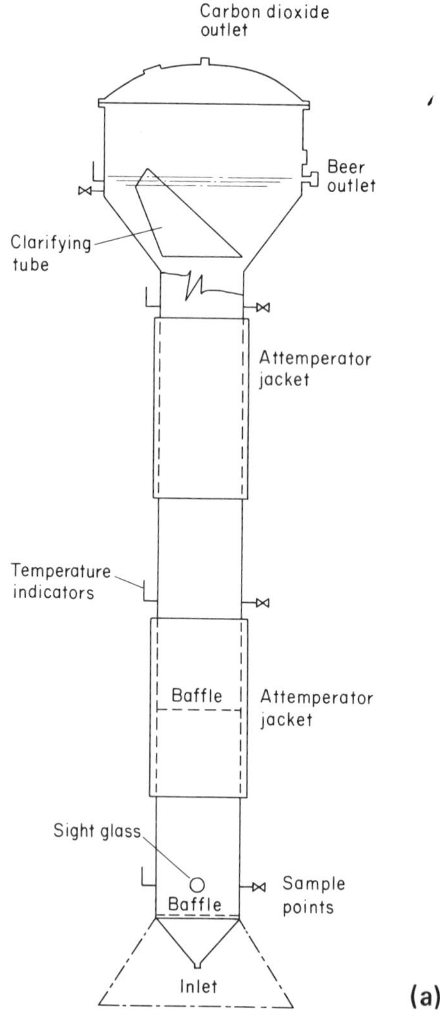

Fig. 2(a). The APV tower fermenter used to produce charging wort in the manufacture of malt vinegar. This figure shows a diagram of the fermenter.

which, if exceeded, causes a complete wash-out of this type of yeast; and (3) flocculent—fermentation limited; these yeasts attain concentrations of 25 to 40% (w/w) and are retained in a tower at all gravities and up to relatively high volumetric efficiencies. At a certain volumetric efficiency the expected attenuation of the wort can no longer be achieved. The yeast strain most suitable in the

Fig. 2(b). A photograph of a fermenter (one metre in diameter) being installed in a British vinegar brewery.

Fig. 3. Photograph of a laboratory glass-tower fermenter containing a culture of a flocculent type 3 yeast. This type of yeast is used for fermenting wort to produce malt-vinegar charging wort.

tower for vinegar brewing must be capable of resisting variable conditions (particularly washout); thus a type 3 strain is preferred (Fig. 3). Strains of *Sacch. diastaticus* and of *Sacch. cerevisiae* with these characteristics have been used. These behave well in having a limitation efficiency of about 4.0 with a wort of original gravity 1.050 before the effluent gravity rises to unacceptable limits. Final gravities of about 1.002 (0.5° Balling) have been obtained from worts with an original gravity of 1.060 (15° Balling). This is perhaps higher than the extent of attenuation desired in vinegar fermentations but, as pointed out by Ault *et al.* (1969), providing air is not introduced into the tower other than that dissolved in the wort, yeast growth is decreased markedly. The concomitant saving in sugar utilization can amount up to an equivalent of 2 to 4° gravity (1° Balling) giving, therefore, a final gravity equivalent to 0.998. The unboiled wort initially provided a sterility problem, but it can be collected at 80°C and held hot before being pasteurized in the tower. Nevertheless remarkable tolerances can be obtained in the vinegar fermentation, as previously stated, although technical sterility is desirable in tower operation.

The charging wort was often stored, for up to 12 weeks, in large wooden vats where acetic-acid and lactic-acid bacteria (even wild yeasts, secondary yeasts or mycoderma) continue to remove the remaining fermentable sugars, lowering the acidity (acetic acid desirable, lactic acid not so) and clarifying the charging wort ready for acetation. It has been shown, however, that storage is both unnecessary and sometimes even deleterious, and acetation of the charging wort immediately presents no problems (Jones and Greenshields, 1969, 1970a).

D. Acetification or Acetation

a. *Commercial.* Production of acetic acid by fermentation is often quoted as one of the oldest continuous systems. However, if Ricica's (through Málek and Fencl, 1966) definition of continuous fermentation is examined, and also the equipment used in acetification compared with Herbert's (1961) classification of continuous processes, it can be clearly seen that, although equipment similar to these types has been used for acetification, it has only been used in

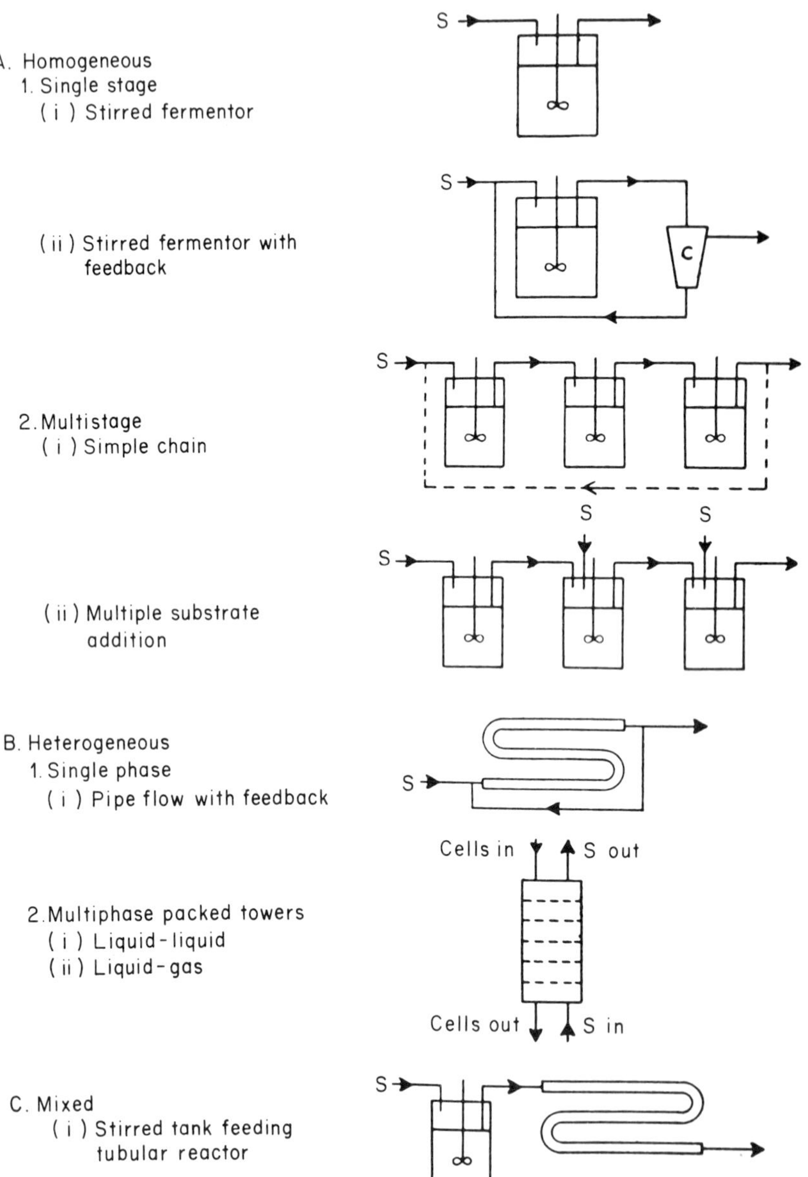

Fig. 4. Classification of continuous-fermentation processes, according to Herbert (1969). Most of these processes have been used to manufacture vinegar.

semicontinuous or batch operation (Fig. 4). It is more likely that effluent disposal is perhaps the oldest truly continuous system.

As previously stated, the souring of alcoholic drinks is a process which has been observed for thousands of years and one which manufacturers have tried to prevent. It comes as no surprise, therefore, that the oldest way to make vinegar was the 'let-alone'

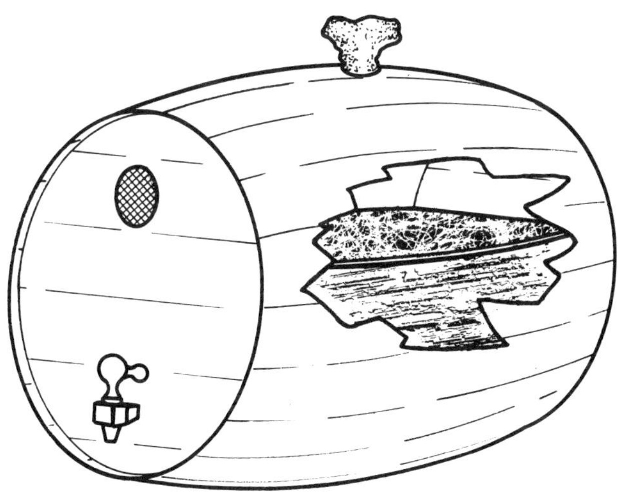

Fig. 5. The 'Slow' or 'Orleans' process for the manufacture of wine vinegar. The film or pellicle of *Acetobacter* spp. floats on the surface, and may be supported by a wooden raft. The vent hole is covered with a mesh to prevent ingress of flies and other organisms.

process. It was certainly known that air was necessary for acetification, but the microbiological action was not understood until the studies of Pasteur in 1862. From his work it was appreciated that the mother, or skin, which grew on the souring wine was a mat of bacteria which can form pellicles. The characteristic form of these is dependent on the nature of the bacteria and the environmental conditions.

It is obvious, therefore, that the earliest industrial manufacture of vinegar consisted of allowing the pellicle skin to form on the surface of wine and beer stored in partially filled casks. Casks with a 50-gallon (200 l) capacity were kept about two-thirds full and, when the fermentation had occurred after about five weeks, the vinegar (10–15 l) was siphoned out from under the pellicle. This volume, some 10% of the total, was then replenished with fresh wine or beer again under the pellicle, and the process repeated at weekly intervals. This, the Slow process as it became called, was often carried out in the open and the casks laid out in fields for rotational sampling much like the Solera sherry system (Fig. 5). Pasteur (1868) suggested that the films could be supported on floating wooden rafts since they often sank in the liquid and the culture had to be restarted. At a later date, particularly in the Orleans district of France, the casks were stored in warmed cellars which speeded up acetification. The Slow process thus became known as the Orleans process. The first recorded account of this method appeared from Richard in 1670. A modification to this process, which gave rise to future developments, was provided by the Dutch technologist H. Boerhaave who realized that the extent of acetification varied with the amount of surface area exposed to the air. He modified the Orleans process by adding pomace (the residue from grape presses) to the casks. The more 'modern' process, finally adopted by most vinegar factories, came after 1823 and was started by the German Schutzenbach, who initially drilled holes in the Boerhaave generators and introduced other types of porous material; this came to be known as the German process. However, Kastner also conducted experiments to improve the Boerhaave process. The exact contributions of Boerhaave and Kastner and the subsequent work of Schutzenbach are the subject of much confusion and speculation in the literature. The large number of small casks was replaced by a few large wooden vats filled with materials such as birch twigs, grape twigs, wood shavings or pomace. The alcoholic wash was continually recirculated and sparged over this material, whilst air was drawn up through the mass (Fig. 6). The types of material used as generator packing are almost too numerous to mention; they include materials such as coke, rattan, charcoal, ceramics, plastics, pomace, various types of wood such as beechwood, fir, cypress, redwood, oak, willow and certain pines in the form of shavings, twigs and segments. Choice of material was to a large extent a function of availability and economics. Usually, beechwood shavings have been preferred, the generator itself being in a variety of sizes and different woods including redwood, cypress, fir and oak. These lasted for up to 100 years and their use can be extended if the vessel is carefully cut down and reformed. This system greatly speeded up the process, cutting down the duration of a weeks-long process to days, and making it less cumbersome and labour intensive. It became known as the 'Quick' process or 'trickling'

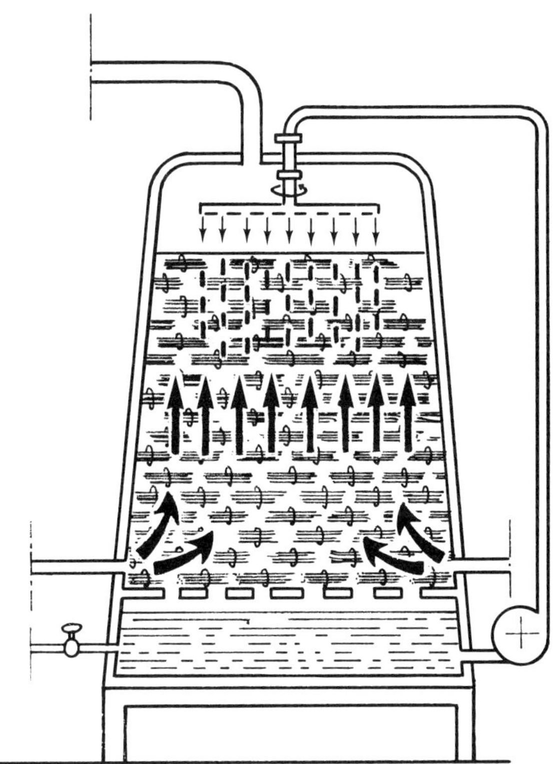

Fig. 6. Diagram showing the Quick or trickling process for the manufacture of vinegar. Two of these generators may be used in tandem in some plants. The packing material may be beech twigs tied in bundles.

process (Shimwell, 1952), and it has provided the basic acetator design for well over a century. There are many of these generator units still operating in the U.K. (White, 1961, 1966, 1971). Although this method was a great improvement, it has certain disadvantages in that it tends to select those *Acetobacter* spp., such as *A. xylinum*, which are slime-forming, and these gradually coat the packing materials. Conversion of alcohol to acetic acid in this process was rarely greater than 75–80%, often falling to 60% or less in time, necessitating replacement of the packing material every few months, although runs of up to a year were usual with careful management. Rolling generators have also been used commercially, but these require long acetification times and consume much power. Some modern manufacturers run two acetators in tandem, and thus gain time and efficiency by shortening the lag phase of growth. Conversion efficiencies of 95% or greater are possible with this system although the volumetric efficiency is lower than that experienced with submerged techniques.

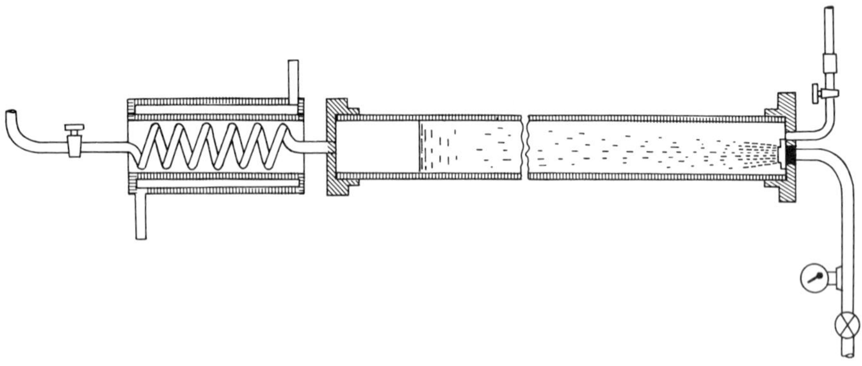

Fig. 7. A diagram of Shimwell's submerged pressure tower acetifier. From Shimwell (1955).

The possibility of using completely continuous techniques came with the appreciation of submerged fermentation as demonstrated and used in the antibiotics and baker's yeast industries. The earliest work using submerged techniques was largely ignored because no practical application came out of it. However, Hromatka and Ebner (1949, 1950, 1951a, b, c) inoculated white wine, in submerged aerated culture, with acetic-acid bacteria, and obtained successful acetification without packing materials. However, interrupting the aeration caused an almost instantaneous death of the bacteria under conditions of high acidity. This discovery was followed by the Enekel *et al.* patent (1953) which led to the development of the Frings generators. This work used aerated towers containing packing materials and was not restricted to the production of acetic acid. In

1955, British Vinegars Ltd., in a patent with Shimwell (1955), also utilized this principle but counted on an increased pressure and high-aspect ratio to enhance the aeration (Fig. 7). These vessels were towers with aspect ratios of 18:1 on the laboratory scale (7.5 x 5.0 cm diam) and a suggested 4:1 on the commercial scale (50 x 12.5 cm diam). However, their operation appeared to be batch or semi-continuous as did the original Frings patent (Els, 1955) although

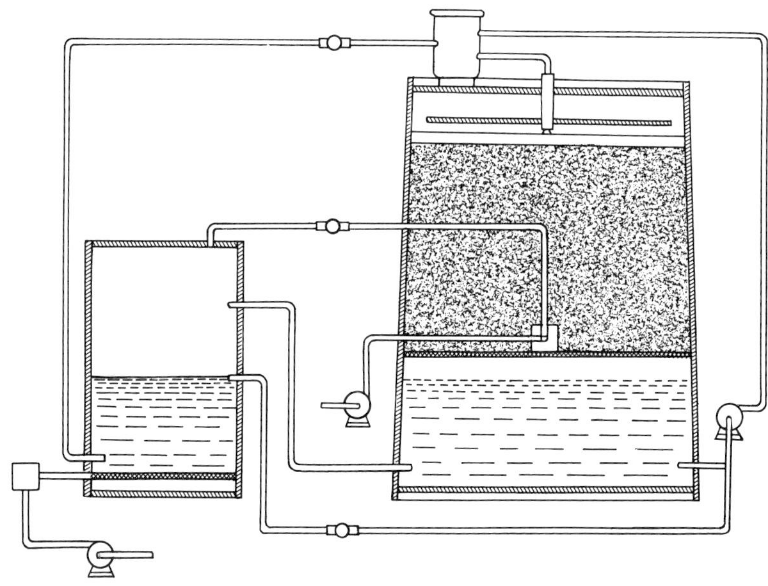

Fig. 8. Diagram of a Fessel generator which led to the development of the Frings acetifier. From Els (1955).

these (Fessel) generators used packing materials (Fig. 8). Perhaps the first truly continuous acetation claim may be that of Simonin and Bernard's device of multiple tower-shaped vessels in sequence which were bubble-aerated by submerged nozzles of special design, and patented in 1958. They claimed 92% conversion in 24 hours at 32°–35°C. The jet-aeration principle and recirculation in submerged culture (mostly in tanks of low aspect ratio) was also adopted by Richardson (1961, Fig. 9) and White (1961; Fig. 10). In the latter's patent there was the safeguard that aeration did not stop in any way since, in the highly acid conditions prevailing in submerged culture, the bacteria would be killed in only 30 seconds (White, 1970). White's system (Fig. 11) has been operated continuously on a commercial basis, but full details of its operation have not been published other than that 98% conversions are achieved.

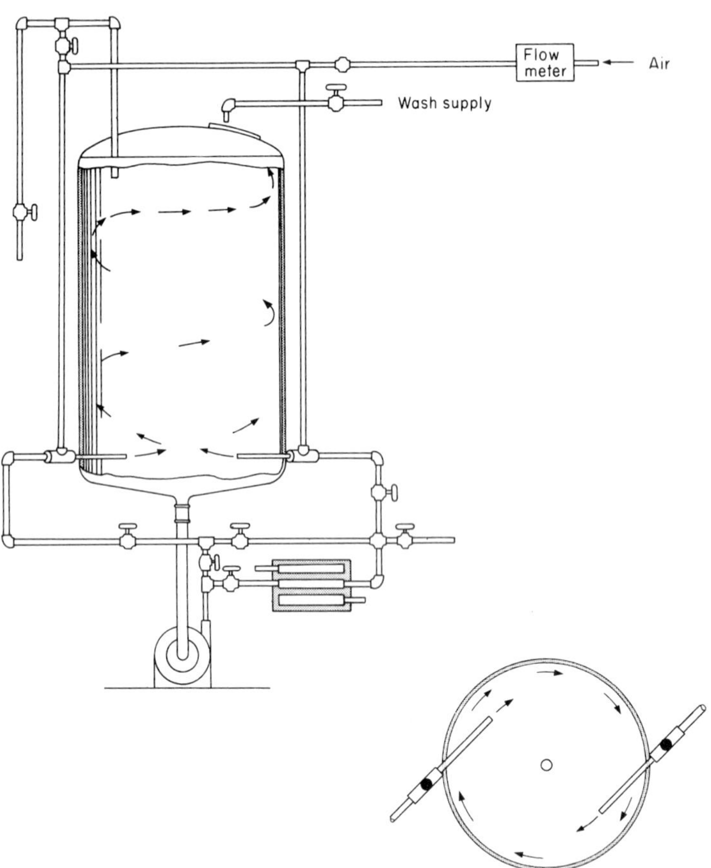

Fig. 9. Diagrams of top and sectional views of a jet-aeration acetifier. In this submerged fermentation system, air is injected tangentially into the base of the fermenter thereby creating a vigorous stirring action and increasing oxygen transfer into the culture. From Richardson (1961).

More widespread, however, are the modified Frings generators (Frings, 1968) which are semicontinuously operated vortex stirred tanks, for which minor modifications can make them almost continuous, giving efficiencies of up to 0.5 with conversions of 96–98%

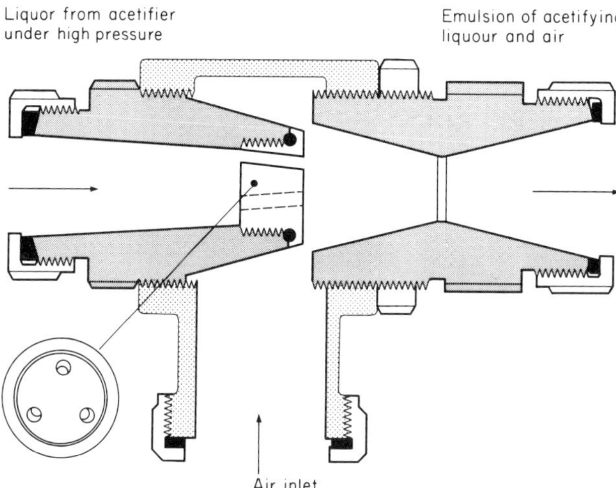

Fig. 10. Diagram of an aeration device used in a continuous submerged-culture acetifier. From White (1961).

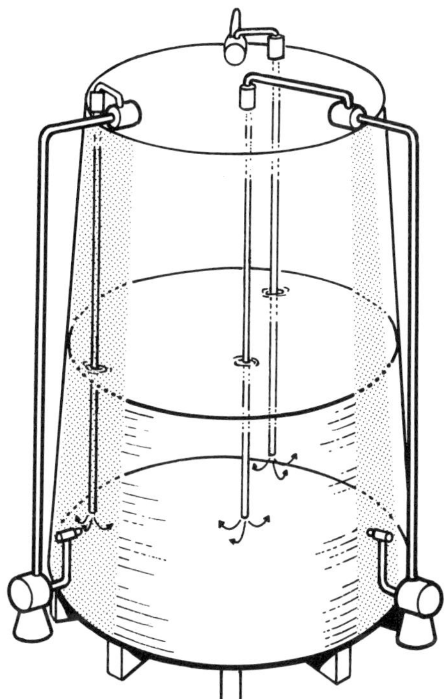

Fig. 11. A sectional diagram of the fermenter designed by White (1970) for submerged-culture acetification. Three aeration devices are used in case of mechanical failure.

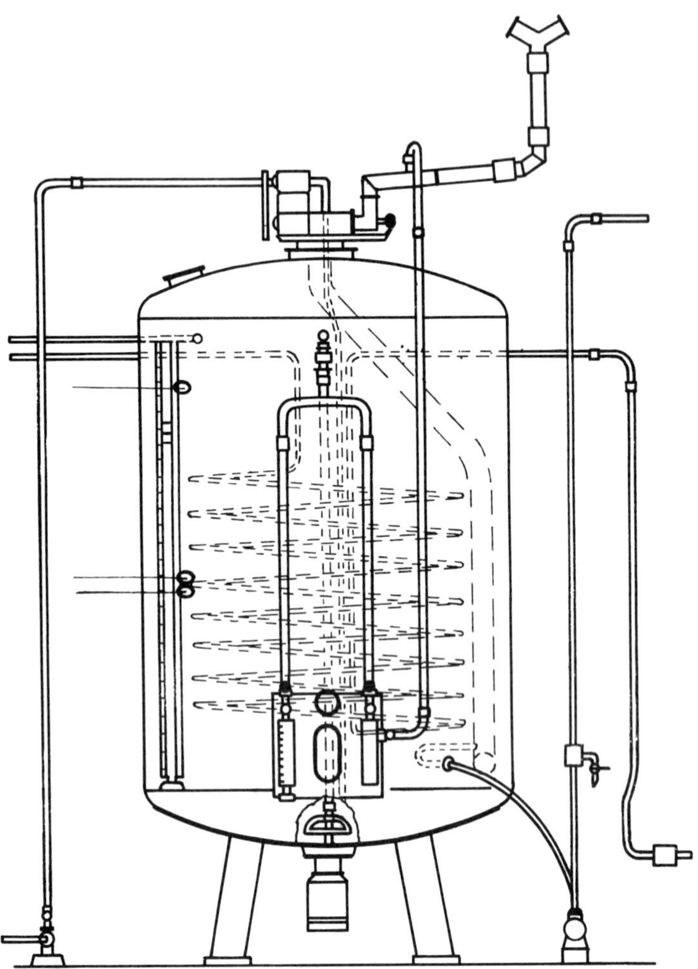

Fig. 12. Diagram of a section through a Frings generator fermenter used for the manufacture of vinegar. The fermenter, which can be used semicontinuously or continuously, employs vortex stirring.

(Fig. 12). These fermenters are fully automated for analyses and operation. Since their inception, the Frings generators have become standard equipment for acetation throughout the World and are probably responsible for most of the World's vinegar, some 350 units producing about 500 million litres of vinegar (10%, v/v, acetic acid) per year. They are capable of operating on the most diverse alcoholic washes at alcohol concentrations from 2 to 12% (w/v) and are capable of operation on a pure alcohol-salts medium for which specially adapted strains of acetic-acid bacteria have been developed. They are free from susceptibility to infection by the slime or film-producers, such as *A. xylinum*, and maintain remarkably consis-

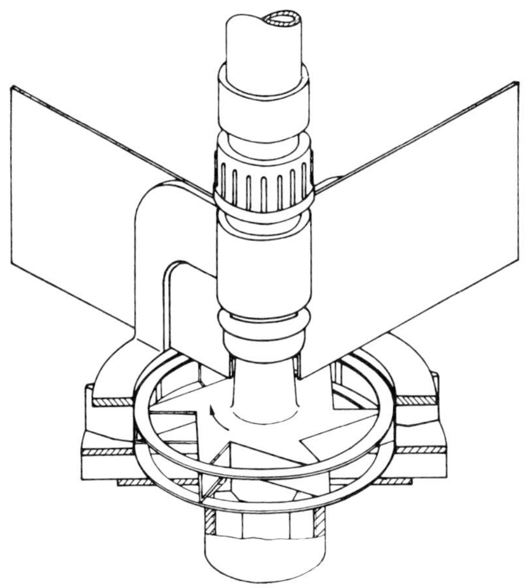

Fig. 13. Axonometric view of the cavitator or agitator used with the Frings generator. The cavitator consists of a turbine and a non-rotating stator. The turbine is designed as a hollow body with openings which are arranged radially and open against the direction of rotation. The openings are shielded by vertical sheets. The turbine sucks liquid from above and below, and mixes it with air sucked in through the openings. The suspension is thrown through the stator towards the circumference of the tank. An upper and lower ring on the turbine help to direct and regulate the air–liquid suspension. The stator consists of an upper and lower ring which are connected by vertical sheets inclined at about 30° towards the radius. The stator also serves to direct the suspension into all parts of the tank. The turbine is driven by a motor with an elongated shaft mounted to the bottom of the fermentation tank.

tent performance for extended periods with low maintenance requirements (Figs. 13, 14 and 15).

A fully continuous process for acetation already exists commercially (Greenshields and Smith, 1974). This is a hybrid tower-stirred tank with the air dispersed and the contents of the tank stirred by a high-speed stirrer at the bottom of the tower as in a conventional stirred tank reactor. The air is injected beneath the stirrer, and this is broken into bubbles. It is also further assisted in this by severe agitation against baffles placed on the walls up the tank (Fig. 16). High volumetric efficiencies with 98% conversion are claimed using air, and operating at high temperatures 40°C (105°F) with specially selected strains of *Acetobacter* spp. adapted for the purpose. It has been claimed that this system is only limited by foam formation which appears with almost non-Newtonian characteristics.

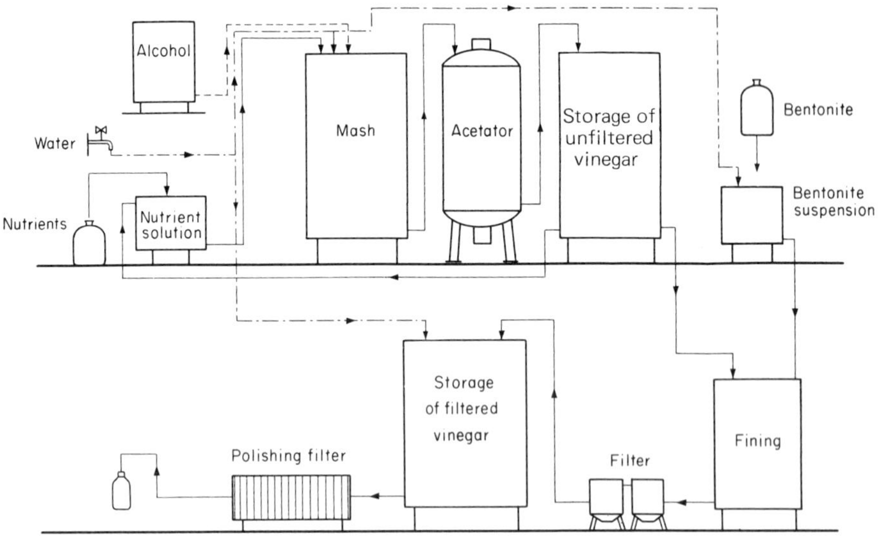

Fig. 14. Flow sheet for production of vinegar using the Frings acetator.

A modification of the tubular reactor for aerobic processes has also been applied to the acetification of alcoholic wash both in isolation and in conjunction with an A.P.V. tower fermenter used to produce the charging wort (Jones, 1970; Greenshields, 1972; Fig. 17). Initial work with M. Kimmett, and subsequently and more extensively with D. D. Jones, showed that successful continuous acetification could be achieved with 98% conversion of washes containing 4 to 8% (w/v) alcohol. A volumetric efficiency of 0.8 was obtained with air, and one of up to 1.8 with oxygen at

Fig. 15. Charts from the alkograph controller on the Frings acetator. The alkograph measures the concentration of residual alcohol during acetification. Signals from the recorder are used to control the rate of media feed and of product output. Chart (a) is from a semicontinuously operated generator, and shows the characteristic 'saw tooth' curve. Chart (b) is from a continuously operated generator.

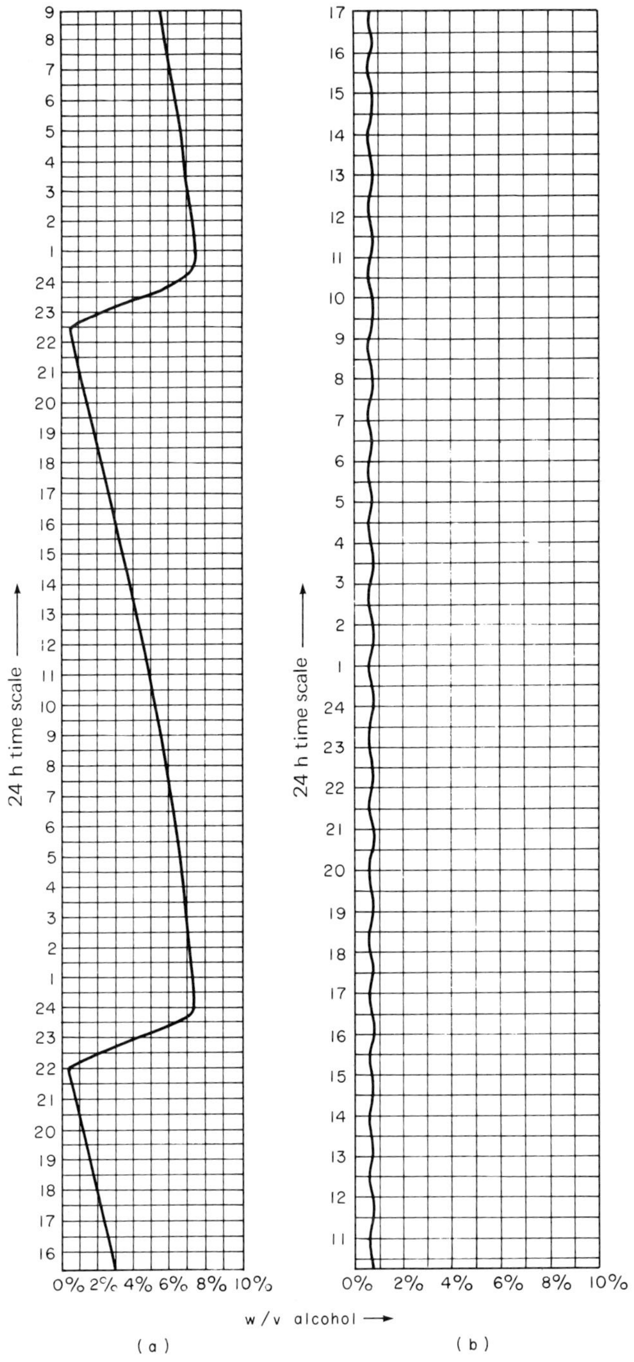
24 h time scale
24 h time scale
0% 2% 4% 6% 8% 10%
0% 2% 4% 6% 8% 10%
w/v alcohol
(a)
(b)

30°C–35°C over periods of 10 weeks using a mixed culture of *Acetobacter* spp. from Quick process generators. The tower had an aspect ratio of about 12:1 and consisted of 7.5 cm diameter glass tubes with special separation devices. The aeration, which was accomplished by sintered glass disks, was found to be limited to 0.5 volumes per volume per minute, mainly due to excessive frothing, although modern methods of mechanical defoaming alter this situation. The main advantages of this system are that the fermenter can

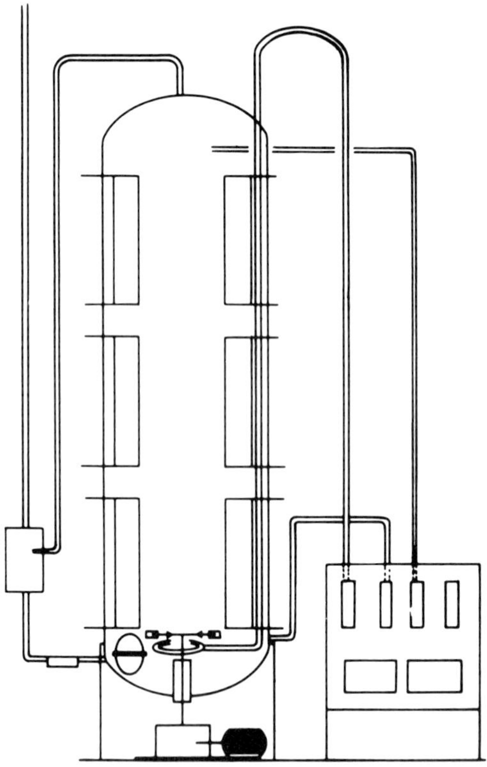

Fig. 16. Diagram of a hybrid stirred-tank tower acetifier for fully continuous submerged acetification in the commercial production of malt vinegar.

be simply and cheaply constructed (in plastic) since there are no moving parts, it is a fully continuous system and stable, and when in steady-state requires little attention and maintenance.

A pilot-scale unit in polypropylene-celmar re-inforced with fibre glass has been constructed for commercial use. The fermenter is 5.0 cm in diameter by some 6 m tall on the tube section, and with an expansion chamber of some 1.3 m diameter by 1.8 m tall and has a working volume of some 3,000 l. Aeration is accomplished by a stainless-steel or plastic perforated plate covering the cross-section of the tower and holding up the liquid. The charging wort or alcoholic wash is fed in continuously at the bottom, and a partially hetero-

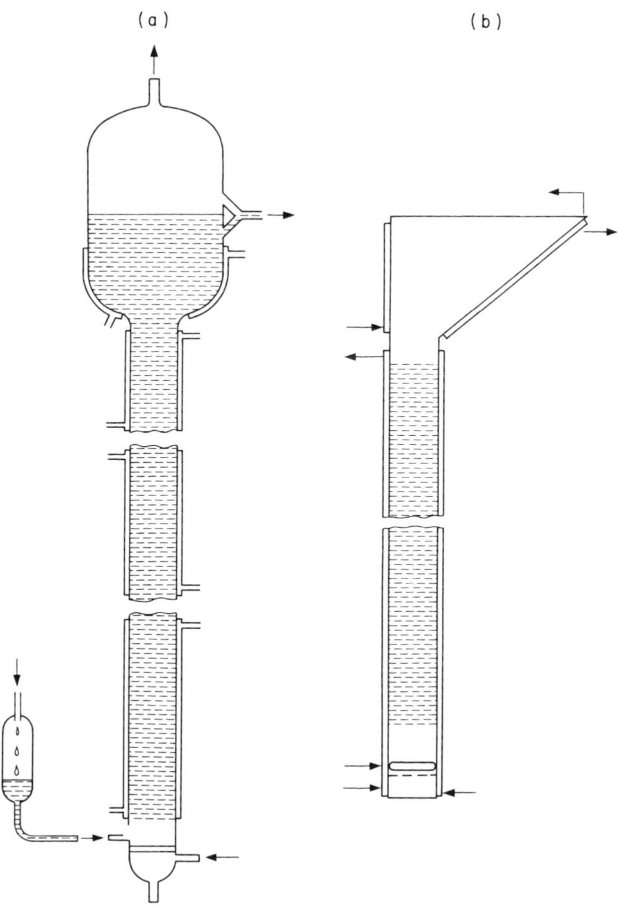

Fig. 17. Diagrams showing sectional views through (a) the Greenshields tower acetifier (from Greenshields, 1972); and (b) the A. P. V. tower fermenter (from Royston, 1963).

geneous system ensues, the aeration accomplishing the necessary agitation, but the aspect ratio giving rise to some degree of heterogeneity. A Y-tube separation system behind a baffle plate provides a quiet area free from the rising gas where the fermented acetic acid can overflow. The *Acetobacter* spp. are not as flocculent as yeast, but aggregate sufficiently to cause relatively high concentrations, and thus high reaction rates, in the tower. The unit can produce up to

450,000 litres of malt vinegar (5%, w/v acetic acid per year). To date, it has successfully been operated continuously for production of malt vinegar for four months with little attention or maintenance, and currently is being developed for spirit vinegar. A range of temperatures have been used, from 20 to 40°C, and a range of air

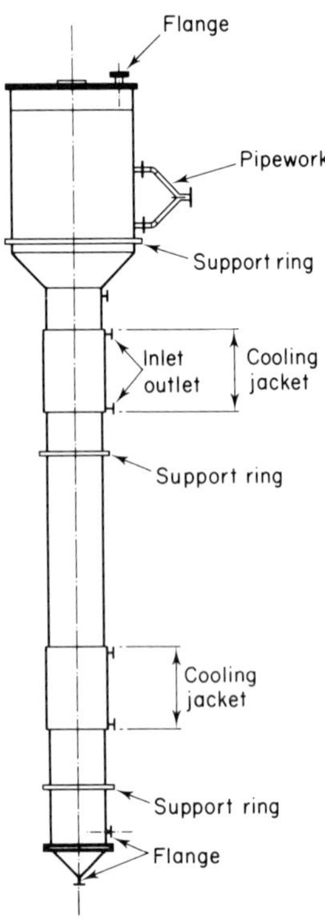

Fig. 18. Diagram of a tower acetifier of a type installed in a U.K. malt-vinegar factory for fully continuous acetification.

flows of 0.3 to 0.5 (v/v)/min (superficial gas velocities of 1 to 3 cm/sec). The culture *Acetobacter* sp. was taken from a Frings generator and requires up to one month to adapt completely to the new conditions as is indicated by the conversion efficiencies and volumetric efficiency of the system rising progressively, whilst the alcohol losses diminish.

The vinegar produced by all three of these methods, i.e. batch, semi- or fully continuous, differs little in flavour, this property

Fig. 19. Photograph of a pilot-scale production tower acetifier, of the type described in Fig. 18, being installed in a U.K. malt-vinegar factory. The polypropylene unit is shown without the air distribution plate attached.

depending more on the raw materials used in the charging-wort manufacture rather than the processing, particularly the acetification. Remarkable similarity in the composition of gross volatiles is found in most malt vinegars. Against a background of 4 to 6% acetic acid, it is a hard task even for the expert to tell differences (although continuous systems tend to increase the acetoin content of vinegar.

b. *Domestic.* Le Fevre (1924) described a convenient method of making vinegar for the household or farm essentially based on the Orleans process, and adaptable to acetification of any alcoholic liquor. A barrel or wine tun holding 25–45 litres is made into a 'converter' as follows. The barrel is supported on its side with the bunghole up. A hole (5 cm diameter) is bored slightly above the centre in one end of the barrel and another in the other end just under the stave containing the bunghole. The holes are covered with a plastic screen or a well-varnished fine wire screen. A spigot is fitted in place. A piece of rubber or plastic tubing, 1.2 cm in diameter, is attached to a glass or plastic funnel. The tubing should be long enough nearly to touch the opposite side of the barrel when the funnel is placed in the bungholes. The funnel is held in the bunghole by a wad of cotton or clean dry cloth. Unpasteurized vinegar (about 1.5 l) containing the mother of vinegar is poured into the converter through the funnel. The yeast-fermented alcoholic liquor is run in until the surface of the liquid is nearly level with the centre air hole. The converter is allowed to remain in a warm place at 30–35°C until acetification is more or less complete. The film which forms on the surface should not be disturbed. When a sufficient acidity has been produced, the vinegar is drawn off through the spigot, about 1.5 l being left in the barrel. A new supply of alcoholic liquor is added through the funnel. Transfers should be made slowly in order not to disturb the film which rises to the surface and starts the fermentation quickly. The converter can be drained and recharged every two or three months.

c. *Spirit and concentrated vinegar.* This account has been limited to a consideration of vinegar manufacture from alcohol derived directly and immediately by fermentation, particularly using natural carbohydrate materials. However, it is important to include discussion of the vinegar called spirit or distilled vinegar. Spirit vinegar should be distinguished from distilled vinegar since the latter type is manufactured by distillation of a finished vinegar such as malt or cider vinegar. It will contain, apart from acetic acid, other volatiles from the primary and secondary fermentations. These volatiles may also be enhanced by the use of herbs in the distillation to give flavoured vinegars suitable for a specialized condiment use for pickling.

Production of distilled or concentrated vinegar was developed in the 1960s to overcome problems of dilution in the pickling industry and also to overcome the problem of high transportation costs. The crucial question was why it should transport a product that is 90% water when it can be more cheaply concentrated without loss and diluted after shipping.

Apart from continuous distillation, another method used commercially is based on the principle of sluch-freezing and then centrifuging ice from the slush to produce a concentrate. The melted ice is re-used for production of vinegar since it can contain 2.5 to 6.0% acid, which is the original vinegar concentration. The slush is melted by precooling incoming vinegar through a scraped-surface heat exchanger.

Spirit vinegar is yet another commodity, a relatively recent innovation depending on the use of pure alcohol for acetification. There are probably two reasons for its development; firstly the requirement of vinegar where there was no ready access to a suitable alcoholic wash, and secondly the requirement of pickle and sauce factories for vinegar with higher concentrations of acetic acid than otherwise available. Considerable dilution occurs in pickling and sauce manufacture, and colour may play an important role in the product presentation. Initially, molasses (cane and beet) were used to produce alcohol as a base product by fermentation with specially selected strains of rum yeasts (*Schizosaccharomyces pombe* and related strains). Fractional distillation gave a spirit 6–7% overproof (see Volume 1, pg. 35). The spirit is transported to the vinegar factor where it is denatured by addition of spirit vinegar (often to overcome tax problems). Salts are added, usually ammonium sulphate, nitrate or phosphate, sodium or potassium phosphates and citrates together with a suitable supply of B-group vitamins either as dried yeast and/or yeast extracts. Successful acetation can be accomplished at a concentration of 10 to 12% (w/v) alcohol although, initially, an adaptation of the bacterial culture is necessary before the process is running satisfactorily (long lag periods of up to three weeks can occur). Various manufacturers have developed cultures suitably adapted for the purpose, and reports of strains capable of acetifying up to 15% (w/v) alcohol are current in the industry. High temperatures have to be avoided to prevent losses due to evaporation either of the alcohol or the acetic acid.

In the early 1950s, the availability of cheap oil and large plants capable of cracking petroleum to give byproducts led to the use of ethylene for production of ethanol. The demand for ethanol other than as a beverage has rapidly risen over the last half century. The alcohol began to be used for spirit manufacture. However, this has run counter to the accepted idea of the definition of vinegar and gradually, as the tide of opinion has risen against the use of oil products for human consumption on the basis of their possible carcinogenic properties, pressure has been exerted to stop the use of synthetic alcohol for this purpose. This only applied to some countries where the definition and food regulations for vinegar are law and strictly enforced. Nevertheless, studies have been initiated to differentiate between vinegar made from synthetic alcohol from that made from fermentation (biogenic) alcohol. It has been shown that biogenic alcohol, and acetic acid made from it, have a measurable radio-activity and the ^{14}C activity in ethanol and acetic acid from quite different origins were measured using liquid scintillation techniques. This allowed researchers to determine the origin of the spirit vinegar (Kaneko *et al.*, 1973; Hildebrandt *et al.*, 1969). However, although the lower ^{14}C radio-activity in fossil carbon can be detected from the higher concentration in photosynthetic carbon, it is not possible to distinguish acetic acid in vinegar derived from fermentation alcohol from that produced by the destructive distillation of wood.

At the moment, therefore, it is left to definition and law in the particular country concerned together with public opinion pressure on the manufacturers to ensure that vinegar is made from acceptable raw materials. However, it is the author's opinion that, in the future, this pressure will result in, firstly, a World-wide acceptable definition of vinegar as a natural product made by fermentation means from natural raw materials, and secondly that sufficient power will be given to regulations and control agencies to ensure this situation.

E. Production and Trading of Vinegar

Vinegar is produced and used in considerable amounts in almost every country of the World. Even where it is not produced, such is the requirement that it represents a large import. Unfortunately the commodity trade statistics rarely include vinegar as a separate item (say, for example, from wine) and it is difficult therefore to determine quantities. Moreover, since in countries where there is a ready availability of suitable alcohol raw material for acetification, vinegar is home produced; again the exact quantities of vinegar produced are difficult to ascertain.

The United Nations F.A.O. trade statistics has established the following facts: (1) the U.K. Food Standards Committee estimates that vinegar production in the U.K. is of the order of 90,000,000 litres per year (1974); (2) European Economic Community Agricultural Statistics (January, 1974) estimate that production of wine vinegar amounts to 830,000 hl for 1970/71 or 0.5% of the total wine production. Data on import of vinegar by major countries are given in Table 1.

An independent World-wide survey by the author elicited the following information (Table 2) from embassy, consular and trade delegation and department sources. Sometimes such information was not directly available but reference was given to internal trade

Table 1

Imports of various vinegars by major countries in recent years

Vinegar	Country		Import (proof gallons) in the year		
			1969	1970	1971
Malt	Canada		8,519	7,205	30,559
	United Kingdom		41,708	106,116	96,770
	Other		1,605	4,230	—
		Total	51,832	117,551	127,329
Wine	France		17,954	33,936	22,974
	Italy		18,091	19,468	18,986
	Japan		64,655	76,155	112,500
	West Germany		4,996	8,152	7,835
	Other		16,459	9,160	27,907
		Total	122,155	146,871	190,202

Table 2

Survey of World vinegar production

COUNTRY	Vinegar type	Commercial quantity per annum (1×10^{-6})	Domestic quantity per annum (1×10^{-6})	Type and quantity imported per annum (1×10^{-6})	Usage of synthetic acetic acid	Date of information	Source
AUSTRIA	Spirit Wine	12–14	—	None	No statistics	July 1974	Fachverband D.Nahrungs'u.Genuss-Mittelindustrie Osterreichs
BELIZE	None					July 1974	Belize Chamber of Commerce, Belize City, Belize
BRAZIL	Wine Spirit Cane	6	12	—	—	Jan. 1974	Antonio Borin S/A, Sao Paolo, Brazil
	Wine Spirit Cane molasses	—	—	—	Forbidden	Jan. 1974	Palhina S/A, Sao Paolo, Brazil
	Wine Fruit Spirit	—	—	—	No	Jan. 1974	Fabrica Trianon de Bebidas Lida, Rio de Janeiro, Brazil
CANADA	Spirit Cider Malt Wine	55.4 8.6 1.3 Small amount	Type not specified 0.2	Wine and spirit less than 0.9	None although a little in pickling fish in the Maritimes	July 1974	Statistics Canada, Canadian High Commission, London, U.K.
CHILE	Spirit Commericial Wine Cider Beer	1.5 4% acetic acid Not defined		None	None	July 1974	Division Alcoholes y Vinas, S.A.B., Chile

Country	Raw material						Source
CHINA	Statistics not available						Chinese Embassy, London, U.K.
COSTA RICA	Guineos (type of banana)	Yes Quantities not specified	Yes	Wine	Yes for food preservation	Oct. 1974	Confederacao Nacional da Industra, Rio de Janeiro, Brazil
CYPRUS	Wine or strengthened wine	0.5	0.15	Bottled malt vinegar	Yes for pickling and increasing acidity of fermented vinegar	Sept. 1974	Vine Products Commission, Cyprus
DENMARK	Spirit Cider Wine Malt	132 25–30 14 little (1972)	—	—	—	1974	Knut Rosen, Danish Distilleries, Asta, Denmark
ARAB REPUBLIC OF EGYPT	Alcohol from cane molasses	18 10% acetic acid	6 6.25% acetic acid	None	None	Aug 1974	Egyptian General Organisation for Food Industries, Cairo, Egypt
FINLAND	Spirit	1.97	—	Wine	None	1973	State Alcohol Monopoly (Alko), Rojamâki Factories, Finland
WEST GERMANY	Brandy Wine Fruit Herb Lemon	110.5	—	0.80 All types	Yes; 23.743,500 kg	July 1973	Verband der Deutschen Essig Industrie a.V. Bonn, Rhein, West Germany
HONG KONG	Rice Beans Spirit	126,760 kg	No statistics	448,550 kg Vinegar from China (? rice)	Yes quantity not known	1973	Commerce and Industry Department, Hong Kong

Table 2—*continued*

COUNTRY	Vinegar type	Commercial quantity per annum (1×10^{-6})	Domestic quantity per annum (1×10^{-6})	Type and quantity imported per annum (1×10^{-6})	Usage of synthetic acetic acid	Date of information	Source
IRAQ	Dates Grape }	70	—	None	Yes 10^6 l	1973	Iraq Embassy, London, U.K.
ITALY	Wine	0.50 (10% acetic acid)	No statistics	0.050 Wine	Forbidden	July 1974	Italian Embassy, London, U.K.
JAMAICA	Cane	0.90	Unspecified	0.255 distilled cider, malt & spirit	None	Aug. 1974	Jamaica National Export Corporation, Kingston, Jamaica
JAPAN		220 1972		Vinegar and substitutes 0.168 1972		Feb. 1974	Japan Trade Centre, London, U.K.
	Rice Spirit Malt Cider Wine }	144 2,000				1974	Kewpie Jozo Company Co. Ltd., Tokyo, Japan
MALAWI	None		None	0.033 Spirit malt	None	July 1974	Ministry of Trade, Industry and Tourism, Blantyre, Malawi
NEW ZEALAND	Malt Cider Spirit Herb Wine Honey }	—	None	Not specified; 0.012	Yes	July 1974	New Zealand High Commission, London, U.K.

PAKISTAN	Fruit wastes Wine	0.045	—	None	Yes; 400 cwt	Aug. 1974	Embassy of Pakistan (Trade Division), London, U.K.
PORTUGAL	Wine	7	—	0.065 Malt	Yes 300,000	Jan. 1975	Portuguese Embassy, Commercial Attache, London, U.K.
RHODESIA	No information due to political situation					1974	Ministry of Commerce & Industry, Salisbury, Rhodesia
SINGAPORE	None	—	None	Not specified 675,000 l (1972)	Yes,	July 1974	Singapore Embassy, London, U.K.
SWITZERLAND	Wine Cider Spirit Whey	25 (1.3 × 10^6 kg acetic acid)	None	None	Yes; 350,000 kg	Aug. 1974	Interessengemeinschaft der Schweiz Garungs-Essen-Industrie (ISG) Solothurn, Switzerland
THAILAND	Spirit Rice Sugar cane Pineapple Corn	0.25	—	(type not specified) 0.050 (Vinegar, 1972)	1,556 tonnes (1971)	July 1974	Thai Chamber of Commerce, Bangkok, Thailand (Applied Scientific Research Corporation of Thailand)
TRINIDAD AND TOBAGO	Tropical fruits Spirit	0.50 (1973)	Not known	Table vinegar 0.30 (1973)	Yes 470,000 l (1973)	July 1974	Export Promotion Office Ministry of Industry and Commerce, Port of Spain, Trinidad and Tobago
TURKEY	Wine	3	7	None	Yes in mustard and pickling	July 1974	Union of Chambers of Commerce Industry and Commodity Exchanges of Turkey (Odalar Birligi), Ankara, Turkey
UGANDA	Vinegar not manufactured					July 1974	High Commission of the Republic of Uganda, London, U.K.

Table 2—*continued*

COUNTRY	Vinegar type	Commercial quantity per annum (1×10^{-6})	Domestic quantity per annum (1×10^{-6})	Type and quantity imported per annum (1×10^{-6})	Usage of synthetic acetic acid	Date of information	Source
UNITED KINGDOM	Malt Spirit Cider Wine Herb	100 (1974)	None	Wine quantity not known	Yes	Jan. 1975	U.K. Federation of Vinegar Manufacturers, London, U.K.
U.S.S.R.	No statistical information					1974	Sojuzchimexport Moscow, U.S.S.R.
U.S.A.	Malt Wine Cider Spirit Sugar (corn)	770		Rice Cider Malt Wine 1.4 (1971)	Yes	April 1974	U.S. Embassy, London, U.K.
YUGOSLAVIA	Spirit Wine	20 (10%)	20 (6.7%	None	Yes 15×10^6 l 10% acetic acid	Jan. 1975	Prof. S. N. Ban

departments, statistics bureaux and special import-export documents. When there was no official survey or statistics available, industrial firms involved in vinegar manufacture were contacted on recommendation. Occasionally the information was classified or not available for political reasons. Although all countries of the World were requested to take part in the survey, it can be seen that only some thirty countries replied and many of these did not return quantitative data. A complete World picture of vinegar manufacture is not therefore possible as was found by the U.N.F.A.O. initial investigation.

However, examination of the available quantitative data from 15 of the countries and suitable extrapolation can give some idea of vinegar usage. Based on population, a range of some 0.2 to 38 l per person per annum of vinegar is utilized by some 670 million people. However, two countries, Japan and Denmark, contribute disproportionately to this ratio (22 and 38 l per person per annum, respectively). Excluding these gives an average ratio of 2.1 l per person per annum for 572 million people. Using this figure on a World population of 3,700 million gives a possible World usage of some 8,000 million litres per annum. Another way to arrive at a similar figure is to examine the ratios for each type of country, and apply this figure to the various types of countries in the World. Figures of 6 to 10,000 million litres per annum are obtained, which is in reasonable agreement with the first estimate. Vinegar has a restricted usage as a commodity and its use depends to a large extent on population and the diet of the country concerned; thus application of the calculated ratios of usage would be a reasonable estimate. Some Third World countries do not either make vinegar or even use a vinegar other than that of accidentally soured-wine. The value of vinegar as a preservative is yet to be discovered or made available to them. Therefore it is envisaged that the use of vinegar World-wide will increase, at least by a factor of two, not including the natural increase due to population.

In all returns of the survey, vinegar was directly used as a condiment, in pickling, in sauce or food preparation and manufacture. In countries where considerable amounts of vinegar were manufactured by fermentation, use of acetic acid from synthetic sources as vinegar or a type-vinegar is prohibited, although frequently allowed to be used to increase acidity in pickling or in sauce or food manufacture. In other countries where little or no fermentation vinegar was made, synthetic acetic acid or distilled vinegars were imported as a condiment vinegar. Two countries with notable differences were Egypt and Thailand. In both, fermented vinegar was distilled to make acetic acid and use of acetic acid from synthetic sources was allowed for all purposes particularly to increase acidity in pickling. Probably the hot climates and the dilution that can occur in pickling make it necessary to use high concentrations of acetic acid in such processes. A number of countries entirely prohibit the use of synthetic acetic acid, and it is believed that, in the future as the technology of vinegar manufacture disseminates throughout the World, this will become the general case. Certainly, fermented vinegar can be made with a concentration of up to 14–15% (w/v) acetic acid, and further concentration to give a distilled or concentrated vinegar will increase this concentration and provide the required concentration of acetic acid for all food processing.

The United Nations Food and Agriculture Organisation is now engaged in surveys of this type, particularly of commodity foods, to determine the World situation. The disastrous harvests of 1973, and the now critical size of the World population and its predicted increase, have clearly highlighted one of the major problems of the future of mankind on planet earth. The importance of the problem clearly culminated in the Rome Conference where the current nutritional problems are clearly highlighted. The effects were shown to be most serious with nations of the Third World with indications that up to 40% of the World population will suffer from poor nutrition. This problem has become more dramatic since its influence does not rest here, and for the first time the rich and more industrialized and sophisticated nations are feeling its effect. They now cannot ignore the problem either directly or politically. Food supplies are now very much their concern to the extent that, in the next decade, they too will feel the pinch and poorer nutrition could be their lot.

This leads to an understanding that food storage and preservation apart from food supplies will take on considerable significance, and it is here that vinegar (acetic acid) will play a major role. Third World nations will need to be introduced to the value of vinegar in food preservation. Plants for manufacturing vinegar, on both a large and small scale (village-level technology), will be required. In the industrialized nations more information will be needed on vinegar and on its use to preserve food. The use of different types of carbohydrate raw materials, particularly those either presently wasted in food processing or not currently considered as tractable, will have to be considered, and the necessity of natural processes for the production of vinegar will become more important as the pressure for energy-raw materials increases. The use of synthetic raw materials derived from non-renewable petroleum products should be avoided and the use of the renewable carbohydrates encouraged. The manufacture and use of vinegar, therefore, has a considerable future. Despite its slow growth, the long tradition of this industry with its special art and science coupled with its use of modern science and fermentation technology will have a yet greater influence on mankind than its humble origin and concept of manufacture the 'let alone' principle would suggest.

F. Biochemistry

1. *Nutritional Requirements of* Acetobacter *spp.*

Whilst, in commercial manufacture, wine, beer, cider or other alcoholic liquids have provided adequate media for acetic-acid bacteria, an investigation of the nutritional requirements of these bacteria has revealed a more complex situation. Moreover it has provided some basis, although by no means exact, for the classification of these and related micro-organisms. Extension of this information has given a valuable insight into the preparation of properly balanced commercial media for the manufacture of spirit vinegar.

The ability of certain acetic-acid bacteria to utilize ammonium salts as a sole source of nitrogen when either alcohol or acetic acid was the carbon source led Hoyer (1898) to develop a synthetic medium for these organisms. It consisted of ammonium phosphate (0.1 g), primary potassium phosphate (0.1 g), magnesium phosphate (0.1 g), sodium acetate (0.1 g), ethanol (3 ml) and 100 ml distilled water. Janke (1916) called these the 'haplotrophic' acetic-acid bacteria and differentiated them from the 'symplotrophic' acetic-acid bacteria with more complex organic nitrogen requirements. Vaughn (in Bergey's Manual, 6th ed.) used the ability to grow in Hoyer's medium in the classification of the acetic-acid bacteria, differentiating *A. aceti* from *A. xylinum* and *A. rancens* on this basis. However, Leifson (1954) found that *A. aceti*, and four strains of *A. xylinum* which produce leathery pellicles, grew well in Hoyer's medium and claimed that the distinction was meaningless. Frateur (1950) modified Hoyer's medium to consist of ammonium sulphate (1 g), potassium dihydrogen phosphate (0.9 g), dipotassium hydrogen phosphate (0.1 g), magnesium sulphate hydrated (0.25 g), 95% ethanol (30 ml) and 0.5 ml of (1%, w/v) aqueous solution of ferric chloride (hydrated) made up to one litre with distilled water, and claimed that it gave better results. Shimwell (1957) concluded from his experiments using this modified Hoyer medium that an inability to grow did not necessarily mean an inability to utilize ammonium salts but rather of the utilization of an ammonium salt with ethanol as the sole source of carbon. In these experiments, *A. xylinum* grew well in the modified medium. Most *Acetobacter* spp. are auxotrophic for ethanol but prototrophic for glucose, although others have opposite properties. With the exception of *A. gluconicus*, even strains which did not grow in a glucose-salts medium were capable of growing in the complete medium proposed by Dunn *et al.* (1947).

The assimilation of ammonium salts by acetic-acid bacteria in synthetic media, therefore, depends on the nature of the carbon source used and, with some species, on the presence of amino acids, vitamins and purines as cofactors. Brown and Rainbow (1956) determined the nutritional patterns of 28 strains of acetic-acid bacteria on lactate and glucose after extensive experiments on the utilization of carbon sources. Two groups, one with a predominantly lactate metabolism (lactaphilic group) and the other with a predominantly glucose metabolism (glycophilic group), were distinguished.

Asai (1968) has studied the assimilation of glucose and ethanol by 25 strains of *Acetobacter* representing 11 species, and 29 strains of *Gluconobacter* representing 13 species grown, in modified Vaughn's medium containing vitamins, amino acids and ammonium sulphate. Strains which assimilated ethanol (Brown and Rainbow's lactaphilic group) were found to belong to the genus *Acetobacter*. Only one *Acetobacter* strain (*A. capsulatus*) utilized glucose as the sole source of carbon. Those strains which assimilated glucose (glycophilic group) were found solely in the genus *Gluconobacter*. No *Gluconobacter* strains examined could utilize ethanol as the sole source of carbon. It was concluded that, although the selective assimilation of ethanol or glucose is a characteristic of *Acetobacter* or *Gluconobacter* spp., this property should not be considered as a criterion for distinguishing the two genera since there are *Acetobacter* strains which do not assimilate ethanol and *Gluconobacter* strains which do not assimilate glucose. Many of these strains could grow following the addition of both glucose and ethanol; some strains, however, did not grow at all even in the presence of both carbon sources.

A more confused situation exists with regard to the requirements for vitamins and amino acids, and no final conclusions can be made. The inability to describe strains accurately, based on nutritional requirements, and the ability of the acetic-acid bacteria to mutate fairly freely, contribute to this dilemma. Completely defined media will support growth of almost all strains, but there are exceptions even to this generalization.

2. *Oxidative Powers*

The reaction termed the 'acetic acid fermentation' was first noted by Pasteur (1868) who recognized the conversion of alcohol (wine) to acetic acid under aerobic conditions. He ascribed it to the action of acetic-acid bacteria, and this has remained the characteristic ever since. According to Henneberg (1926) the highest ethanol concentration that can be converted by a particular species is 11% by volume, but it is well known in the industry that spirit vinegar can be readily made from 12% by weight of alcohol. It is known (C. Ebner, personal communication) that up to 14–15% by weight of alcohol can be converted, but it is probable that this is by specially adapted mixed cultures. At the lower end of the scale, 2% by weight can be converted and, in Japan, such dilute vinegars are frequently made from rice alcohol washes (A. Mori, personal communication).

Tanaka (1938) studied oxidation of various alcohols and aldehydes by *Acetobacter* spp. by measuring oxygen uptake by resting cells, and Table 2 gives values for the relative activity to alcohol oxidation at pH 6.0.

Addition of such compounds to cultures of commercial acetic-acid bacteria growing on alcohol wash derived from malt gave different activities (Jones, 1970) but the pattern was the same. This is to be expected since most commercial cultures are mixed and adapted to the particular chemical and physical environment in which they occur. During acetification, oxidation of ethanol by *Acetobacter* spp. is thought to occur by a two-stage oxidation system through acetaldehyde (Hoyer, 1898; Henneberg, 1891) to acetic acid (Neuberg and Nord, 1919); Neuberg and Nord verified the reaction by trapping the acetaldehyde with sulphite:

$$CH_3.CH_2OH + [O] \longrightarrow CH_3.CHO + H_2O$$
$$\text{acetaldehyde}$$

$$CH_3.CHOH.OH + [O] \longrightarrow CH_3.COOH + H_2O$$
$$\overline{CH_3CHO + O_2 \longrightarrow CH_3.COOH + H_2O}$$

This is a highly exothermic reaction, as is well understood in industry since almost all acetators have to be cooled. The reaction can occur over the temperature range 20–44°C with an optimum at 30–32°C, but the reaction is slow at 25°C or below and overoxidation can occur above 40°C.

A second mechanism or pathway has been proposed by Neuberg and Windisch (1925) involving a dismutation (Neuberg, 1928) in which two molecules of acetaldehyde form one molecule of acetic acid and one molecule of ethanol. This was discovered under anaerobic conditions with *A. ascendens*, *A. pasteurianus* and *A. xylinum*, but it proceeds under aerobic conditions up to 50%, the remaining half of the acetaldehyde being oxidized to acetic acid:

$$2CH_3.CHO + H_2O \longrightarrow CH_3.CH_2OH + CH_3.COOH$$
$$\text{acetaldehyde} \qquad \text{ethanol} \qquad \text{acetic acid}$$

Thus 75% of the acetaldehyde is converted to acetic acid and 25% to ethanol:

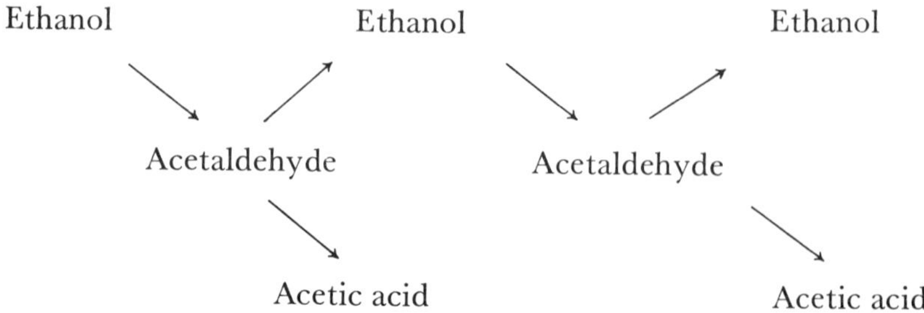

Table 3

Oxidation of alcohols and aldehydes by *Acetobacter* spp. After Tanaka (1938)

Substrate	Species		
	Acetobacter peroxydans	*Acetobacter rancens*	*Acetobacter aceti*
Acetaldehyde	473	536	405
Allyl alcohol	740	720	600
n-Amyl alcohol	600	640	850
Isoamyl alcohol	150	63	70
Benzaldehyde	—	17	8
Benzyl alcohol	35	2	3
n-Butyl alcohol	750	740	927
Isobutyl alcohol	140	61	60
sec-Butyl alcohol	84	8	7
tert-Butyl alcohol	0	0	0
Isobutyraldehyde	231	272	170
n-Cetyl alcohol	2	0.2	0.3
Cinnamaldehyde	17	15	5
Cinnamic alcohol	25	22	30
Citronellal	19	15	2
Citronellol	31	13	22
Crotonaldehyde	263	110	50
Dimethyl propyl alcohol	0	0	0
Ethyl alcohol	1000	1000	1000
Ethylene glycol	200	35	40
Formaldehyde	140	130	20
Furfural	134	235	200
Furfuryl alcohol	250	128	130
Glyceraldehyde	30	36	—
Glycerol	15	17	10
n-Hexyl alcohol	350	256	420
Methyl alcohol	25	8	7
Methyl heptyl alcohol	12	3	2
Methyl butyl alcohol	22	4	2
Phenylacetaldehyde	—	—	—
Phenylethyl alcohol	39	6	8
Phenylpropyl alcohol	45	37	17
n-Propyl alcohol	700	766	960
Isopropyl alcohol	48	9	8
n-Propionaldehyde	497	842	700
Saligenin	4	3	1
Salicylaldehyde	—	1	4
Vanillin	8	14	6

Figures show relative activity compared to that for ethyl alcohol = 1000.

Wieland (1913) demonstrated production of acetic acid in the absence of oxygen, providing hydrogen acceptors such as methylene blue were added. Since this reaction is not inhibited by cyanide or carbon monoxide, it was concluded that acetic acid was formed by the action of dehydrogenases. Moreover, Wieland and Bertho (1928) found that, under fully aerobic conditions, dismutation activity is low and thus they proposed that it is an unlikely pathway in normal acetification situations. Jones and Greenshields (1969, 1971), examining the volatile compounds in commercial vinegars, also concluded that dismutation could only be minimal under the highly aerobic conditions of submerged aeration, low ethanol concentrations and low pH values found in modern acetators. Moreover, little acetaldehyde could be found at the low pH values either during or after the process and, with the low residence times, they considered the main pathway of Wieland and Bertho (1928) the likely route. White (1970) suggests that, in his continuous acetator, since interruption of air for only 30 seconds can kill the acetic-acid bacteria almost completely, a build up of acetaldehyde within the cells is responsible. This occurs, he proposes, because of the difference in the speed of the reactions in the Wieland mechanism and the lack of a suitable hydrogen acceptor when the air is cut off.

However, according to Simon (1930), the ratio of dismutation to direct oxidation is dependent on environmental pH value, with dismutation decreasing under acidic conditions, as is found in commercial manufacture (pH 2.0 to 3.0). Nevertheless, Bertho and Basu (1931) state that dismutation is independent of pH value.

3. Production of Other Compounds

Whilst oxidation of alcohol to acetic acid characterizes *Acetobacter* spp., various other oxidations by these bacteria are of particular significance, since some of the compounds which form suitable substrates for these bacteria can only be prepared with great difficulty by chemical methods. Those species which oxidize such substrates completely to carbon dioxide and water have little industrial value, but those which only partially oxidize the substrate may be of considerable importance.

A good example of this phenomenon is *A. suboxydans* which is well adapted for industrial use for it mostly brings about an incomplete oxidation of sugars, alcohols and acids even when supplied with large amounts of oxygen sufficient to oxidize the complete substrate. *Acetobacter xylinum*, the sorbose bacterium, also brings about an incomplete oxidation of certain substrates but, in the presence of large quantities of oxygen, may oxidize the substrates completely to carbon dioxide and water. In the list of substrates oxidized by *A. suboxydans* (Table 4), the products in most instances may also be produced by other species of the genus *Acetobacter*.

Table 4

List of substrates oxidized, and products formed, by
Acetobacter suboxydans

Substrate	Product
2,3-Butylene glycol	Acetylmethylcarbinol
Erythritol	L-Erythrulose
Ethylene glycol	Glycollic acid
D-Gluconic acid	D-2-Oxogluconic acid
D-Gluconic acid	D-5-Oxogluconic acid
D-Glucose	D-Gluconic acid
Glycerol	Dihydroxyacetone
D-Mannitol	D-Fructose
Perseitol	Perseulose
α-Propylene glycol	Acetol
D-Sorbitol	L-Sorbose

Oxidation of mannitol to fructose was conducted on an industrial scale during the 1950s, but the process was discontinued because of the high cost of the raw material (mannitol) and other economic considerations. Preparation of fructose by enzymic isomerization on an economic scale will now make this fermentation most uneconomic.

A modern industrial fermentation utilizing *Acetobacter suboxydans* for preparing dihydroxyacetone from glycerol amounting to some U.S.$500,000 (Peppler, 1967) has only recently been mounted on a commercial scale, despite the fact that this conversion was known in the late Nineteenth Century and the usefulness of the product as a pharmaceutical known in the early part of the Twentieth Century. Although manufacture of dihydroxyacetone depends on efficient chemical purification with carbon treatment, resin treatment and evaporation, the fermentation process has a high conversion efficiency (95–97% yield) and a pure product (0.2% ash).

The sorbose fermentation is of special interest since L-sorbose is used in synthesis of L-ascorbic acid (vitamin C). The substrate (sorbitol) is a cheap raw material produced by electrolytic or catalytic hydrogenation of glucose. It was shown by the U.S. Tariff Commission in 1964 that annual production of ascorbic acid in all of its forms was some three million kilograms from which it can be seen that the production of sorbose must be at least as high. Again *A. suboxydans* is used, and 20 to 30% solutions of sorbitol plus accessory nutrients with vigorous aeration and agitation allow almost quantitative conversion and subsequent 70% yield after chemical purification to a crystalline product.

Perseulose, a rare ketoheptose, gluconic acid and 5-oxogluconic acid, acetylmethylcarbinol or acetoin, L-erythrulose and D-tartaric acid have all been prepared industrially by *Acetobacter* species. Recently there have been reports (Asai, 1968; Peppler, 1967) that antifungal, antibacterial and antitumour cell compounds can be prepared from acetic-acid bacteria. Certainly it is well known that certain of these species have the capacity to produce anti-yeast factors but whether these are specific organic chemicals or viral factors has yet to be clarified.

Finally, in this context of products other than acetic acid formed by *Acetobacter* spp., the formation of cellulose can not be overlooked since it has been used to prepare membranes for osmometry (Masson *et al.*, 1946). However, since cellulose is available so readily and cheaply in a variety of forms from the Vegetable World, it is unlikely that the microbiological process will be developed to any great extent. Brown in 1886 first reported the phenomena when studying *A. xylinum* grown in a nutrient medium, and many reports of other species doing the same have been published since (Asai, 1968). From these reports, two forms of cellulose have been recognized, namely a leathery, hard and tough pellicle ('hard cellulose') and a soft, slimy pellicle ('soft cellulose'; Shimwell and Carr, 1958). *Acetobacter xylinum* can produce both forms while other *Acetobacter* spp. produce only the soft cellulose. The former species also has the property of mutating to give a non-cellulosic form.

Hard cellulose has a fibre size of some 25 nm with a degree of polymerization of $2\text{–}6 \times 10^3$ and of all the microbiologically synthesized polysaccharide is pure cellulose. Pure-cellulose producers are apparently limited to the genus *Acetobacter* and give celluloses which closely resemble cotton cellulose in X-ray analysis. Soft cellulose, whilst resembling plant cellulose, generally has an amorphous structure although a unit cell can be calculated. Low degrees of polymerization of about 600 units have been found.

In the author's experience, in commercial practice, the two forms of cellulose are usually found to be produced by *A. xylinum*. The hard cellulose often referred to as the fibrous form appears in bottled or casked vinegars when they have been insufficiently filtered, pasteurized or salt-treated, or when they have been in contact with dirty plant. In warm weather, growth can be rapid and the pellicle forms in layers one on top of the other (a 'mother') until the total mass is too heavy to float and sinks into the main liquid. Here it is frequently mistaken by the public as being of animal origin and gives rise to appropriate complaints. In days before pasteurization or filtration of vinegar, the public were used to decanting vinegar to remove the 'mother' as it was called.

Soft cellulose (often referred to as colloidal cellulose) frequently forms in vinegar storage vats and can appear in considerable quantities (usually only seen when the vats are infrequently cleaned) having the appearance of 'tripe'. Various forms of the cellulose cause Quick generators to slow down and become inefficient, since the twigs or packing material get covered in slime (probably of soft cellulose). Whether this soft cellulose is produced entirely to *A. xylinum* is debatable since mixed cultures are used in vinegar manufacture and they mutate freely. Cellulose production appears to be rare in submerged culture either in the Frings generator, the tower fermenter or the Yates hybrid stirred-tower. Indeed, after prolonged continuous or semicontinuous operation, cellulose-producing *Acetobacter* spp. disappear from the whole plant and give a pronounced advantage for this type of operation.

G. Analysis of Vinegar and its Characteristics of Composition

1. *Characteristics*

Although specifications for vinegar obviously vary from country to country, the U.N. F.A.O./W.H.O. Codex Alimentarus Commission chose the Spanish specification to serve as an example and give the following details:

> Vinegar has to be limpid with a characteristic flavour and colour, without sediments or other perceptible alterations.
>
> Vinegar should contain the characteristic substances according to its origin which have not been transformed as a result of its production. The total acidity calculated as acetic acid anhydride should not be less than 50 g per litre (ranging from 3–6% in other countries).
>
> The dry extract without sugars should be not less than 10 g per litre.
>
> Total ash content should be not less than 1 g per litre.
>
> The alcohol content should not be more than 1%.
>
> The total residue of sulphur dioxide should not be more than 50 mg/kg.

The specifications given in the different countries for several varieties of vinegar include other data on contaminants and additives and requirements for limits on concentrations of compounds which are typical for certain vinegars.

a. *Additives.* To improve the storing properties of the product and its appearance, some specified substances may be added to vinegar, for example, salt.

b. *Colourants.* Colouring compounds may be added, including caramel (e.g. Spain, Austria, United Kingdom, France and the Netherlands), Cochineal (France), Orchil (France) and colourants from red wine 'marc' for red-wine vinegar (Austria).

c. *Other additives.* The Netherlands permits addition of formic acid in acetic acid solution (not more than 0.5 g/100 ml).

2. Composition

From the technological point of view, there are three categories of vinegar, namely natural fermented vinegar, acetified distilled vinegar, and artificial vinegar. The first type contains fermentation by-products from the alcoholic fermentation and acetification, while the second type contains only those from the distillation and acetification. Artificial vinegar contains none of these compounds.

The composition of natural fermented vinegar is given in Table 5. The composition of three different types of vinegar may be distinguished by paper-partition chromatography, gas–liquid chromatography and by ultraviolet absorption spectroscopy. The presence

Table 5

Chemical composition of cider vinegars

From Joslyn, M.A., 1970. *In* 'Encyclopedia of Chemical Technology', (R. E. Kirk and D. F. Othmer, eds.), vol. 21, pp. 254–269. Interscience Publishers. John Wiley and Sons Inc., New York.

Constituent	Concentration		
	Average	Maximum	Minimum
Total acid (%, w/v, as acetic acid)	4.94	7.96	3.29
Total solids (%, w/v)	2.54	4.52	1.37
Non-sugar solids (%, w/v)	1.90	2.89	1.26
Reducing sugars in solids (%, w/v)	19.6	45.0	5.6
Total ash (%, w/v)	0.367	0.52	0.20
Alkalinity of water-soluble ash (ml 0.1 N acid[a])	35.7	56.0	21.5
Ash in non-sugar solids (%, w/v)	18.8	26.5	11.2
Soluble phosphate (mg P_2O_5[a])	17.3	39.9	6.7
Insoluble phosphate (mg P_2O_5[a])	29.3	64.2	15.1
Alcohol (%, v/v)	0.35	2.0	0.03
Glycerol (%, w/v)	0.30	0.46	0.23
Polarization (direct; $^\circ$V[b])	−1.46	−3.6	−0.2
Polarization (indirect; $^\circ$V[b])	−1.69	−3.1	0.0

[a] Content from 100 ml vinegar.

[b] $^\circ$V indicates one degree Vanzkescale, and equals 0.34657° angular rotation, with sodium D light.

of acetoin distinguishes natural from artificial vinegars, while malic acid can be found only in cider vinegar. Acetone is normally detectable in vinegars made from ethanol. A larger concentration of residual extract is obtained from undistilled vinegars. Methods of analysis for vinegars in the United Kingdom (Table 6) include certain procedures which may be used to differentiate vinegars.

Table 6

Analytical values for various types of vinegar

Type of vinegar	Value for			
	Oxidation number	Alkaline oxidation number	Iodine number	Ester value
Malt	360	34	64	14
Distilled malt	850–1,000	—	800–1,050	30–36
Wine	1,350–2,400	100–180	400–850	70–220
Spirit	90–225	75–23	6–40	0–16
Artificial	1–16	0.8–8.0	0–250	0–14

3. Analysis

Standard methods of analysis have been established for vinegars in many countries of the World. In the United States of America, methods have been established for organoleptic examination, and determination of total solids content, total ash, soluble and insoluble ash, soluble and insoluble phosphate, total acids and alcohol.

In the United Kingdom, the following analyses have been used and are derived from Edwards and Nanji (1938), Whitmarsh (1942) and the recommendations of the Association of Official Analytical Chemists in the U.S.A.

a. *Total acid.* Dilute 10 ml of vinegar with recently boiled and cooled distilled water until it appears only slightly coloured. Titrate against 0.5 N alkali and calculate the result as a percentage of acetic acid.

b. *Non-volatile acid.* Measure 10 ml of vinegar into an evaporating dish, evaporate just to dryness on a water bath, add 5–10 ml of distilled water and repeat until at least five evaporations have been made. Take up in about 200 ml of recently boiled and cooled distilled water and titrate with 0.1 N alkali. Express the result as percentage acetic acid.

c. *Volatile acid.* Percentage volatile acid (as acetic acid) = (1)–(2).

d. *Total solids.* Measure 10 ml of vinegar into an evaporating dish, evaporate on a bath of boiling water for 30 min and then dry for precisely 30 min in a water oven at 100°C. Cool in a desiccator and weigh. Return the result as a g dry matter per 100 ml of vinegar.

e. *Oxidation value.* This value is defined as the number of millilitres of 0.01 N $KMnO_4$ used by 100 millilitres of vinegar in 30 min under the described standard conditions. In a 400 ml distillation flask place a quantity of sample (the example quoted is for malt vinegar, for other vinegars see Whitmarsh, 1942) such that on dilution to 25 ml the total acid content (as acetic acid) is 4.0%. Add sufficient distilled water to bring the volume to 75 ml, add a pinch of pumice and distil *slowly* until 60 ml of distillate have collected (measuring cyclinder). Cork or rubber stoppers should be covered with tin or aluminium foil. This distillate is also used for determination of the iodine value (see f. below).

Pipette 25 ml of the distillate into a 200 ml glass-stoppered bottle, add 10 ml of dilute (1 to 3) H_2SO_4 and then 10 ml of 0.1 N $KMnO_4$. Stand for 30 min at 18°C, add 5 ml of 10% KI and titrate against 0.02 N thiosulphate. A blank experiment on 25 ml of distilled water is also necessary. The oxidation value is given by the expression 19.2 (A–B) where A and B are the volumes of thiosulphate used for the blank and sample, respectively.

f. *Iodine value.* This is defined as the number of millilitres of 0.01 N iodine absorbed by 100 millilitres of vinegar under the standard conditions described. Pipette 25 ml of distillate (see e. above) into a 200 ml glass-stoppered bottle. Just neutralize to litmus paper with 10 N KOH, then add 10 ml of N KOH followed by 10 ml of 0.1 N iodine. Stand at room temperature in the dark for 15 min, add 10 ml of diluted (1 to 3) H_2SO_4 and titrate against 0.02 N thiosulphate. A blank experiment on 25 ml of distilled water using the same quantities of all reagents is also necessary.

g. *Ester value.* This is defined as the number of millilitres of 0.01 N KOH required to saponify the esters contained in 100 millilitres of vinegar under the described standard conditions. Measure 100 ml of vinegar into a 400 ml distilling flask, add a pinch of pumice and distil slowly until 30 ml of distillate are collected. Add a few drops of phenolphthalein to the distillate and then N KOH until just pink. Add 0.2 N HCl drop by drop until the colour is just discharged; then add 10 ml by pipette of 0.1 N KOH and saponify by heating under reflux for 2 h. Cool, add a few spots of phenolphthalein, and titrate against 0.02 N HCl. A blank experiment with 30 ml of water and 10 ml of 0.1 N KOH heated under reflux as before is necessary. The ester value equals 2 (A − B), where A and B are, respectively, the volumes of 0.02 N HCl used in the blank and sample titrations. An authentic sample of malt vinegar gave the following analysis:

Total acidity	4.98%
Non-volatile acidity	0.20%
Volatile acidity	4.78%
Total solids	2.29%
Oxidation value	1,195

IV. SUMMARY

Vinegar manufacture or the double fermentation due to the action of yeast to form alcohol and the acetic-acid bacteria to give acetic acid represents one of the best examples of art and science combined in Man's activities. Certainly, despite the restrictive economics of the industry, it has been a valuable crucible for many of the developments in fermentation technology and has attracted the intellectual power of technologists and academics alike. The contribution to the biochemistry of microbial activity and the understanding of metabolism is not to be underestimated. The full potential of the fermentative activity of *Acetobacter* spp. has not yet been realized either directly to form vinegar or indeed to prepare other useful biochemicals. In the preface to Pepplers' book (1967) 'Microbial Technology' he states: 'Fermentation industries, like the micro-organisms they depend upon grow in stages from infancy through maturity and then decline'—this indeed was true but now no longer. The existence of the traditional fermentation industries will undoubtedly be assured, but Man will depend more and more on microbial activities for survival on this planet and perhaps others. It is no understatement to say that much remains to be done which will contribute to our understanding and use of microbial activities, and *Acetobacter* spp. will play a valuable part in this work.

The final word can be best given by the quotation which Professor Asai chose to open his book on the 'Acetic Acid Bacteria': 'Dear God, what marvels there be in so small a creature'.

V. ACKNOWLEDGEMENTS

The author wishes to acknowledge the following people and associated companies for supplying valuable information to enable this article to be written. In addition, he acknowledges the valuable help and encouragement which gave rise to the tower acetator, and the associated research and development which made it possible to become a commercial reality. Individual acknowledgements go to M. C. A. Greenshields, Frances Greenshields, M. Philips, P. M. Nicholson, M. Farrar, Dr. C. Ebner, M. Wright, Jean Yates, D. T.

Shore, Dr. D. D. Jones, Dr. H. A. Conner, M. Kimmitt, Mrs. O. Deeley, The A.P.V. Company Ltd., Crawley, Sussex, British Vinegars Co. Ltd., Penistone Brewery, Derbyshire, Beecham Food and Drinks, Barbourne Brewery, Worcestershire, The Hammonds Group of Companies, Barrowford Brewery, Lancashire, Man Vinegar Co. Ltd., Burntwood, Staffs., H.P. Sauce Ltd., Aston Cross, Birmingham, Frances Harmon Ltd., London, Frings G.m.b.H., Bonn, West Germany.

REFERENCES

Adams, F. (1849). 'Genuine Works of Hippocrates', vol. 1. The Sydenam Society, London.

Allgeier, R. J., Nichol, G. B. and Conner, H. A. (1974). *Food Product Development* June, July, August.

Asai, T. (1935). *Journal of the Agricultural Chemistry Society of Japan* 11, 686.

Asai, T. (1968). 'Acetic Acid Bacteria. Classification and Biochemical Activities'. University of Tokyo Press, Tokyo, Japan.

Asai, T., Iizuka, H. and Komagata, K., (1964). *Journal of General Microbiology* 10, 95.

Ault, R. G., Hampton, A. N., Newton, R. and Roberts, R. H. (1969). *Journal of the Institute of Brewing* 75, 260.

Beijerinck, M. W. (1900). *Proceedings of the Academy van Wetenschappen Amsterdam* 2, 495.

Bergey, D. H., Breed, R. S. and Murray, E. G. D., (1939). 'Bergey's Manual of Determinative Bacteriology', 5th ed. The Williams & Wilkins Company, Baltimore.

Breed, R. S., Murray, E. G. D. and Smith, N. R. (1957). 'Bergey's Manual of Determinative Bacteriology', 7th ed. The Williams and Wilkins Company, Baltimore.

Bertho, A. and Basu, K. P. (1931). *Liebigs Annalen* 485, 26.

Boerhaave, H. (1732). *Elementa chemicae. Lugduni Batovorum* 2, 179.

Brooks, R. O. (1927). 'Critical Studies in the Legal Chemistry of Foods'. Reinhold Publishing Corporation, New York.

Brown, A. J. (1886). *Journal of the Chemical Society* 49, 172, 432.

Brown, G. D. and Rainbow, C. (1956). *Journal of General Microbiology* 15, 61.

Cohen, D. (1872). *Beitrage zur Biologie der Pflanzen* 1, Heft 2, 127.

De Ley, J. (1961). *Journal of General Microbiology* 24, 31.

Desmazières, L. (1826). *Annales des Sceances Naturelles* 10, 42.

Dunn, M. S., Shankman, S., Camien, M. H. and Block, H. (1947). *Journal of Biological Chemistry* 168, 1.

Dussance, H., (1871). 'A General Treatise on the Manufacture of Vinegar'. H. C. Baird & Co., Philadelphia.

Edwards, F. W. and Nanji, H. R. (1938). *Analyst* 63, 410.

Els, H. (1955). British Patent 731,804.

Enenkel, R., Enenkel, A., Maurer, R. and Enenkel, G. (1953). British Patent 686,849.

Fetzer, W. R. (1930). *Food Industries* 2, 489.

Frateur, J. (1950). *La Cellule* 53 (3), 288.

Frateur, J. (1950). *La Cellule* 53, (3), 333.

Frings, G.m.b.H. (1968). British Patent 1,101,560.

Greenshields, R. N. (1972). British Patent 1,253,059.

Greenshields, R. N. and Smith, E. L. (1971). *Chemical Engineer* C. E. 249,182.

Greenshields, R. N. and Smith, E. L. (1974). *Process Biochemistry* 9(4), 11.

Greenshields, R. N., Yates, Jean, Sharpe, P. and Davies, T. M. C. (1971). *Journal of the Institute of Brewing* 78, 236.

Hansen, E. C. (1879). *Comptes Rendus des Traveaux du Laboratoire Carlsberg* 1, 49, 96.

Harden, A. (1911). 'Alcoholic Fermentation'. Longmans, London.

Henneberg, W. (1891). *Zentralblatt fur Bakteriologie, Parasitenkunde, Infektionskrankheiten und Hygiene (Abteilung III)*, 933.

Henneberg, W. (1926). 'Handbuch der Garungs Bakteriologie'. Bd. 2. 190. Paul Parcy, Berlin.

Herbert, D. (1961). *Society for Chemical Industries Monograph* 12, 21.

Hildebrandt, F. M., Nichol, G. B. and Kahn, J. H. (1969). *U.S.I. Vinegar Newsletter* 56.

Holland, P. (1964). 'The Natural History of C. Plinius Secundus', Book 10. McGraw Hill Co., New York.

Hoyer, D. P. (1898). Dissertation: University of Leiden, Waltmann, Delft, Holland.

Hromatka, O. and Ebner, H. (1949). *Enzymologia* 13, 6, 369.

Hromatka, O. and Ebner, H. (1950). *Enzymologia* 14, 2, 96.

Hromatka, O. and Ebner, H. (1951). *Enzymologia* 15, 2, 57; 31, 134; 6, 337.

Huber, E. (1927). *Deutsche Essigindustrie* 31, 1, 12, 28.

Janke, A. (1916). *Zentralblatt fur Bakteriologie, Parasitenkunde. Infektionskrankheiten und Hygiene (Abteilung III)*. 45, 1.

Jarvis, D. C. (1959). 'Folk Medicine'. Henry Holt and Co., New York.

Jones, D. D. (1970). MSc Thesis: The University of Aston in Birmingham.

Jones, D. D. and Greenshields, R. N. (1969). *Journal of the Institute of Brewing* 75, 457.

Jones, D. D. and Greenshields, R. N. (1970a). *Journal of the Institute of Brewing* 76, 55.

Jones, D. D. and Greenshields, R. N. (1970b). *Journal of the Institute of Brewing* 76, 235.

Jones, D. D. and Greenshields, R. N. (1971). *Journal of the Institute of Brewing* 77, 160.

Kaneko, T., Ohmori, S. and Masai, H. (1973). *Journal of Food Science* 38, 350.

Kat v Diment, (1950). *I.K.B.* 34.

Klopper, W. J., Roberts, R. H., Royston, M. G. and Ault, R. G. (1965). *Proceedings of the European Brewing Convention Congress, Stockholm*, p. 238.

Knieriem, W. von and Mayer, A. (1873). *Lander. Versuchsstation Mitt* 16, 305.

Kutzing, D. T. (1837). *Journal of Praktikale Chemische* 11, 385.

Le Fevre, E. (1924). *Making Vinegar in the Home and on the Farm, U.S. Department of Agriculture Farmers Bulletin*, 1424.

Leifson, E. (1954). *Antonie van Leeuwenhoek* 20, 102.

Lemery, M. S. (1720). 'A Course of Chemistry', 4th English ed. From 11th French edition, 6.

Ludwig, F. (1898). *Zentralblatt fur Bakteriologie, Parasitenkunde, Infektionskrankheiten und Hygiene (Abteilung III).* 4, 867.

Málek, I. and Fencl. Z. (1966). 'Theoretical and Methodological Basis of Continuous Culture of Micro-organisms.' Academic Press, London.

Masson, C. R., Menzies, R. F. and Cruickshank, J. (1946). *Nature, London* 157, 74.

Maule, A. P. and Greenshields, R. N. (1970). *Process Biochemistry* 5 (2), 39.

Maule, A. P. and Greenshields, R. N. (1971). *Process Biochemistry* 6 (7), 28.

Ministry of Agriculture, Fisheries and Food (1971). *Food Standards Committee Report on Vinegars.* H.M.S.O. London.

Mitchell, C. A. (1926). 'Vinegar: its Manufacture and Examination'. Charles Griffin and Co. Ltd., London.

Naegeli, C. (1857). *Beiträge Zeitung* 760.

Neuberg, C. (1928). *Biochemische Zeitschrift* 199, 1928.

Neuberg, C. and Nord, E. F. (1919). *Biochemische Zeitschrift* 96, 158.

Neuberg, C. and Windisch, F. (1925). *Biochemische Zeitschrift* 166, 454.

Ochs, I. L. (1950). *Journal of the American Medical Association* 142, 1361.

Orla-Jensen, S. (1909). *Zentralblatt fur Bakteriologie, Parasitenkunde, Infektionskrankheiten und Hygiene (Abteilung III).* 2 Abt. 22, 312.

Pasteur, L. (1862). *Comptes Rendus Hebdomadaire des Séances de l'Academie des Séances* 54, 265.

Pasteur, L. (1868). 'Études sur le Vinaigre'. Masson and Sons, Paris, France.

Peppler, H. J. (1967). 'Microbial Technology'. Reinhold Publishing Corporation, London, U.K.

Persoon, C. H. (1822). *Mycologia Europea* 1, 96.

Rainbow, C. (1966). *Wallerstein Laboratories Communications* 25, 9.

Rainbow, C. and Mitson, G. W. (1953). *Journal of General Microbiology* 9, 371.

Richard, E. (1670). *Philosophical Transactions of the Royal Society* 5, 2002.

Richardson, A. C. (1961). British Patent 865,508.

Royston, M. G. (1963). British Patent 929,315.

Scopes, A. W. (1962). *Journal of General Microbiology* 28, 69.

Shimwell, J. L. (1952). British Patent 781,584.

Shimwell, J. L. (1955). British Patent 727,039.

Shimwell, J. L. (1957). *Journal of the Institute of Brewing* 63, 44.

Shimwell, J. L. and Carr, J. G. (1958). *Journal of the Institute of Brewing* 64, 477.

Shimwell, J. L. (1959). *Antonie van Leeuwenhoek* 25, 49.

Shimwell, J. L. and Carr, J. G. (1964). *Nature, London* 201, 1051.

Shore, D. T. and Royston, M. G. (1968). *Chemical Engineer*, 99.

Simon, E. (1930). *Biochemische Zeitschrift* 224, 253.

Simonin, R. F. and Bernard, M. (1958). British Patent 805,698.

Spillan, D. and Edmonds, C. (1895). 'The History of Rome', (Titius Livius), Book 21, 2. Harper and Brothers, New York, U.S.A.

Tanaka, K. (1938). *Journal of Science. Hiroshima University Series B. Div. 2. Vol. 3.* 125.

The Holy Bible, Revised Standard Version, John 19:29; Ps 69:31; Prov. 10:26; 25:20; Matt. 27:48; Mark 15:36; Luke 23:36.

The Jewish Encyclopaedia (1906). 12 Talmud-Zweifel. Funk and Wagnalls Co., New York.

Thompson, (1931). *Journal of the Royal Asiatic Society.*

Thompson, R. T. (1852). *Liebegs Annalen* 83, 89.

Tosic, J. and Walker, T. K. (1946). *Journal of the Society of Chemical Industry* **115**, 180.

United Nations Food and Agriculture Organisation (1974). Codex Alimentarius Commission. Alinorm 74/32.

U.S. Department of Agriculture, Food and Drugs Administration (1936). Service and Regulatory Announcements, No. 2. Rev. 5, November.

Vinegar Brewers' Federation, U.K. (1952). 'The Legal History of Vinegar.'

White, J. (1961). British Patent 878,949.

White, J. (1966). *Process Biochemistry* **1**(3), 139.

White, J. (1970). *Process Biochemistry* **5**(10) 54.

White, J. (1971). *Process Biochemistry* **6**(5), 21.

Whitmarsh, J. A. (1942). *Analyst* **67**, 188.

Wieland, H. (1913). *Berichte der Deutschen Chemische Gesellschaft* **46**, 3335.

Wieland, H. and Bertho, A. (1928). *Liebigs Annalen* **467**, 95.

Wustenfeld, H. (1930). 'Lerbuch der Essigfabrikation'. Paul Barey, Berlin.

Zeidler, A. (1896). *Zentralblatt fur Bakteriologie, Parasitenkunde, Infektionskrankheiten und Hygiene (Abteilung III)* **2**, 406.

5. Production of Nucleotides by Micro-Organisms

ARNOLD L. DEMAIN

Department of Nutrition and Food Science, Massachusetts Institute of Technology, Cambridge, Mass. 02139, U.S.A.

I. INTRODUCTION

Although nucleotides show various pharmacological effects of potential medical and nutritional interest, and are of use in the preparation of synthetic oligo- and polynucleotides, the greatest interest in their industrial production is based on their ability to enhance the flavour of foods (Kuninaka *et al.*, 1964; Margalith, 1974). Flavour enhancement is a property of three purine ribonucleoside 5′-monophosphates, namely (in order of decreasing potency) guanylic acid (GMP), inosinic acid (IMP) and xanthylic acid (XMP). The other natural purine 5′-ribonucleotide, adenylic acid (AMP), does not exhibit flavour potentiation. Also inactive are 2′ and 3′ isomers, free purine bases, nucleosides and pyrimidine nucleotides. Thus only 6-hydroxypurine

This is Contribution No. 2542 from the Department of Nutrition and Food Science, Massachusetts Institute of Technology, Cambridge, Massachusetts 02139, U.S.A.

ribonucleoside 5'-monophosphates (as well as the corresponding deoxyribonucleotides) are active (Fig. 1). In combination with monosodium glutamate (MSG), a synergistic flavour enchancing effect is observed. On a weight basis, nucleotides are more potent than MSG by several orders of magnitude.

Production of purine 5'-ribonucleotides by fermentation was reviewed several years ago (Demain, 1968a) but many new findings

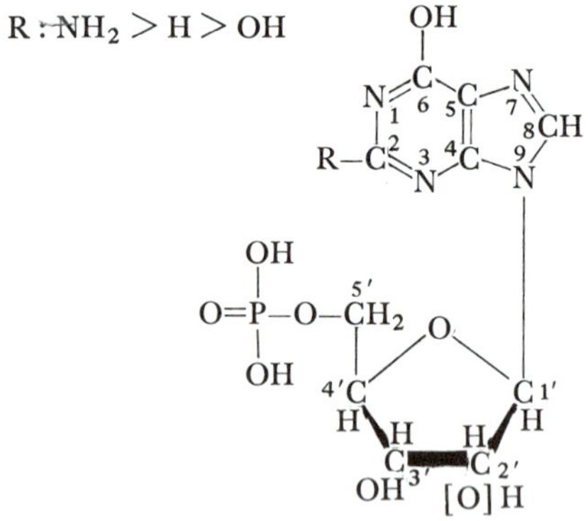

Fig. 1. General structure of 6-hydroxypurine ribonucleoside 5'-monophosphates and the corresponding deoxyribonucleoside 5'-monophosphates.

have been reported since then. As was the case with the earlier developments, research in this area continues to be almost an exclusively Japanese accomplishment. Due to the overwhelming economic importance of purine 5'-ribonucleotides, production of pyrimidine ribonucleotides will not be discussed in this chapter. It should be understood, however, that pyrimidine 5'-ribonucleotide biosynthesis and regulation are well understood (for a review, see O'Donovan and Neuhard, 1970). Furthermore, deregulated mutants (Jund and Lacroute, 1970; O'Donovan and Gerhart, 1972) and biosynthetic processes (Škodová *et al.*, 1969; Kawamoto *et al.*, 1970; Nakayama and Tanaka, 1971) have been described. Also beyond the scope of this review are bioconversions leading to production of sugar nucleotides (Tochikura *et al.*, 1972).

II. BIOSYNTHESIS OF PURINE 5'-RIBONUCLEOTIDES

The pathway of purine ribonucleotide biosynthesis is virtually identical in all living systems. Steps on the pathway are shown in Fig. 2. In the first specific step of purine ribonucleotide bio-synthesis, 5'-phosphoribosyl pyrophosphate (PRPP), derived from ATP and ribose 5-phosphate, is aminated with glutamine to form 5'-phosphoribosylamine (PRA). By a series of seven known enzymic steps, PRA is converted to 5-amino-1-(5'-phosphoribosyl)-imidazole 4-carboxamide (AICAR). The next steps lead to IMP which is the first purine ribonucleotide formed. This pivotal intermediate is the precursor of both AMP and GMP. In step 11 (Fig. 2) IMP is converted to adenylosuccinate (S-AMP), aspartate acting as the succinyl donor and GTP as the energy source. Adenylosuccinate is then converted to AMP in step 13 by adenylosuccinate lyase, a bifunctional enzyme which also catalyses step 8 which occurs prior to IMP. Inosinic acid is oxidized by IMP dehydrogenase using NAD^+ as a cofactor to form XMP (step 12). The latter is aminated by XMP aminase to GMP (reaction 14). In this reaction, ATP is the energy source and, depending on the organism, the amino donor is either glutamine or ammonia. Both AMP and GMP can be used for synthesis of nucleic acids after being converted to their respective di- and triphosphates. Alternatively, GMP can be reduced to IMP in a reaction catalysed by GMP reductase (reaction 15 which utilizes NADPH as an electron donor) and AMP can be converted to IMP by deamination (step 16) or to AICAR by a series of steps. These reactions permit a cyclic regeneration of intermediates common to both purine nucleotides from an accumulated excess of either. Most of the reactions shown in Fig. 2 are essentially irreversible. All of the purine ribonucleotide interconversions occur via their 5'-isomers. For the remainder of this chapter, it is to be understood that any ribonucleotide mentioned has the phosphate group in the 5'-position.

III. REGULATION OF BIOSYNTHESIS OF NUCLEOTIDES

With such a complex series of biosynthetic reactions, it is essential to the economy of the cell that there be regulation of the rates of the

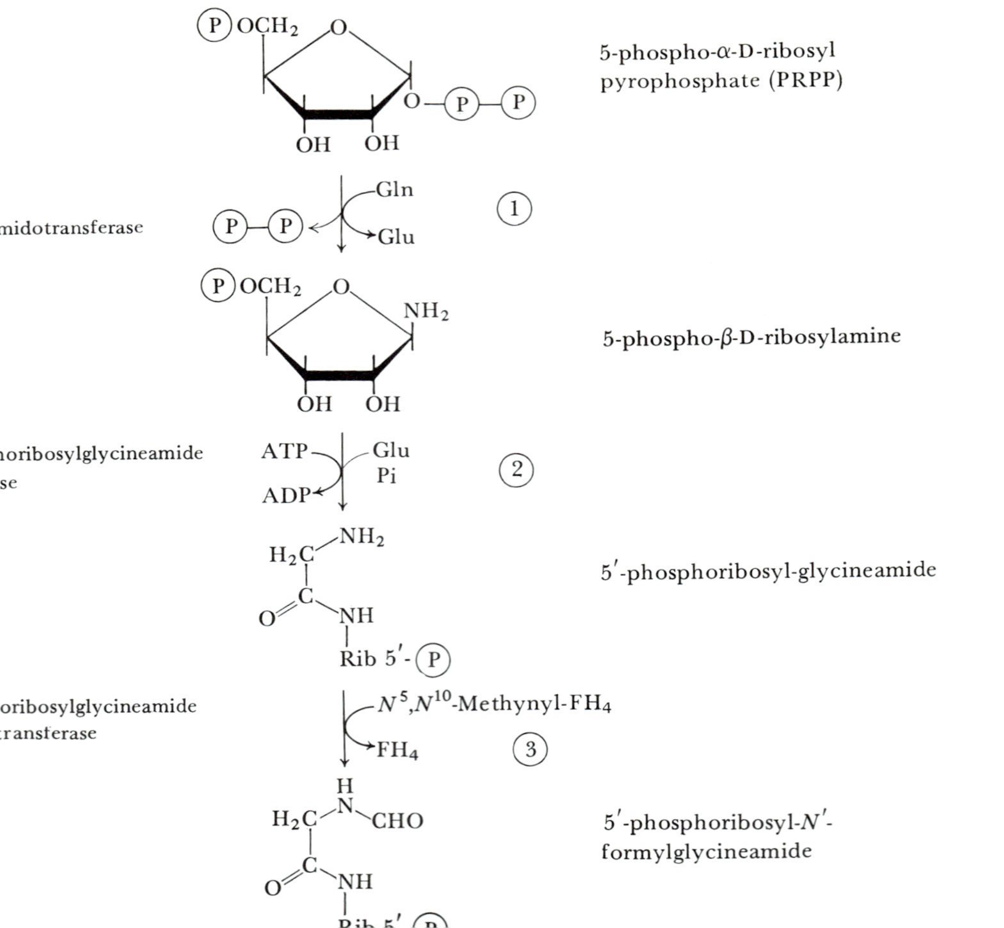
5-phospho-α-D-ribosyl pyrophosphate (PRPP)
PRPP amidotransferase
Gln
Glu
1
5-phospho-β-D-ribosylamine
phosphoribosylglycineamide synthase
ATP
ADP
Glu
Pi
2
5′-phosphoribosyl-glycineamide
Rib 5′-
phosphoribosylglycineamide formyltransferase
N⁵,N¹⁰-Methynyl-FH₄
FH₄
3
5′-phosphoribosyl-N′-formylglycineamide

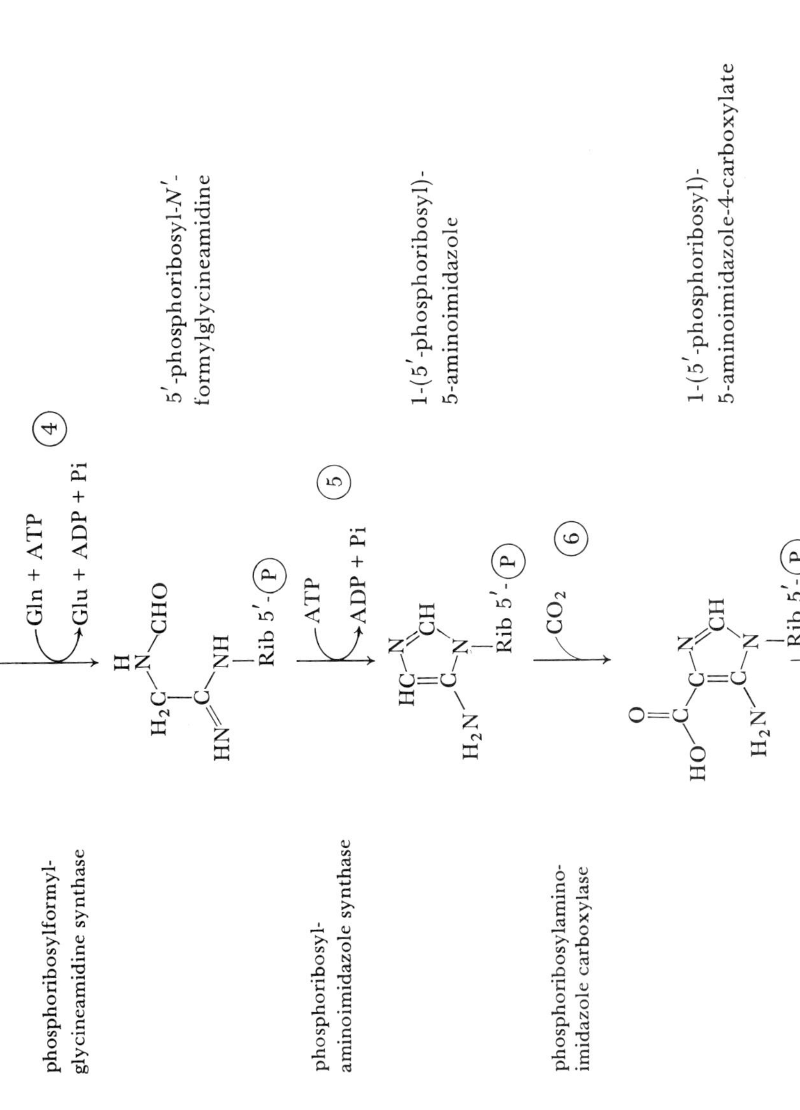

Fig. 2. Reactions leading to biosynthesis of purine ribonucleotides. Ⓟ indicates a phosphate residue. Continued on page 192.

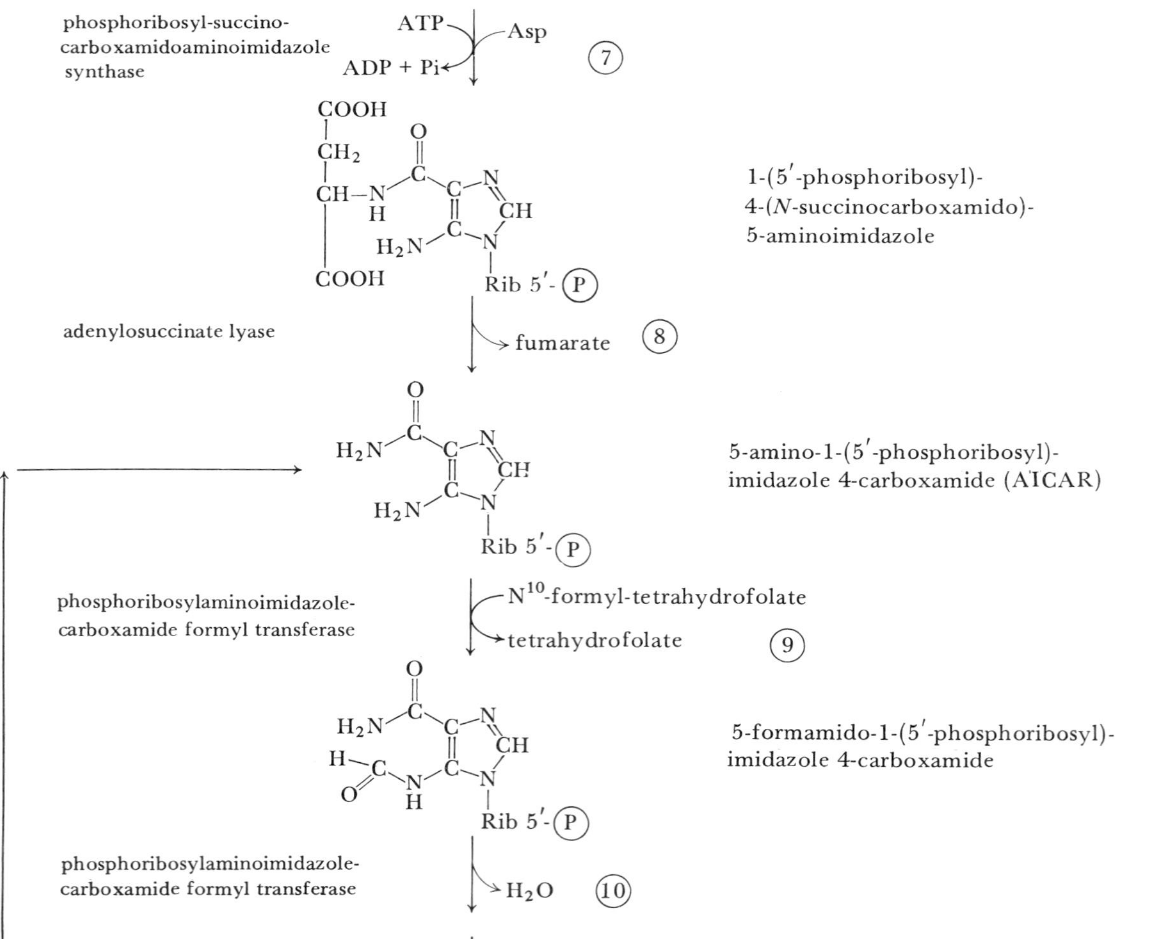
phosphoribosyl-succino-
carboxamidoaminoimidazole
synthase
ATP
Asp
ADP + Pi
7
COOH
CH₂
CH—N
H
H₂N
COOH
Rib 5'-(P)
1-(5'-phosphoribosyl)-
4-(N-succinocarboxamido)-
5-aminoimidazole
adenylosuccinate lyase
fumarate
8
H₂N
H₂N
Rib 5'-(P)
5-amino-1-(5'-phosphoribosyl)-
imidazole 4-carboxamide (AICAR)
phosphoribosylaminoimidazole-
carboxamide formyl transferase
N¹⁰-formyl-tetrahydrofolate
tetrahydrofolate
9
H₂N
H
Rib 5'-(P)
5-formamido-1-(5'-phosphoribosyl)-
imidazole 4-carboxamide
phosphoribosylaminoimidazole-
carboxamide formyl transferase
H₂O
10

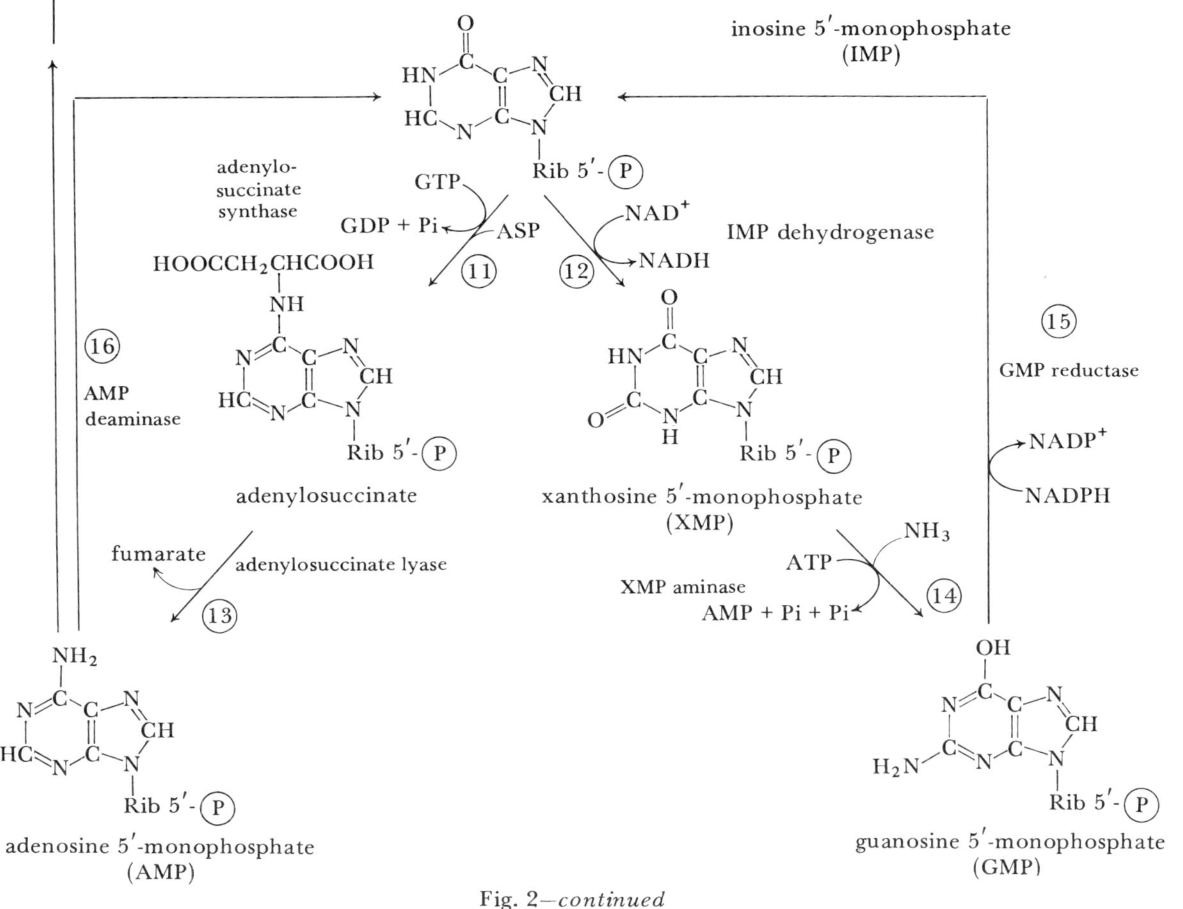

Fig. 2—*continued*

individual reactions. This is accomplished by negative feedback inhibition of enzyme action and repression of enzyme synthesis. In feedback inhibition, the *action* of the first enzyme of the pathway or of a branch is inhibited by the end product. In feedback repression,

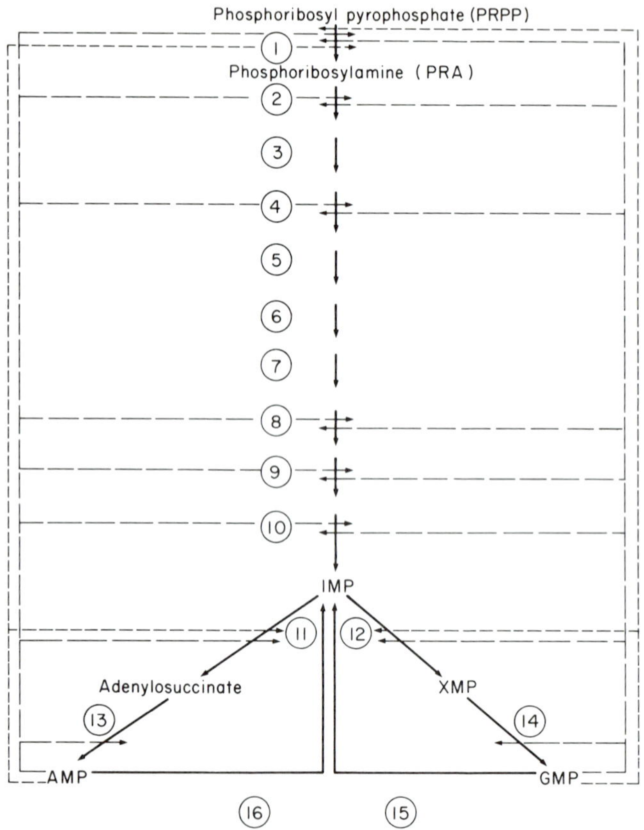

Fig. 3. Composite feedback control scheme for regulation of synthesis of purine nucleotides in a number of micro-organisms. Short dashed lines indicate operation of feedback inhibition of enzyme action, long dashed lines of repression of enzyme synthesis, and thick arrows indicate the reactions on the biosynthetic pathway. Also indicated are the numbers assigned to enzymes that catalyse reactions on the pathway.

the *formation* of one or more of the enzymes on the pathway is inhibited by a derivative of the end product.

Figure 3 is a composite feedback control scheme for several micro-organisms which have been examined. Minor differences between the regulatory mechanisms in these micro-organisms are not shown in the figure. Usually both AMP and GMP exert feedback repression and inhibition on the common pathway and on their respective branches. An important control site is step 1, catalysed by PRPP amidotransferase. In *Klebsiella aerogenes*, AMP and GMP each inhibit action of the enzyme completely at high concentrations and cause a synergistic inhibition at low concentrations (Nierlich and Magasanik, 1965). The situation is different in *Bacillus subtilis* where AMP alone causes complete inhibition but GMP is only weakly inhibitory (Shiio and Ishii, 1969). Synthesis of the amidotransferase, as well as many of the other enzymes of the common pathway, is also repressed by AMP and GMP. Another important control site is IMP dehydrogenase (step 12) which is usually inhibited and repressed by GMP. Step 11 (catalysed by adenylosuccinate synthase) is usually repressed and inhibited by AMP.

IV. PURINE RIBONUCLEOSIDE OVERPRODUCTION BY *BACILLUS SUBTILIS*

Bacillus subtilis has a great ability to overproduce purine ribonucleosides when the control mechanisms described in Section III are eliminated by various manipulations. The manipulations are of two major types. In the first, auxotrophic mutants are obtained and fed growth-limiting concentrations of the end-product requirement so that the intracellular levels of feedback inhibitors and repressors can be lowered. In the second method, constitutive mutants which are no longer subject to feedback repression or desensitized mutants (resistant to feedback inhibition) are selected by resistance to toxic purine analogues.

The regulatory controls existing in *B. subtilis* have been summarized in an excellent paper by Ishii and Shiio (1973). The main points (Fig. 4) are as follows: (1) PRPP amidotransferase is feedback inhibited by AMP but not significantly by GMP and other guanine derivatives; (2) synthesis of PRPP amidotransferase and the other enzymes of the common pathway is repressed by AMP and GMP which act co-operatively at low concentrations (Sato and Shiio, 1970). In the actual experiments, adenine and guanosine are added to the culture since the intact nucleotides cannot be taken up;

(3) IMP dehydrogenase is strongly feedback inhibited (Ishii and Shiio, 1968; Wu and Scrimgeour, 1973) and its synthesis repressed by GMP (Sata and Shiio, 1970); (4) synthesis of adenylosuccinate synthase is repressed by AMP but the enzyme is not significantly feedback inhibited *in vivo*.

Because of these controls (Fig. 5a, b), specific auxotrophs which lack the ability to synthesize adenylosuccinate synthase accumulate inosine (Ishii and Shiio, 1973). The amount of adenine added to the medium is kept low so that high levels of AMP, which could inhibit the activity of PRPP amidotransferase, are not built up inside

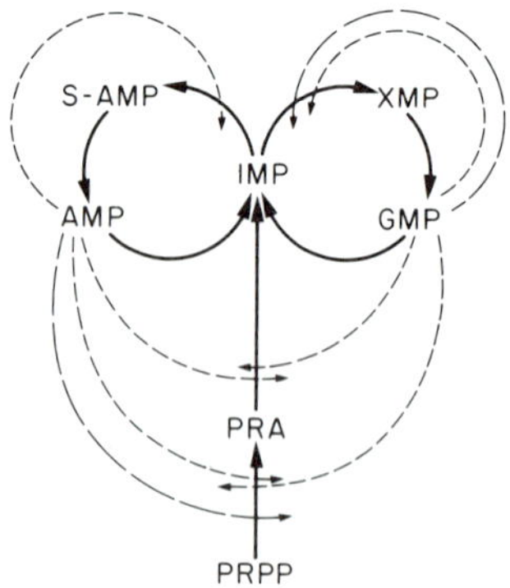

Fig. 4. Scheme showing feedback regulation of synthesis of purine nucleotides in *Bacillus subtilis*. Continuous lines indicate reactions on the biosynthetic pathway, short dashed lines operation of repression of enzyme synthesis, and long dashed lines feedback inhibition of enzyme action.

the cells. Because of the strong inhibition and repression of IMP dehydrogenase by GMP, the accumulation does not proceed past IMP into the GMP branch of the pathway. Certain strains of *B. subtilis* excrete as much as 12 grams of inosine per litre of culture (Momose and Shiio, 1969). Such organisms are still subject to a certain degree of GMP repression of synthesis of the enzymes of the common path. To minimize the severity of this regulation, the adenine auxotrophs have been further mutated to delete the ability to synthesize IMP dehydrogenase. The adenine-xanthine double auxotrophs show a two-fold increase in specific activity of some common-path enzymes and accumulate up to 15 g inosine per litre under conditions of limiting adenine and xanthine (or guanosine; Fig. 5c).

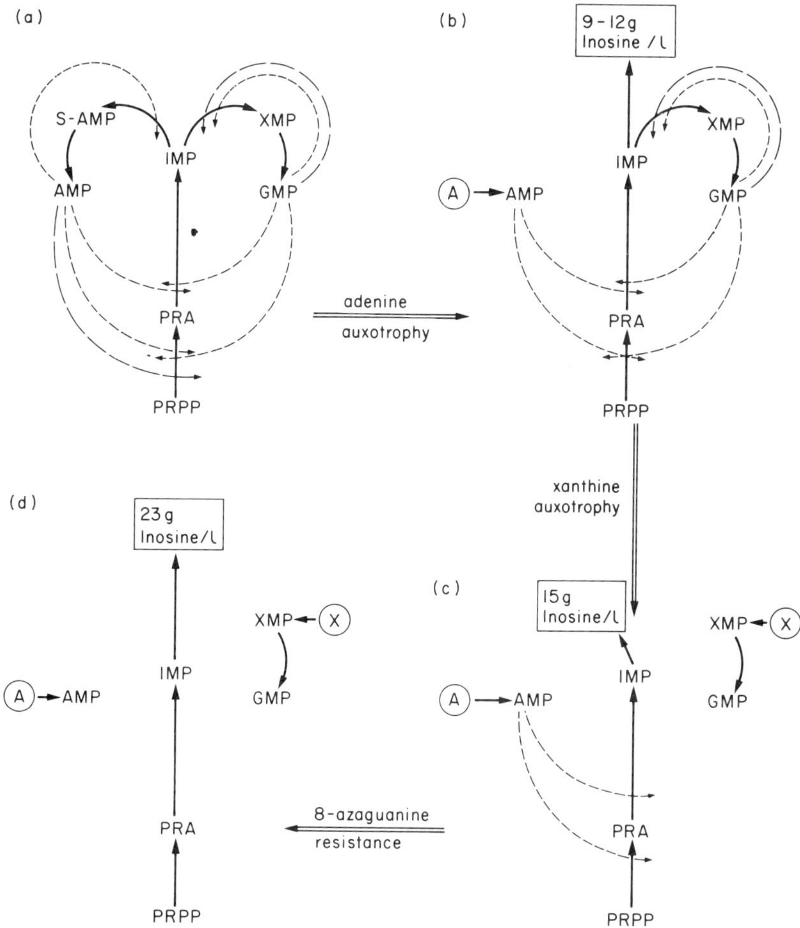

Fig. 5. Development of the inosine fermentation with *Bacillus subtilis*. The concentration of excreted product is indicated in the box. A required nutrient is circled. Figure 5(a) indicates the behaviour of the starting strain which lacked AMP deaminase and GMP reductase. Figures 5(b), (c) and (d) describe the activities of successive mutants.

A means of eliminating AMP repression of the common-path enzymes is to mutate the adenine-requiring auxotroph and select for resistance to 8-azaguanine. Such double mutants produce up to 18 g of inosine per litre (Shiio and Ishii, 1971) accompanied by increased specific activities of the common-path enzymes (Ishii and Shiio, 1972). These double mutants, like their parent, must be grown under conditions of adenine limitation to prevent inhibition of the activity

　　　　　　　　　　　　　　　　　　　　　　　　　A. L. DEMAIN

of PRPP amidotransferase. A further increase in inosine productivity can be obtained by combining adenine auxotrophy, xanthine auxotrophy and azaguanine resistance (Sasajima *et al.*, 1970). Such triple mutants accumulate up to 23 g inosine per litre (Fig. 5d).

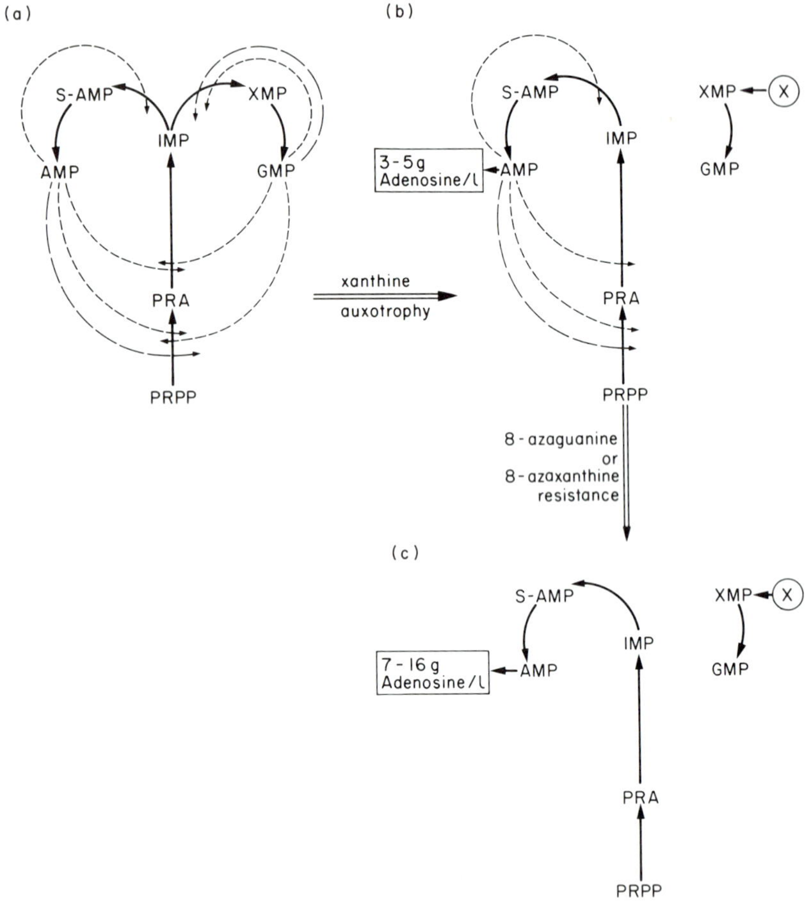

Fig. 6. Development of the adenosine fermentation with *Bacillus subtilis*. See the legend for Fig. 5 for an explanation of the figure.

Xanthine auxotrophs (lacking IMP dehydrogenase) accumulate adenosine (3–5 g/l) when grown with limiting concentrations of xanthine or guanosine (Ishii and Shiio, 1973). Accumulation of adenosine rather than inosine is due to the weak *in vivo* feedback inhibition of the activity of adenylosuccinate synthase by AMP in *B. subtilis* (Fig. 6a, b). The accumulation is due to a bypassing of GMP repression of the common enzymes (the addition of excess guanosine represses synthesis of the common-path enzymes down to wild-type levels and prevents adenosine accumulation) but AMP inhibition and repression of PRPP amidotransferase synthesis keeps the accumulation at a rather low level. To increase adenosine production, genetic derepression by isolation of azaguanine-resistant mutants can be carried out. Such double mutants produce nearly 7 g adenosine per litre (Fig. 6c). Another xanthine-requiring auxotroph, resistant to 8-azaxanthine, produces 16 g/litre (Haneda *et al.*, 1971). Although adenosine is of no use in producing flavouring agents, it can be used to prepare ATP, a widely used biochemical. A yeast process has been devised (Tanaka and Hironaka, 1972) in which 1.6 g of crystalline sodium salt of ATP is obtained from each gram of adenosine.

Guanine-requiring auxotrophs, which lack XMP aminase, produce about 5 g of xanthosine per litre when grown under conditions of guanosine limitation (Fig. 7a, b; Ishii and Shiio, 1973). Under such conditions (Fig. 7b), GMP repression of synthesis of the common-path enzymes and of IMP dehydrogenase, as well as GMP inhibition of activity of this enzyme, are bypassed. Of interest is the metabolic flow into the GMP arm of the pathway rather than into the AMP branch, i.e. a lack of adenosine accumulation. The reason is that *B. subtilis* contains 10–60 fold more IMP dehydrogenase than adenylo-succinate synthase when grown under derepressed conditions. The K_m values for both enzymes with respect to IMP are the same. The low concentration of xanthosine accumulated (5 g/l) is due to AMP inhibition of the activity of PRPP amidotransferase and AMP repression of synthesis of the common-path enzymes. This can be partially relieved by a second mutation to adenine auxotrophy resulting in accumulation of 10 g xanthosine and 5 g inosine per litre (Fig. 7c). Further derepression resulting from mutation to aza-guanine resistance results in the production of 18 g xanthosine per litre and 6 g inosine per litre (Fig. 7d). In both the double and triple auxotrophs, the relative proportions of xanthosine and inosine produced can be markedly changed by the relative concentrations of adenine and guanosine included in the medium.

Appreciable accumulation of guanosine is dependent upon: (1)

adenine auxotrophy to eliminate AMP feedback inhibition of PRPP amidotransferase activity; (2) removal of GMP reductase; and (3) bypassing the inhibitory effect of GMP on the activity of IMP

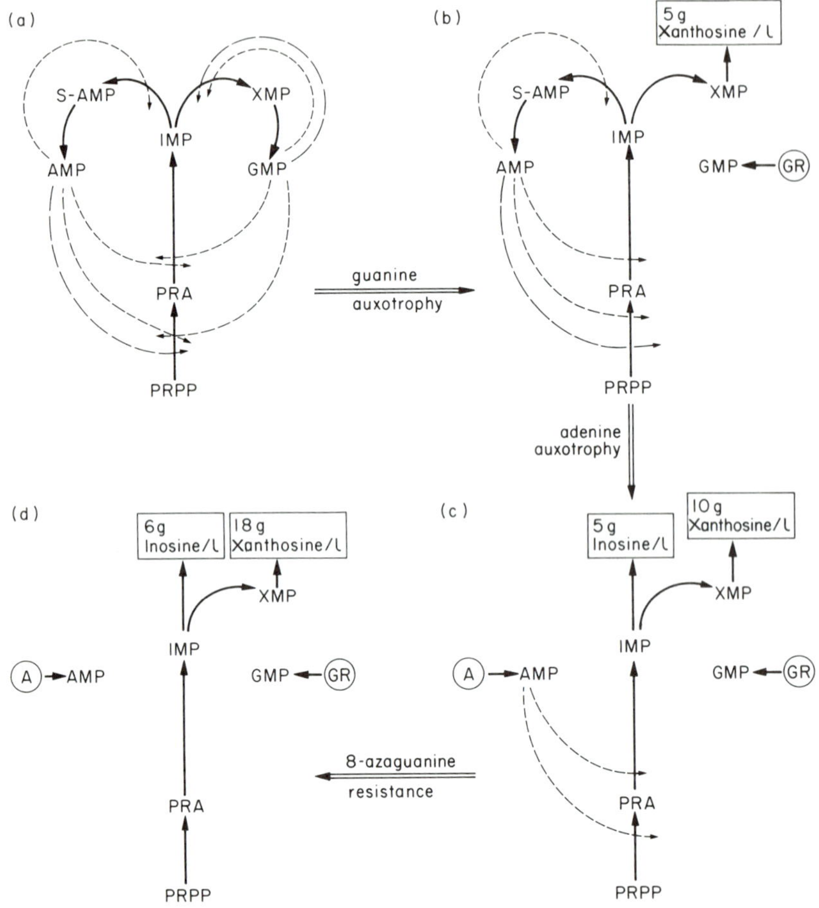

Fig. 7. Development of the xanthosine fermentation with *Bacillus subtilis*. See the legend for Fig. 5 for an explanation of the figure.

dehydrogenase. This was demonstrated by Konishi and Shiro (1968) who reported that an 8-azaguanine-resistant, GMP reductase-deficient, adenine-requiring auxotroph accumulated 4 g guanosine and 3 g inosine per litre of culture. In the studies of Momose and Shiio (1969), IMP dehydrogenase in a GMP reductase-deficient adenine-requiring auxotroph was desensitized by mutation to 8-azaxanthine resistance. About 4 g guanosine per litre were accumulated by these *B. subtilis* mutants in addition to 9 g inosine per litre.

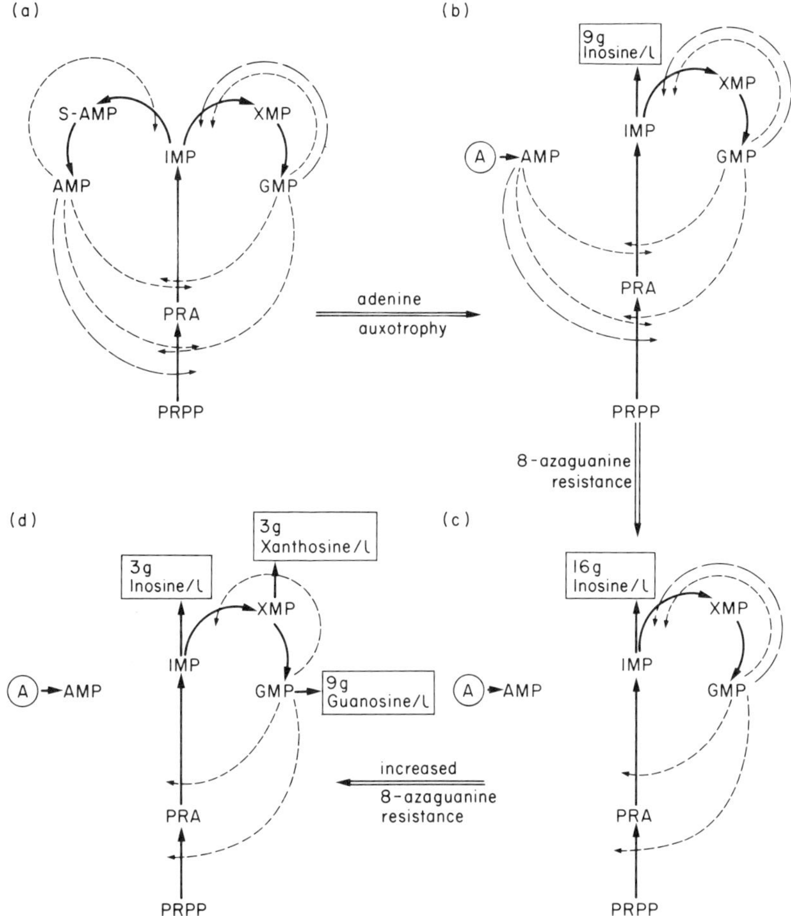

Fig. 8. Development of the guanosine fermentation with *Bacillus subtilis*. See the legend for Fig. 5 for an explanation of the figure.

More recent work by this group utilizing resistance development first to low and then to high concentrations of 8-azaguanine (Fig. 8) has resulted in a mutant producing 9 g guanosine, 3 g xanthosine and 3 g inosine per litre (Shiio, 1974). Another group (Komatsu *et al.*, 1972) mutated *B. subtilis* to auxotrophy for adenine, 8-azaxanthine resistance and GMP reductase deficiency, and obtained 4 g gaunosine, 3 g inosine and 2 g xanthosine per litre in addition to low concentrations of the respective purine bases. Further mutation to partial deficiency of the riboside-hydrolysing activity eliminated production of the free bases, riboside production amounted to 9–11 g guanosine, 0–2 g inosine and 1–3 g xanthosine per litre (Komatsu and Kodaira, 1973).

Nogami *et al.* (1968) described a procedure involving successive mutations to adenine auxotrophy, and thence to GMP reductase deficiency, sensitivity to adenine inhibition, and finally resistance to adenine inhibition. The final mutant produced (per litre) 5 g guanosine, 3 g inosine and 3 g hypoxanthine. The type of deregulation accomplished by the two mutations involving adenine inhibition was not explained by the authors, but probably involves IMP dehydrogenase. Since ATP is an inhibitor of the activity of IMP dehydrogenase under certain conditions (Ishii and Shiio, 1968), mutation to adenine sensitivity could have been one involving elimination of most (but not all) of the IMP dehydrogenase activity; in other words, with only a low level of activity, the inhibitor (ATP) which could be formed from adenine could be expected to inhibit growth. Mutation to adenine resistance probably resulted in a return to a high activity of IMP dehydrogenase but with the protein genetically modified so that it was no longer sensitive to feedback inhibition by GMP.

In addition to the direct fermentations already described for *B. subtilis*, various bioconversions have been reported for this species. Some of these are:

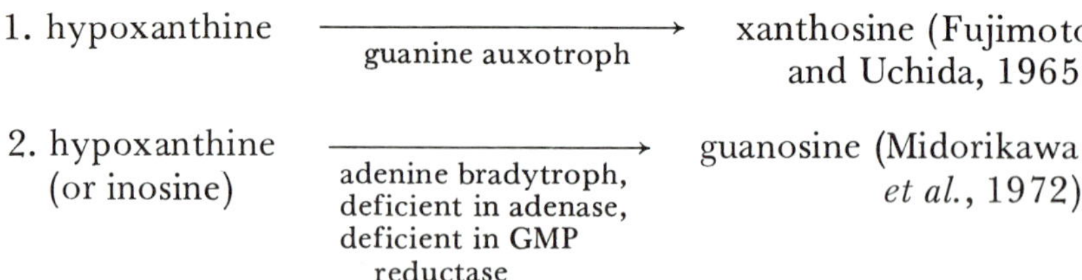

It should be evident from the data presented in this section of the chapter that *B. subtilis* has performed well in the accumulation of purine ribonucleosides. Much work has been done on these processes because ribonucleosides can be chemically phosphorylated to the more economically important purine 5′-ribonucleotides. Ribonucleosides are undoubtedly the breakdown products of internally accumulated purine 5′-ribonucleotides which are apparently unable to pass out through the permeability barrier of the bacterium unless the phosphate group is removed (Demain, 1968b). One strain of *B. subtilis*, however, has been described which can excrete intact nucleotides under certain conditions (Akiya *et al.*, 1972). This strain requires adenine and, in addition, is deficient in 5′-nucleotidase as a result of two mutations. Under normal growth conditions at 30°C it excretes about 1 g IMP per litre. However, when grown at an elevated temperature (40°C), excretion amounts to 3 g/litre. When the strain was further mutated to auxotrophy for guanine, 1–2 g XMP per litre were produced while IMP was not detectable in the culture filtrate.

V. PURINE NUCLEOTIDE OVERPRODUCTION BY *CORYNEBACTERIUM GLUTAMICUM*

Corynebacterium glutamicum and other members of the group of glutamic acid bacteria differ from most micro-organisms in their ability to excrete ribonucleotides after elimination of feedback controls. Thus, adenine-requiring auxotrophs of *C. glutamicum* excrete IMP when grown under conditions of adenine limitation. Further mutational elimination of the capacity to synthesize IMP dehydrogenase yields adenine-xanthine double auxotrophs which show increased IMP formation since intracellular accumulation of both AMP and GMP can be limited (Demain *et al.*, 1965; Misawa *et al.*, 1969a). Accumulation of as much as 13 g of IMP per litre has been reported (Furuya *et al.*, 1968).

Guanine-requiring mutants or adenine–guanine double auxotrophs accumulate XMP (Misawa *et al.*, 1964; Demain *et al.*, 1965). Culture-filtrate concentrations are of the order of 7 g of XMP per litre (Misawa *et al.*, 1969b). As one would expect, excess guanine inhibits XMP production after conversion of the base to GMP which in turn represses the synthesis of and inhibits action of IMP dehydrogenase.

Development of a direct fermentation to yield GMP has been more difficult than with the intermediates IMP and XMP. One of the main obstacles is the regulation of the activity of IMP dehydrogenase by GMP. A number of years ago, a mutant accumulating a low concentration (1 g/l) of GMP was obtained by mutation of an adenine-requiring auxotroph of *C. glutamicum* to adenine–xanthine double auxotrophy followed by a second mutation back to just the adenine requirement (Demain *et al.*, 1966). This double mutation of the gene coding for IMP dehydrogenase presumably desensitized the enzyme to feedback inhibition. Subsequent development of the guanosine fermentation in *B. subtilis* (see Section IV, pg. 201) suggests that the important requirements in a bacterium to be used in an economical GMP fermentation are: (a) nucleotide excretion such as in *C. glutamicum*; (b) removal of ability to synthesize adenylo-succinate synthetase (i.e. adenine auxotrophy) to eliminate feedback inhibition of the activity of PRPP amidotransferase by AMP; (c) successive resistance mutations to 8-azaguanine and/or 8-azaxanthine to desensitize IMP dehydrogenase and eliminate repression of synthesis of the common-path enzymes and IMP dehydrogenase; and (d) genetic removal of the gene coding for GMP reductase. The existence of such an organism has not yet been reported.

The feedback controls exerted by GMP on its own biosynthesis can also be overcome by environmental manipulations such as addition of anion-exchange resins or dioxane to the medium (Schwartz and Margalith, 1973). Dioxane acts by forming quantitatively a complex with GMP and water as discovered by Vitali *et al.* (1965). Using the resin technique, Schwartz and Margalith (1973) were able to recover 4 g IMP, 2 g XMP and 1 g GMP from a one-litre culture of an azaguanine-resistant mutant of *Streptomyces* sp.

VI. SALVAGE SYNTHESIS OF PURINE NUCLEOTIDES

While investigating what appeared to be a direct IMP fermentation by an adenine-requiring auxotroph of *Brevibacterium ammoniagenes*, Nara *et al.* (1967) noticed that, during the early part of the fermentation, only hypoxanthine was excreted. The excreted purine then disappeared as extracellular IMP was detected. Investigation showed that the phenomenon involved intracellular accumulation of IMP due to adenine auxotrophy, degradation to inosine and then to hypoxanthine, excretion and finally a remarkable extracellular 'salvage synthesis' of IMP from the released hypoxanthine.

It was later found that several bases could be converted to their respective ribonucleotides by intact cells of wild-type *B. ammonia-genes* (Nara *et al.*, 1968a) as long as certain requirements were met. These requirements were high concentrations of inorganic phosphate and Mg^{2+}, addition of thiamin and pantothenate, and a growth-limiting concentration of Mn^{2+} (Nara *et al.*, 1968b). Under normal conditions in a medium containing a lower concentration of phosphate, none of the above requirements is necessary for growth. When the high concentration of phosphate required as a nucleotide precursor is used, however, growth does not occur unless Mn^{2+}, thiamin, pantothenate and a high Mg^{2+} concentration are added (Nara *et al.*, 1969a, b). Under these abnormal growth conditions, the concentration of Mn^{2+} is crucial to extracellular salvage synthesis, i.e. it must be low. If so, it triggers nucleotide formation by altering the composition of cellular lipids and increasing the rate of exit of intracellular compounds (Furuya *et al.*, 1970). Accompanying the change in permeability are inhibition of DNA synthesis, decrease in cell viability, increase in cell mass, and a shift in cellular morphology from normal small rods and ellipsoidal cells to elongated and swollen forms (Furuya *et al.*, 1968; Oka *et al.*, 1968). The cells leak into the medium ribose 5-phosphate, PRPP, 5-phosphoribose pyrophosphokinase, several purine nucleotide pyrophosphorylases and presumably also ATP. Thus, addition of a purine allows the following reactions to take place extracellularly:

$$\text{ribose 5-phosphate} + \text{ATP} \xrightarrow[\substack{\text{5-phosphoribose} \\ \text{pyrophosphokinase}}]{Mg^{2+},\ PO_4^{3-}} \text{PRPP} + \text{AMP}$$

$$\text{PRPP} + \text{added purine} \xrightarrow[\substack{\text{purine nucleotide} \\ \text{pyrophosphorylase}}]{Mg^{2+},\ PO_4^{3-}} \text{5}'\text{-purine ribonucleotide} + \text{PPi}$$

Using this organism, hypoxanthine is converted to IMP (15 g IMP/litre) with a 96% conversion of hypoxanthine used (Misawa and Kinoshita, 1970). Guanine is converted to GMP, GDP, GTP, and adenine is converted to AMP, ADP and ATP (Tanaka *et al.*, 1968). Production of the di- and triphosphates of guanosine and adenosine is due to an additional leakage of nucleotide phosphotransferases (Misawa *et al.*, 1969c). Phosphate donors are ATP, GTP, CTP and PRPP. The ratio of mono-, di- and triphosphates can be controlled by addition of Hg^{2+}, Zn^{2+} and Ag^+ which inhibit the action of nucleotide phosphotransferases (Komuro *et al.*, 1969) and by addition of xylene or other solvents which increase excretion of these enzymes and thus increase production of the di- and triphosphates (Misawa *et al.*, 1969d). *Brevibacterium ammoniagenes* can also be used for preparing nicotinic acid mononucleotide by adding nicotinic acid or nicotinamide; nicotinamide adenine dinucleotide (NAD^+) is formed when adenine is also added. Nicotinamide adenine dinucleotide is produced at concentrations of 2 g/l (Nakayama *et al.*, 1968).

The beneficial effect of Mn^{2+} limitation in extracellular salvage synthesis can be obtained also by addition of antibiotics or cationic surface-active agents to log-phase cells (Nara *et al.*, 1969c). Mutants have been isolated which produce nucleotides even in the presence of high concentrations of Mn^{2+} (Furuya *et al.*, 1969; Kato *et al.*, 1971). The biochemical change elicited by mutation to Mn^{2+} insensitivity is unknown.

A combination of permeability alteration and mutation has been used to effect bioconversion of XMP to GMP by *B. ammoniagenes* (Furuya *et al.*, 1971). Limitation of Mn^{2+} increased permeability so that XMP could penetrate the cells. Mutation to resistance to decoyinine, an inhibitor of XMP aminase, resulted in conversion of the XMP to GMP, presumably by derepressing XMP aminase. Also accumulated were GDP, GTP and guanine. A similar mutant with lower 5′-nucleotidase activity converted XMP to the guanine nucleotides without any degradation to guanine (Furuya *et al.*, 1972). Guanosine monophosphate was also produced directly by use of a mixed culture combining this strain with an XMP-accumulating mutant.

REFERENCES

Akiya, T., Midorikawa, Y., Kuninaka, A., Yoshino, H. and Ikeda, Y. (1972). *Agricultural and Biological Chemistry* **36**, 227.

Demain, A. L. (1968a). *Progress in Industrial Microbiology* **8**, 35.

Demain, A. L. (1968b). *Biotechnology and Bioengineering* **10**, 291.

Demain, A. L., Jackson, M., Vitali, R. A., Hendlin, D. and Jacob, T. A. (1965). *Applied Microbiology* **13**, 757.

Demain, A. L., Jackson, M., Vitali, R. A., Hendlin, D. and Jacob, T. A. (1966). *Applied Microbiology* **14**, 821.

Fujimoto, M. and Uchida, K. (1965). *Agricultural and Biological Chemistry* **29**, 1150.

Furuya, A., Abe, S. and Kinoshita, S. (1968). *Applied Microbiology* **16**, 981.

Furuya, A., Abe, S. and Kinoshita, S. (1969). *Applied Microbiology* **18**, 977.

Furuya, A., Abe, S. and Kinoshita, S. (1970). *Agricultural and Biological Chemistry* **34**, 210.

Furuya, A., Abe, S. and Kinoshita, S. (1971). *Biotechnology and Bioengineering* 13, 229.

Furuya, A., Okachi, R., Takayama, K. and Abe, S. (1972). *Abstracts, Fourth International Fermentation Symposium, Kyoto*, pg. 192.

Haneda, K., Hirano, A., Kodaira, R. and Ohuchi, S. (1971). *Agricultural and Biological Chemistry* 35, 1906.

Ishii, K. and Shiio, I. (1968). *Journal of Biochemistry, Tokyo* 63, 661.

Ishii, K. and Shiio, I. (1972). *Agricultural and Biological Chemistry* 36, 1511.

Ishii, K. and Shiio, I. (1973). *Agricultural and Biological Chemistry* 37, 287.

Jund, R. and Lacroute, F. (1970). *Journal of Bacteriology* 102, 607.

Kato, F., Furuya, A. and Abe, S. (1971). *Agricultural and Biological Chemistry* 35, 1061.

Kawamoto, I., Nara, T., Misawa, M. and Kinoshita, S. (1970). *Agricultural and Biological Chemistry* 34, 1142.

Komatsu, K. and Kodaira, R. (1973). *Journal of General and Applied Microbiology* 19, 263.

Komatsu, K., Haneda, K., Hirano, A., Kodaira, R. and Ohsawa, H. (1972). *Journal of General and Applied Microbiology* 18, 19.

Komuro, T., Nara, T., Misawa, M. and Kinoshita, S. (1969). *Agricultural and Biological Chemistry* 33, 1018.

Konishi, S. and Shiro, T. (1968). *Agricultural and Biological Chemistry* 32, 396.

Kuninaka, A., Kibi, M. and Sakaguchi, K. (1964). *Food Technology* 19, 29.

Margalith, P. (1974). *Science Progress, Oxford* 61, 443.

Midorikawa, Y., Akiya, T., Kuninaka, A. and Yoshino, H. (1972). *Agricultural and Biological Chemistry* 36, 1529.

Misawa, M. and Kinoshita, S. (1970). *Paper presented at the Annual Meeting of the American Society for Microbiology, Boston*, U.S.A.

Misawa, M., Nara, T., Udagawa, K., Abe, S. and Kinoshita, S. (1964). *Agricultural and Biological Chemistry* 28, 688.

Misawa, M., Nara, T. and Kinoshita, S. (1969a). *Agricultural and Biological Chemistry* 33, 514.

Misawa, M., Nara, T., Udagawa, K., Abe, S. and Kinoshita, S. (1969b). *Agricultural and Biological Chemistry* 33, 370.

Misawa, M., Nara, T. and Kinoshita, S. (1969c). *Agricultural and Biological Chemistry* 33, 521.

Misawa, M., Nara, T. and Kinoshita, S. (1969d). *Agricultural and Biological Chemistry* 33, 532.

Momose, H. and Shiio, I. (1969). *Journal of General and Applied Microbiology* 15, 399.

Nakayama, K. and Tanaka, H. (1971). *Agricultural and Biological Chemistry* 35, 518.

Nakayama, K., Sato, Z., Tanaka, H. and Kinoshita, S. (1968). *Agricultural and Biological Chemistry* 32, 1331.

Nara, T., Misawa, M. and Kinoshita, S. (1967). *Agricultural and Biological Chemistry* 31, 1351.

Nara, T., Misawa, M. and Kinoshita, S. (1968a). *Agricultural and Biological Chemistry* 32, 561.

Nara, T., Misawa, M. and Kinoshita, S. (1968b). *Agricultural and Biological Chemistry* 32, 1153.

Nara, T., Misawa, M., Komuro, T. and Kinoshita, S. (1969a). *Agricultural and Biological Chemistry* 33, 358.

Nara, T., Komuro, T., Misawa, M. and Kinoshita, S. (1969b). *Agricultural and Biological Chemistry* **33**, 1030.

Nara, T., Misawa, M., Komuro, T. and Kinoshita, S. (1969c). *Agricultural and Biological Chemistry* **33**, 1198.

Nierlich, D. P. and Magasanik, B. (1965). *Journal of Biological Chemistry* **240**, 358.

Nogami, I., Kida, M., Iijima, T. and Yoneda, M. (1968). *Agricultural and Biological Chemistry* **32**, 144.

O'Donovan, G. A. and Gerhart, J. C. (1972). *Journal of Bacteriology* **109**, 1085.

O'Donovan, G. A. and Neuhard, J. (1970). *Bacteriological Reviews* **34**, 278.

Oka, T., Udagawa, K. and Kinoshita, S. (1968). *Journal of Bacteriology* **96**, 1760.

Tanaka, A. and Hironaka, J. (1972). *Agricultural and Biological Chemistry* **36**, 867.

Tanaka, H., Sato, Z., Nakayama, K. and Kinoshita, S. (1968). *Agricultural and Biological Chemistry* **32**, 721.

Tochikura, T., Kawai, H., Kawaguchi, K., Mugibayashi, Y. and Ogata, K. (1972). *In* "Fermentation Technology Today", (G. Terui, ed.), pp. 463–471. Society of Fermentation Technology, Osaka.

Sato, H. and Shiio, I. (1970). *Journal of Biochemistry, Tokyo* **68**, 763.

Sasajima, K., Nogami, I. and Yoneda, M. (1970). *Agricultural and Biological Chemistry* **34**, 381.

Schwartz, J. and Margalith, P. (1973). *Biotechnology and Bioengineering* **15**, 85.

Shiio, I. (1974). *Paper presented at 2nd International Symposium on Genetics of Industrial Microorganisms*, Sheffield.

Shiio, I. and Ishii, K. (1969). *Journal of Biochemistry, Tokyo* **66**, 175.

Shiio, I. and Ishii, K. (1971). *Journal of Biochemistry, Tokyo* **69**, 339.

Škodová, H., Šolínová, H., Škoda, J. and Dyr. J. (1969). *Folia Microbiologica* **14**, 145.

Vitali, R. A., Inamine, E. S., Rothrock, J. W. and Jacob, T. A. (1965). *Archives of Biochemistry and Biophysics* **112**, 313.

Wu, T.-W. and Scrimgeour, K. G. (1973). *Canadian Journal of Biochemistry* **51**, 1391.

6. Amino Acids

S. KINOSHITA AND K. NAKAYAMA

Tokyo Research Laboratory, Kyowa Hakko Kogyo Co. Ltd., Tokyo, Japan

I. INTRODUCTION

Following the increasing demand for monosodium glutamate as a flavouring agent, an efficient L-glutamic acid-producing micro-organism, namely *Corynebacterium glutamicum* (synonym. *Micrococcus glutamicus*), was isolated in authors' laboratory (Kinoshita *et al.*, 1957). Following this, much research activity has been focused on microbial amino-acid production. The primary reason for these efforts was the hope to improve the nutritional value of low-cost vegetable proteins by enrichments with essential amino acids. Nowadays, many processes for production of various amino acids have been developed. The total World production of L-glutamic acid is considered to be in excess of $150,000 \times 10^3$ kg per year and it is used mainly as a flavouring agent. L-Lysine has also been produced on a large scale by fermentation and is used mainly as a feed supplement. Its total World production is considered in excess of $15,000 \times 10^3$ kg per year. Other proteinaceous amino acids, except methionine, alanine, glycine and cystine and cysteine, are being produced by fermentation process on a scale less than 10^6 kg per year. L-Methionine and L-alanine can be produced by enzymic processes although the racemic forms of these amino acids are being produced by chemical synthesis.

Once *C. glutamicum* was discovered by screening isolates from Nature, similar efforts lead to the isolation of bacteria producing DL-alanine or L-valine. However, it was found that most wild-type strains isolated from Nature could not produce industrially significant amounts of other amino acids except a few amino acids such as those already referred to. One of the main reasons for this fact is that regulation of cellular metabolism avoids oversynthesis. The existence of these regulatory phenomena was just being clarified at the time of isolation of *C. glutamicum* (Yates and Pardee, 1957; Umbarger and Brown, 1958) and is now well recognized. An auxotrophic mutant which cannot produce the regulatory effector or corepressor (usually the end product or a derivative of the end product) overproduces and excretes the precursor or the related metabolite of a blocked reaction when grown on a limiting supply of the required nutrient. This is the principle of the application of an auxotrophic mutant to the microbial production of amino acids. Active attempts to utilize this phenomenon for the industrial production of microbial metabolites were launched in the 1950s. The first prominent result was obtained in producing α,ϵ-diaminopimelic acid using a lysine-requiring auxotroph of *Escherichia coli* (Casida and Baldwin, 1956). Diaminopimelic acid was decarboxylated to yield L-lysine in the second step. Soon afterwards, a process for L-lysine production was developed with a homoserine-requiring auxotroph of *C. glutamicum* (Kinoshita *et al.*, 1958). Many amino acids are being produced with auxotrophic mutants.

It is obviously useless to accumulate the end product of an unbranched pathway such as arginine and histidine with an auxotrophic mutant. The production of such a metabolite depends on the use of a regulatory mutant. A mutant which has lost some biosynthetic regulation can be obtained by selecting for analogue resistance, and as a prototrophic revertant from the auxotroph having a deficiency in a regulatory enzyme.

To improve the yield of an amino acid, mutants having multiple markers including auxotrophy and analogue-resistance contributing to production of the designated amino acid are selected. Multiple markers also contribute to the yield by stabilizing the productivity against back mutation during fermentation (Nakayama, 1972). With metabolic regulation, the permeability barrier is another mechanism which protects the micro-organism from leaking organic compounds to the environment. It allows cells to retain intermediates and macromolecules necessary for life of the micro-organism. Production of L-glutamic acid by *C. glutamicum* was found to be due mainly to the permeability change induced by limiting the supply of biotin required by the bacterium. Permeability is thus another important factor for amino-acid production.

Although certain specific environmental conditions make it possible to exploit a single enzymic process or a process using a precursor, most amino acids can now be produced by the so-called 'direct fermentation' process, i.e. microbial production from a cheap carbon source by a fermentation process. Examples of each

type and process will be described in the remaining sections of this
chapter.

II. L-GLUTAMIC ACID AND L-GLUTAMINE

A. Glutamic Acid Production from Carbohydrate

Production of glutamic acid and L-glutamine was reviewed recently
(Kinoshita and Tanaka, 1972). Production of glutamic acid from
carbohydrate in high yield is carried out by a group of bacteria
represented by *Corynebacterium glutamicum* (synonym. *Micrococcus
glutamicus*). These bacteria were classified species in different genera;
they include *Corynebacterium glutamicum* (*Micrococcus glutamicus*), *Brevibacterium flavum*, *Brev. lactofermentum*, *Brev. divari-
catum*, *Brev. thiogenitalis*, *Corynebacterium callunae*, *C. herculis*,
Microbacterium ammoniaphilum and others. The guanine + cytosine
(G-C) content of the DNA of these bacteria falls into a narrow range
from 51.2 to 54.4 mole %. Morphological and physiological properties also supported their close affinities. Moreover, they could be
classified into a group of bacteria belonging to the Corynebacteri-
aceae (Abe *et al.*, 1967; Yamada and Komagata, 1972). For convenience we call these bacteria 'glutamic acid bacteria' hereafter.

The special conditions which allow these bacteria to excrete large
amounts of glutamate are a nutritional requirement for biotin and
the lack, or very low content, of α-oxoglutarate dehydrogenase. The
biotin requirement is the major controlling factor in the fermentation. When enough biotin is supplied for optimal growth, the
organism produces lactate. Under conditions of suboptimal growth,
glutamate is excreted.

Under optimal culture conditions, glutamic-acid bacteria convert
about 50% of the supplied carbohydrate into L-glutamic acid with
little formation of by-products. Various carbohydrate materials can
be used as the carbon source. Glucose and sucrose are particularly
suitable. For industrial purposes, hydrolysed starch solutions, cane
molasses and beet molasses are preferred. Other carbon sources such
as acetic acid and ethanol are also used. Carbon sources, such as cane
molasses, with a high content of biotin, are used with the addition of
penicillin during logarithmic growth or of fatty-acid derivatives such

as polyoxyethylene sorbitan mono-oleate (Tween 60) before or during logarithmic growth. Ammonium sulphate, ammonium chloride, ammonium phosphate, aqueous ammonia, ammonia gas and urea have been used as nitrogen sources. Although a large amount of ammonium ion is necessary, a high concentration of it is inhibitory to growth of the organism as well as to production of glutamic acid. Therefore, ammonium ions are added as the fermentation progresses. Ammonia water, or gaseous ammonia, is generally used industrially. Other ions supplied include K^+, Mg^{2+}, Fe^{2+}, Mn^{2+}, PO_4^{3-}, SO_4^{2-} and Cl^-. These are usually supplied by the following inorganic salts (%, w/v): 0.05–0.2 KH_2PO_4, 0.05–0.2 K_2HPO_4, 0.025–0.1 $MgSO_4 . 7H_2O$, 0.0005–0.01 $FeSO_4 . 7H_2O$, 0.0005–0.005 $MnSO_4 . 4H_2O$ and 0.5–4 $CaCO_3$. The most important factor in the medium for the glutamic-acid fermentation is biotin, which is an essential growth factor for glutamic-acid bacteria. The concentration of biotin must be limited to a suboptimal one for growth. The optimal concentration of biotin for the glutamic-acid fermentation depends on the strain, kind and concentration of carbon source, but it is generally somewhat below 5 μg per litre of medium. Some strains require thiamin or cystine in addition to biotin. Certain iron-chelating compounds are necessary for growth of glutamic-acid bacteria, but their presence is not necessary when the carbohydrate is autoclaved simultaneously with other ingredients of the medium because iron-chelating compounds are formed by autoclaving.

The pH value optimal for growth and glutamic-acid production is 7.0–8.0. Continuous feeding of NH_4^+ can adjust the pH value and also supply ammonium ions to the medium. Urea can replace the ammonium ion in the nutrition of those glutamic-acid bacteria which possess urease activity. The optimal value of K_d (the overall co-efficient of oxygen transfer) for the glutamic-acid fermentation is considered to be $3–5 \times 10^{-6}$ (mol O_2) atm^{-1} min^{-1} ml^{-1}. Under conditions of insufficient oxygen, production of glutamic acid is poor and large amounts of lactic acid and succinic acid accumulate, while excess oxygen increases the amount of lactic acid and α-oxoglutaric acid. The optimal temperature for the glutamic-acid fermentation is usually 30°C to 35°C. The time-course of the glutamic-acid fermentation by *C. glutamicum* No. 541 is shown in Figure 1 (Tanaka *et al.*, 1960). The medium used had the following composition (w/v): 10% glucose, 0.25% corn-steep liquor, 0.25% 'NZ-amine'

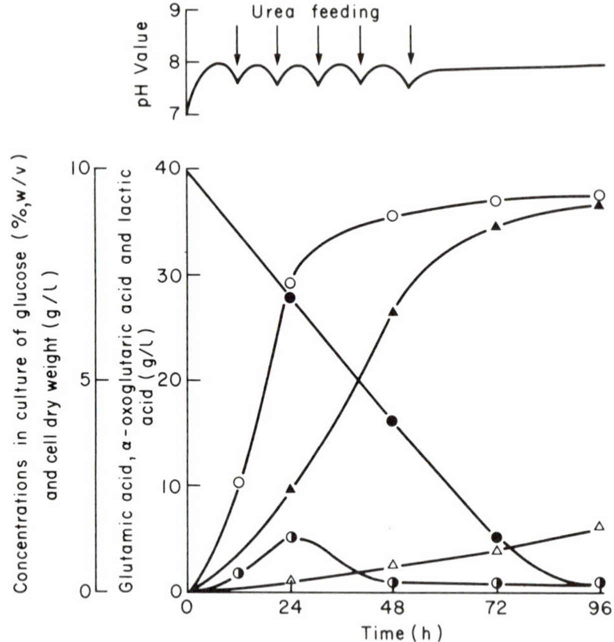

Fig. 1. Time-course of a glutamic acid fermentation by *Corynebacterium glutamicum* no. 541. ○ indicates changes in the concentration of cell dry weight, ● of glucose, ▲ of glutamic acid, △ of α-oxoglutarate, and ◑ of lactic acid. The top line on the figure indicates changes in the pH value of the culture which was maintained at 7.5–8.0 by additions of urea solutions. Also indicated are the times of urea feeding. Reproduced from Tanaka *et al.* (1960).

(enzymic hydrolysate of casein), 0.1% K_2HPO_4, 0.025% $MgSO_4 . 7H_2O$ and 0.5% urea. Fermentation was carried out in shake flasks at 28°C.

B. Glutamic Acid Production from Non-Carbohydrate Materials

The availability of acetic acid at a reasonable price and the waste water problem associated with the use of cane molasses as the carbon source for the glutamic-acid fermentation prompted the search for a process using acetic acid. With *Brev. flavum*, L-glutamic-acid production reached 98 g per litre (48% on the basis of acetic acid) in 48 h (Tanaka *et al.*, 1971). Adding 25 to 250 μg of Cu^{2+} per litre to a medium increased L-glutamate yield from acetate remarkably (Kanzaki *et al.*, 1973). The poor yield of L-glutamate from acetate by copper-deficient cells seems to be due to a decrease in energy supply which was caused by the low efficiency of oxidative phosphorylation (Sugiyama *et al.*, 1973). The productivity of an oleic acid-requiring auxotroph of *Brev. thiogenitalis* was superior to the parent strain (Kanzaki *et al.*, 1972). Brevibacterium sp. B 136 converted ethanol to L-glutamic acid with a yield of 60% (Oki *et al.*, 1968).

A group of bacteria represented by *Nocardia erythropolis* (synonym. *Corynebacterium hydrocarboclastus* (Komura *et al.*, 1973) produce L-glutamic acid from *n*-paraffins in media containing a suboptimal concentration of thiamin. The glutamic acid yield is stimulated by addition of penicillin to an exponentially growing culture with excess thiamin in the medium. Cupric ions stimulate growth and glutamic-acid production from *n*-paraffins by *Arthrobacter paraffineus* (Suzuki *et al.*, 1971). Production of glutamic acid by *A. paraffineus* reached 82 g per litre at 48 h. A penicillin-resistant mutant of *N. erythropolis* produced 84 g of L-glutamic acid per litre (Kobayashi *et al.*, 1971). A glycerol-requiring auxotroph of *C. alkanolyticum* produced 72 g of L-glutamic acid per litre (Kikuchi *et al.*, 1972).

C. Metabolic Pathways Involved in the Biosynthesis of Glutamic Acid from Glucose

The pathway of glutamate biosynthesis from glucose is shown in Fig. 2. The major route, shown with the heavy arrows, involves at least 16 enzymic steps. *Alpha*-oxoglutarate is converted to glutamate by reductive amination. The enzyme catalysing this conversion is the $NADP^+$-specific glutamate dehydrogenase. The presence of this enzyme is essential for glutamate formation. When resting cells are incubated with glucose in the absence of ammonia, α-oxoglutarate accumulates rather than glutamate. The NADPH required for the action of glutamate dehydrogenase is supplied by the preceding isocitrate dehydrogenase reaction. Glutamate dehydrogenase, in turn, provides the $NADP^+$ required for isocitrate dehydrogenase.

The low content of α-oxoglutarate dehydrogenase favours glutamate production. The importance of the lack of this enzyme has been demonstrated with *E. coli* which does not require biotin and is not a glutamate excreter. Even without the biotin requirement, a mutant which lacks α-oxoglutarate dehydrogenase was found to excrete 2.3 g of glutamate per litre while its parent excreted none.

In addition to the EMP pathway, glutamic acid-producing bacteria also use the HMP pathway to convert glucose to three-carbon and

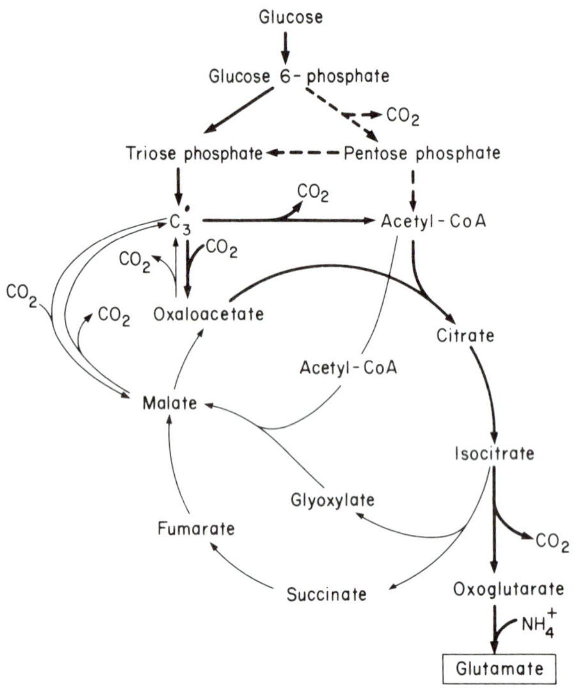

Fig. 2. Metabolic pathways involved in biosynthesis of glutamic acid from glucose. Major routes are indicated by heavy arrows. Thin lines indicate reactions of the glyoxylate cycle, and broken lines reactions of the hexose monophosphate pathway.

two-carbon compounds (shown by broken lines in Fig. 2). These compounds can then feed into the TCA cycle. Estimates of the relative utilization of glucose by the two pathways have shown the predominance of the EMP pathway under fermentation conditions.

When a glutamate fermentation is carried out in the presence of $^{14}CO_2$, the radio-activity is fixed into the α-carboxyl group of glutamate. Two enzymes have been found in glutamate excreters which participate in the fixation of carbon dioxide, namely oxaloacetate carboxylase and the $NADP^+$-linked malic enzyme which catalyses fixation of carbon dioxide to pyruvate to yield malate. Malate is oxidized to oxaloacetate by malate dehydrogenase. The oxaloacetate is converted to citrate. Because of a deficiency in the TCA cycle, glutamic acid-producing bacteria use the glyoxylate pathway shown by thin lines in Fig. 2. Therefore, the two competing reactions of isocitrate are important. The isocitritase reaction is needed during the growth phase for energy production and to produce intermediates for biosynthetic reactions. However, after the growth phase, glutamate production would be best without operation of the isocitritase reaction. This would indicate that optimal conditions for the growth phase and the glutamate production phase should be different. After growth, the ideal fermentation would proceed by the following reaction:

$$C_6H_{12}O_6 + NH_3 + 1.5O_2 \rightarrow C_5H_9O_4N + CO_2 + 3H_2O$$

This represents a 100% molar conversion of 81.7% weight conversion of sugar to glutamic acid. On the other hand, the poorest fermentation would be represented by complete oxidation of glucose as occurs with growth in media containing an optimal concentration of biotin and which results in no conversion. What is actually found lies between these limits, i.e. 50–75% molar conversion by resting cells. This indicates that carbon is being lost as carbon dioxide by reversal of action of the malic enzyme and of oxaloacetate carboxylase during operation of the glyoxylate bypass. Therefore, assimilation of ammonium ions in this organism is almost totally dependent upon action of $NADP^+$-linked glutamate dehydrogenase.

D. Alteration of Permeability and Glutamic Acid Production

Alteration of permeability in relation to glutamic-acid production was reviewed by Demain and Birnbaum (1968). Production of a large amount of glutamic acid in bacteria is attained by the alteration of the permeability barrier. Increased permeability can be induced in glutamic-acid bacteria either by biotin deficiency (Tanaka *et al.*, 1960), oleic acid deficiency in an oleic acid-requiring auxotroph (Kanzaki *et al.*, 1967), glycerol deficiency in a glycerol-requiring auxotroph (Nakao *et al.*, 1970), treatment with fatty acid derivatives (Udagawa *et al.*, 1962; Takinami *et al.*, 1963, 1964; Shiio *et al.*, 1963; Shibukawa *et al.*, 1964a, b) or by addition of penicillin to the growth medium (Sommerson and Phillips, 1962). Biotin deficiency and treatment with fatty-acid derivatives cause an aberration in the normal synthesis and distribution of cellular fatty acids. The cell membrane from such cells contains an abnormal ratio of saturated to unsaturated fatty acids. This abnormality directly correlates with increased permeability of the cell and with the excretion of high concentrations of glutamic acid.

Biotin- and oleic acid-requiring auxotrophs were found to be inapplicable for production of L-glutamic acid from *n*-paraffins. Excretion of L-glutamic acid by addition of penicillin or cephalosporin was accompanied by excretion of both phospholipid and *N*-acetylglucosamine, which are known to be components of the cell membrane and cell wall respectively (Nakao *et al.*, 1973). Based on this observation, a glycerol-requiring auxotroph of *Corynebacterium alkanolyticum* was obtained in which cellular phospholipid synthesis is regulated by the amount of glycerol supplied (Nakao *et al.*, 1970). The mutant produced about 40 g of L-glutamic acid per litre from *n*-paraffins in the culture in the presence of 0.01% of glycerol but in the absence of penicillin (Nakao *et al.*, 1972). Studies with the auxotroph suggest that the permeability of L-glutamic acid is not always controlled by the cellular content of unsaturated fatty-acyl residues, and that phospholipid content controls the membrane permeability to L-glutamic acid (Kikuchi and Nakao, 1973).

Addition of penicillin to log-phase cultures of *C. glutamicum* results in a rapid (97–99.5%) decrease in viability accompanied by a rapid rate of glutamate excretion. Cell mass (measured by optical density) and total cell count remain fairly constant after a slight initial increase. The packed cell volume, however, decreases by 60–80% during the phase of decreasing viability indicating a change in the surface properties of the cells. The cells continue to produce glutamate for 40–50 h after penicillin addition with no lysis throughout the entire fermentation. Apparently, the lack of an overabundance of potent mucopeptidases allows the cells to retain their form and to metabolize as permeable 'resting' entities (Demain and Birnbaum, 1968). During growth in a medium containing a high concentration of biotin, *C. glutamicum* synthesizes glutamate until the cell becomes saturated at 25–35 μg per mg of dry weight. Cessation of production was assumed to be due to some feedback control of glutamate on its own synthesis (Demain and Birnbaum, 1968). When the cells become more permeable, glutamate passed from the cells into the medium, thus relieving the feedback effect and allowing further synthesis.

E. L-Glutamine and N-Acetyl-L-Glutamine

L-Glutamine and N-acetyl-L-glutamine are produced along with glutamic acid by glutamic-acid bacteria. Production of glutamine is increased by maintaining the medium at a weakly acidic pH value and production of N-acetylglutamine is increased by maintaining a neutral to weakly acidic pH value in the medium. Under optimal conditions, a 20% conversion of glucose to glutamine and 12–14% to N-acetylglutamine are observed. High concentrations of ammonium ion retard growth and glutamine production. Increase in the biotin supply and addition of natural nutrients such as corn-steep liquor and meat extract to the medium retard growth and increase glutamine production. These nutrients or Zn^{2+} at a concentration over that required for growth, repress formation of N-acetylglutamine (Nakanishi, 1972). L-Glutamine and N-acetylglutamine are used in the treatment of gastric ulsers.

III. L-LYSINE

A. L-Lysine Production by Fermentation

Microbial production of L-lysine was reviewed by Nakayama (1972). A microbial process for L-lysine production was first developed by a combination of diaminopimelate production by a lysine-requiring auxotroph or a lysine-histidine-requiring double auxotroph of *Escherichia coli* and decarboxylation of the compound by *Aerobacter aerogenes* or wild-type *E. coli*. The process was developed by workers of E. I. de Pont de Nemours and Co. Inc. The yield of diaminopimelic acid reached 24 g per litre. Direct production of L-lysine from carbohydrate was developed first with a homoserine-

or threonine- plus methionine-requiring auxotroph of *Coryne-bacterium glutamicum* by the present authors. The same type process was reported with a homoserine-requiring auxotroph of *Brev. flavum.* The leaky homoserine-requiring auxotroph was recognized as a threonine-sensitive mutant because growth was inhibited by excessive threonine and the inhibition was released by addition of methionine. This phenomenon is due to feedback inhibition of residual homo-serine dehydrogenase by threonine (see Section A, pg. 223). Homo-serine- (or threonine plus methionine)-requiring auxotrophs of other bacteria were also found to produce L-lysine, but the yields were lower than that from the homoserine-requiring auxotroph of coryne-form bacteria. Threonine- or leucine-requiring auxotrophs of *C. glutamicum* produce fairly large amounts of *L*-lysine but they are inferior to the homoserine-requiring auxotrophs. Other auxotrophs of *C. glutamicum* and other bacteria were also inferior to the homoserine-requiring auxotrophs of *C. glutamicum.* The fermen-tation performance of the homoserine-requiring auxotrophs could be stabilized by the use of mutants having a double amino-acid defic-iency, one of which is homoserine. Double auxotrophs, which require in addition to homoserine at least one of the amino acids threonine, isoleucine or methionine for growth, have been found to be highly stabilized showing little tendency to revert to homoserine independence. It is possible not only to prevent reversion of the cultures to a wild-type state, but many of the micro-organisms, being double-mutants in the homoserine pathway, produce lysine in higher yields (Woodruff and Jackson, 1970).

Cane molasses is now generally used as a carbon source in the industrial production of lysine, though other carbohydrate materials, acetic acid and ethanol can be used. The pH value of the medium is maintained near neutrality during the fermentation by feeding ammonia or urea. Ammonia and ammonium salts are generally good nitrogen sources, and urea can be used for organisms having urease activity. An example of a fermentation using cane molasses in a 2 Kl fermentor is as follows. The medium for first seed culture contained 2% glucose, 1% peptone, 0.5% meat extract and 0.25% NaCl in tap water. For the second seed culture, the medium contained 5% cane molasses, 2% $(NH_4)_2SO_4$, 5% corn-steep liquor and 1% $CaCO_3$ in tap water. The fermentation medium contained 20% cane molasses (as glucose) and 1.8% soybean meal hydrolysate (as weight of meal before hydrolysis with 6N H_2SO_4 and neutralization with ammonia water) in tap water. The fermentation was carried out at 28°C. Figure 3 shows the time course of the fermentation using *C. glutamicum* No. 901 (a homoserine-requiring auxotroph), which produced 44 g of L-lysine per litre in 60 h. Foaming in the aerated

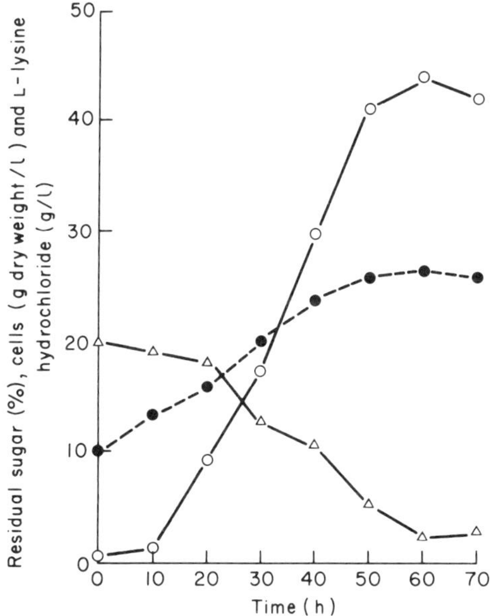

Fig. 3. Time-course of a lysine-producing fermentation using *Corynebacterium glutamicum*. ○ indicates changes in the concentration of L-lysine, ● in the concentration of cells, and △ in the concentration of residual sugar. From Shigeto (1962).

culture can be repressed by addition of proper antifoaming agents. The amount of the growth factors (homoserine or threonine and methionine) should be appropriate for the production of L-lysine. It is supplied in limited amounts and is suboptimal for the growth. The biotin concentration in the medium must generally be greater than 30 μg per litre. Cane molasses usually supplies enough biotin but, when beet molasses or starch hydrolysate is used, biotin must be added. Yields of L-lysine as the monohydrochloride reach 30–40% in relation to the initial sugar concentration.

Coryneform glutamic acid-producing bacteria can utilize acetic acid as a carbon source for growth and lysine production. Figure 4 shows

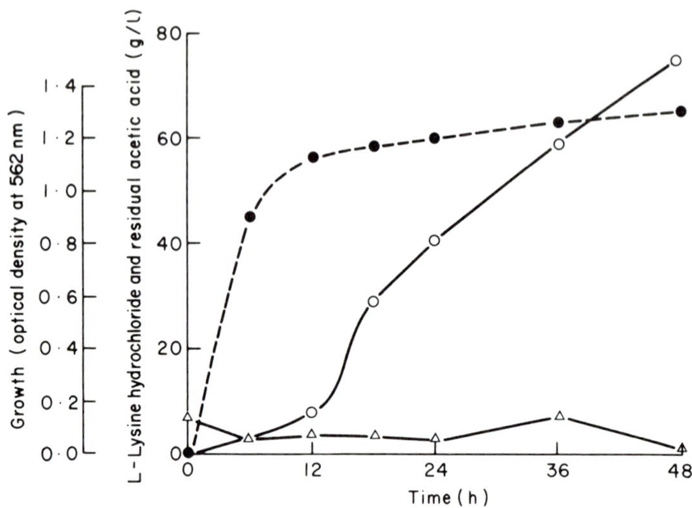

Fig. 4. Time-course of production of L-lysine from acetic acid by *Brevibacterium flavum*. ○ indicates concentration of L-lysine, ● of cell concentration, and △ of residual acetic acid. Based on the data of Tanaka *et al.* (1971).

an example of L-lysine production from acetic acid by a homoserine-leaky (threonine-sensitive) threonine-requiring auxotrophic mutant of *Brev. flavum*. The medium contained 0.7% acetic acid, 0.2% KH_2PO_4, 0.04% $MgSO_4 . 7H_2O$, 0.001% $FeSO_4 . 7H_2O$, 0.001% $MnSO_4 . 4H_2O$, 3.5% hydrolysate of soybean protein, 3.0% glucose, 50 μg biotin per litre, 40 μg thiamin HCl per litre (pH 6.0).

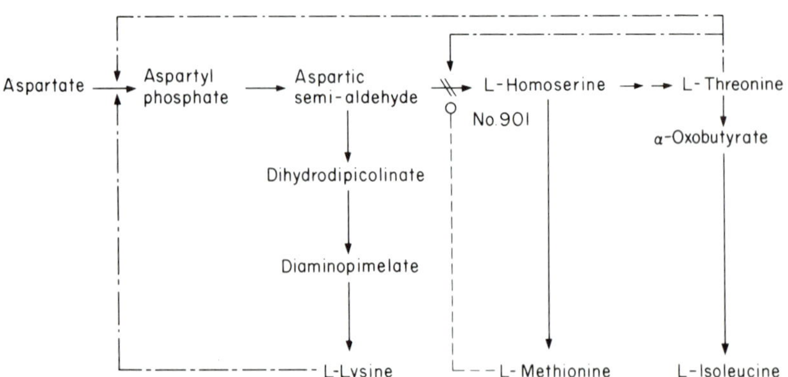

Fig. 5. Regulation of lysine biosynthesis by *Corynebacterium glutamicum*. Dotted lines indicate operation of feedback inhibition, dashed lines of repression of enzyme synthesis.

Fermentation was carried out at 33°C with feeding of a solution of acetic acid. The feeding solution contained 60% acetic acid composed of a mixture of acetic acid and ammonium acetate having molar ratio of 100 : 25, and 3% glucose. The feeding was controlled automatically till the end of the fermentation keeping the pH value of the medium at 7.4. Acetic acid is assimilated constantly throughout the fermentation, and L-lysine yield (as the hydrochloride) at 48 h reached 75 g per litre (29% on the basis of acetic acid and glucose supplied; Tanaka *et al.*, 1971).

A mutant of *Brev. flavum* resistant to S-(β-aminoethyl)-L-cysteine (AEC), a lysine analogue, produced fairly large amounts of L-lysine (Sano and Shiio, 1970). The increase in lysine yield (more than 10%) was obtained using a mutant of *C. glutamicum*, which requires homoserine and leucine and is resistant to AEC. It produced 39.5 g of L-lysine per litre in a medium containing 10% cane molasses (as glucose) while the homoserine- plus leucine-requiring auxotroph produced 34.5 g of L-lysine per litre (Nakayama and Araki, 1973). Some patents have been issued for the process to produce L-lysine from *n*-paraffins.

Regulation of lysine biosynthesis is represented in Fig. 5 (Nakayama *et al.*, 1966). A similar regulatory pattern was also observed in *Brev. flavum* (Shiio and Miyajima, 1969). The blocking of homoserine synthesis at homoserine dehydrogenase results in the release of the concerted feedback inhibition by threonine and lysine on aspartokinase, and the aspartic semi-aldehyde produced proceeds to lysine through the lysine synthetic pathway on which no feedback inhibition is found a situation which differs from that in *E. coli*. Resistance to AEC brought by the desensitization of aspartokinase also releases the concerted feedback inhibition. The conversion of aspartic semi-aldehyde to threonine is feedback inhibited by L-threonine. Thus the overproduced aspartic semi-aldehyde is channelled into L-lysine production.

B. L-Lysine Production from DL-α-Aminocaprolactam

L-Lysine production from DL-α-aminocaprolactam was first studied by Seto (1962). *Aspergillus ustus* hydrolysed only the L-form of α-aminocaprolactam. The remaining D-form was recycled after racemization. However, the yield was low. More recently, a very efficient process has been developed for the conversion (Fukumura, 1974). Incubation of a mixture of 100 ml 10% DL-α-aminocaprolactam (adjusted to pH 8.0 with HCl), 0.1 g acetone-dried cells of *Cryptococcus laurentii* and 0.1 g acetone-dried cells of *Achromobacter obae* nov. sp. with gentle shaking at 40°C for 24 h resulted in the conversion of DL-α-aminocaprolactam to L-lysine in 99.8% yield.

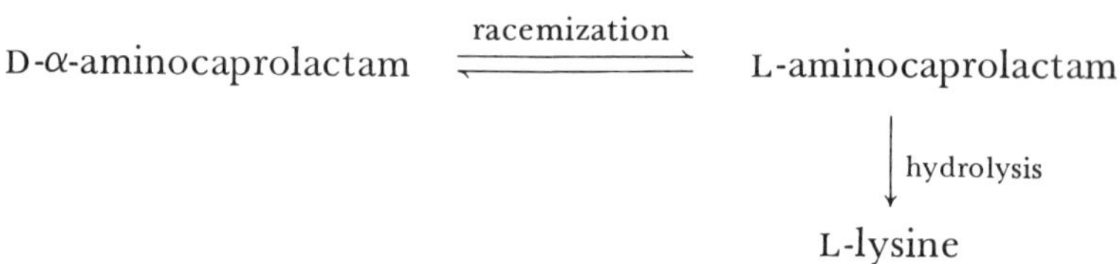

Cryptococcus laurentii produces L-aminocaprolactam hydrolase inductively in a medium containing L-α-aminocaprolactam, glucose and other ingredients. *Achromobacter obae* produces aminocaprolactam racemase using both D- and L-α-aminocaprolactam as an inducer (Sato *et al.*, 1974).

IV. L-THREONINE, L-HOMOSERINE AND L-SERINE

A. L-Threonine

L-Threonine production from L-homoserine has been studied by several groups (Shimura, 1972a), but it was unsuccessful as an industrial process because of the high cost of homoserine. Direct production of L-threonine from a carbohydrate was pioneered by Huang, but its industrial establishment was delayed until Nakayama's group established the process using an *E. coli* auxotroph. Using a diaminopimelate-requiring auxotroph and diaminopimelate-plus-methionine double auxtroph of *E. coli*, Huang obtained L-threonine production at 2–4 g per litre (Huang, 1961). Triple auxotrophs of *E. coli* W, which require diaminopimelate, methionine and isoleucine, and their isoleucine revertants produced L-threonine in increased yields. One of the isoleucine revertants, KY 8280 produced 13.5 g of L-threonine per litre. The cultivation was carried out in 5 litre jar fermenters containing 3 litres of a medium having following composition: 7.5% fructose, 1.4% $(NH_4)_2SO_4$, 0.3% KH_2PO_4, 0.03% $MgSO_4 . 7H_2O$, 2.0% $CaCO_3$ (pH 7.8) A time-course is shown in Fig. 6 (Kase *et al.*, 1971).

The process using the *E. coli* auxotroph has also been developed by Watanabe's group. They isolated a methionine-requiring auxotroph (no. 15) from *E. coli* C-6. It produced 4.3 g of L-threonine per litre in a medium containing 5% (as sugar) cane molasses. A

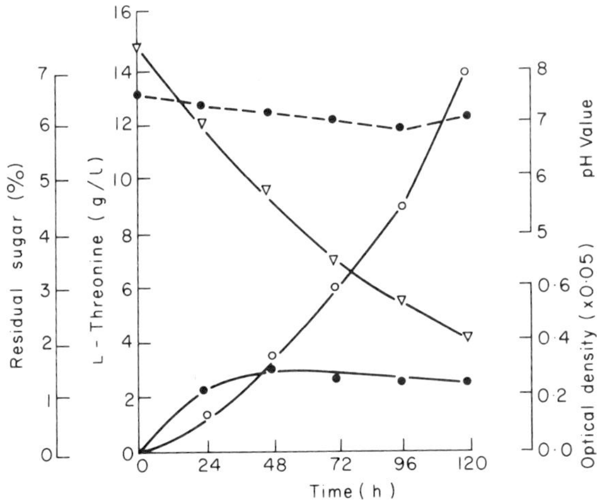

Fig. 6.   Time-course of an L-threonine fermentation by *Escherichia coli* KY 8280. o——o indicates L-threonine concentration, ▽——▽ concentration of residual sugar, ●----● changes in pH value, and ●——● changes in optical density at 660 nm. Reproduced from Kase *et al.* (1971).

methionine- plus valine-leaky double auxotroph (no. 234) derived from no. 15, produced 13.6 g of L-threonine per litre in a medium containing 10% glycerol (Hirawaka *et al.*, 1973). In strain no. 15, the lysine- or methionine-sensitive aspartokinase, which is insensitive to feedback inhibition, was derepressed about five-fold when the auxotroph was cultured in the presence of a limited concentration of methionine (Hirakawa and Watanabe, 1974). L-Threonine production by auxotrophic mutants is also encountered in other members of the Enterobacteriaceae (Kase *et al.*, 1971) and in *Candida guilliermondii* var. *membranaefaciens* (Tsukada and Sugimori, 1971).

Since the regulation of biosynthesis is different from that in *E. coli*, L-threonine production using glutamic acid-producing bacteria, represented by *C. glutamicum*, was unsuccessful (see Section A, pg. 228). L-Threonine production by these bacteria was attained using a regulatory mutant. An α-amino-β-hydroxyvaleric acid (AHV, an analogue of L-threonine)-resistant mutant of *Brev. flavum* BB-82 produced 13.5 g of L-threonine per litre in a medium having the following composition: 10% glucose, 3% $(NH_4)_2SO_4$, 1.5% KH_2PO_4, 0.04% $MgSO_4 . 7H_2O$, 2 p.p.m. Fe^{2+}, 2 p.p.m. Mn^{2+}, 200 μg biotin per litre, 300 μg thiamin HCl per litre, 4 ml Mieki (an HCl-hydrolysate of soybean protein) per litre, 5% $CaCO_3$, pH 7.2 (adjusted with potassium hydroxide; Shiro and Nakamori, 1970). Strain BBM-21, a methionine-requiring auxotroph derived from another AHV-resistant mutant of *Brev. flavum*, produced about 18 g of L-threonine per litre although it showed decreased production of homoserine (Nakamori and Shiio, 1972). Auxotrophs of methionine-requiring strains of *C. glutamicum* resistant to AHV and thialysine produced per litre 14 g and 10 g, respectively, of L-threonine in a medium containing 10% glucose and in a medium containing 5% (as glucose) cane molasses. Another AHV- and thialysine-resistant methionine-requiring auxotroph was found to produce both 9 g of L-threonine per litre and 5.5 g of L-lysine per litre (Kase and Nakayama, 1972). Production of L-threonine with an AHV-resistant mutant was reported with other bacteria including *Proteus rettgeri* (Nakamura and Aida, 1970) and *Corynebacterium acetoacidophilum* (Shiio and Nakamori, 1970).

An example of L-threonine production from acetic acid by an AHV-resistant mutant of *Brev. flavum* was reported (Tanaka *et al.*, 1971). The fermentation medium had the following composition: 0.4% ammonium acetate, 0.41% sodium acetate, 1.0% $(NH_4)_2SO_4$, 0.2% urea, 0.3% KH_2PO_4, 0.04% $MgSO_4 . 7H_2O$, 0.001% $FeSO_4 . 7H_2O$, 0.001% $MnSO_4 . 4H_2O$, 50 μg biotin per litre, 5 mg thiamin HCl per litre, 1.5% hydrolysate of soybean protein, 0.3% glucose (pH 7.2). After inoculation of the seed culture, the fermentation was conducted at 31°C with automatical maintenance of the pH value at 7.7 by feeding a mixed solution of acetic acid and ammonium acetate having a molar ratio of 100 : 17. After 48 h culture, L-threonine production reached 27 g per litre (14% on the basis of acetic acid and glucose).

An isoleucine-leaky auxotroph of *Arthrobacter paraffineus* produced L-threonine and L-valine each at 9 g per litre. Besides these

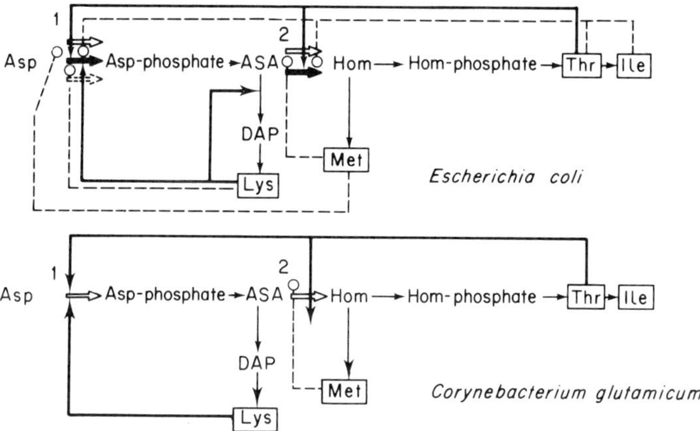

Fig. 7. Regulation of threonine biosynthesis in *Escherichia coli* and *Corynebacterium glutamicum*. Thick lines indicate operation of feedback inhibition of enzyme action, ――― of repression of enzyme synthesis. 1 indicates operation of aspartokinase, 2 of homoserine dehydrogenase. ASA is the abbreviation for aspartate semi-aldehyde, DAP for diaminopimelate, and Hom for homoserine.

amino acids, 2 g per litre each of L-serine and L-leucine were produced (Takayama *et al.*, 1969). Some of the double auxotrophs derived from the isoleucine-requiring auxotroph, and some of their revertants with respect to isoleucine requirement, produced more threonine than the original isoleucine-requiring auxotroph (Kase and Nakayama, 1973). A revertant derived from a methionine plus

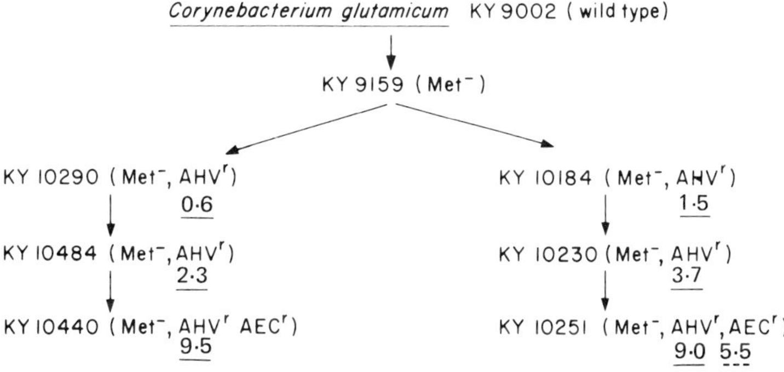

Fig. 8. Genealogy of mutants of *Corynebacterium glutamicum* that produce L-threonine or L-threonine and L-lysine. The figures underlined indicate the concentrations (g per litre) of L-threonine (solid lines) or L-lysine (dashed line) produced in a medium containing 10% (w/v) glucose. AHV is the abbreviation for α-amino-β-hydroxyvaleric acid, and AEC for S-(β-amino-ethyl)-L-cysteine.

isoleucine double auxotroph produced 12 g of L-threonine per litre in seven days incubation with the medium containing 10% *n*-paraffin (C_{12}-C_{14} rich). By-production of valine was decreased in this strain.

Regulation of threonine biosynthesis in *E. coli* and *C. glutamicum* is shown in Fig. 7. *Corynebacterium glutamicum* has one homoserine dehydrogenase which is threonine-sensitive and methionine-repressible, and one aspartokinase which is concerted feedback inhibited by threonine plus lysine. On the other hand, *E. coli* has two homoserine dehydrogenases, one of which is methionine repressible and threonine-insensitive while the other is threonine-sensitive and threonine-repressible. Furthermore, phosphorylation of aspartate in *E. coli* is catalysed by three iso-enzymes of aspartokinase, one of which is repressible by methionine and insensitive to threonine and is a complex with the homoserine dehydrogenase repressible by methionine; the second one is multivalent repressible by threonine and isoleucine and inhibited by threonine, and is a complex with the homoserine dehydrogenase sensitive to threonine; the third aspartokinase is repressible and inhibited by lysine. These regulatory processes explain the contribution of methionine, lysine and isoleucine deficiencies to threonine overproduction in *E. coli*. Isoleucine auxotrophy contributes also in blocking metabolism of the threonine produced. Figure 7 explains also the reason why the auxotrophic mutant of *C. glutamicum* could not overproduce L-threonine.

The genealogy of the threonine-producing strains of *C. glutamicum* is shown in Fig. 8 (Kase and Nakayama, 1974a). The activities of homoserine dehydrogenase in the mutant which produced L-threonine or L-threonine together with L-lysine were slightly less susceptible to inhibition by L-threonine than the activity in the original methionine-requiring auxotroph (KY 9159) from which the threonine producers were derived. The genetical alteration of the enzyme may be at least a cause of L-threonine production in these mutants. The aspartokinase in the threonine-producing mutants, (KY 10484 and KY 10230), which are resistant to AHV and more sensitive to AEC than the parent, were sensitive to concerted feedback inhibition by L-lysine and L-threonine to the same degree as with KY 9159. The aspartokinase activity in KY 10440 was less susceptible to feedback inhibition than KY 10484 or KY 9159. In KY 10251, which produces both L-threonine and L-lysine, L-threonine and L-lysine simultaneously added hardly inhibited the activity of aspartokinase. The difference in L-lysine production between strains KY 10440 and KY 10251 would be mainly due to the difference in the susceptibility of the aspartokinase to concerted feedback inhibition by L-lysine and L-threonine. Furthermore, higher levels of L-threonine production in AHV- and AEC-resistant mutants, as compared with the parent AHV-resistant mutants, could be brought about by the genetically determined desensitization in aspartokinase to the end product inhibition.

B. L-Homoserine

Homoserine production was reviewed by Nara (1972). A threonine-requiring auxotroph of *C. glutamicum* produced 13–15 g of L-threonine per litre and 9 g of L-lysine per litre in a chemically defined medium consisting of 10% glucose, 2% $(NH_4)_2SO_4$, 400 mg L-threonine per litre, 30 μg biotin per litre, 0.1% K_2HPO_4, 0.03% $MgSO_4 . 7H_2O$ and 2% $CaCO_3$. In the presence of excess methionine, the yield of homoserine was lowered and that of lysine increased without any change in the total amounts produced of both amino acids. The effect of methionine explained the low yield in natural media. Excess threonine lowered the yield of both amino acids. Homoserine production by threonine-requiring auxotrophs has been reported also for *E. coli* and *Brev. flavum*. Homoserine production from *n*-paraffins has also been reported in a threonine-requiring auxotroph of *Corynebacterium* sp. KY 4403.

C. L-Serine

Corynebacterium glycinophilum A.T.C.C. 21341, isolated from a putrefied banana and resistant to a high concentration of glycine, produced 10 g of L-serine per litre with a medium containing 2% glycine in a favourable condition (Kubota *et al.*, 1972). An attempt to obtain a mutant which produced L-serine directly from cheap carbon sources was not successful although some mutants produced a small amount of L-serine (Yoshida and Nakayama, 1974). By-production of L-serine by threonine producers was noted (Kase and Nakayama, 1974b). Threonine, homoserine and glycine contributed to L-serine production by some bacteria (Yoshida and Nakayama, 1974; Kase and Nakayama, 1974b). L-Serine production by some bacteria with a medium containing DL-glycerate has been patented (Wada, 1967).

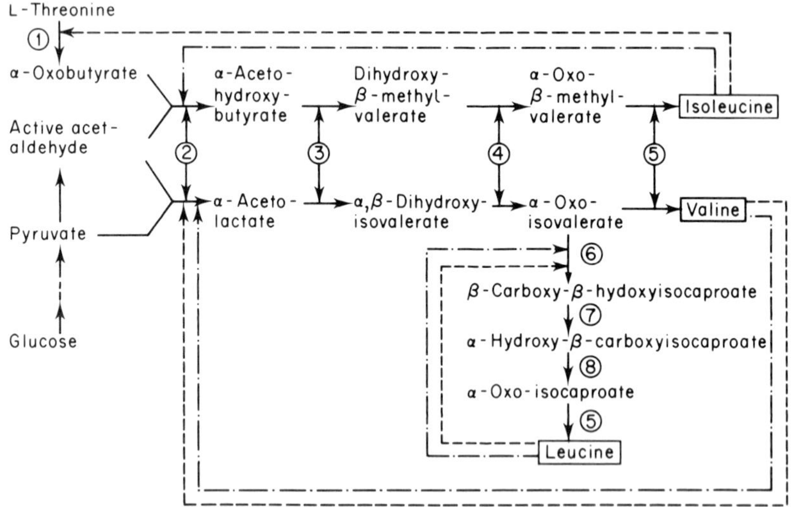

Fig. 9. Regulation of the biosynthetic pathways for isoleucine, valine and leucine. ––– indicates operation of feedback inhibition, –·– of repression of enzyme synthesis. 1 indicates operation of threonine dehydratase, 2 of acetohydroxyacid synthetase, 3 of dihydroxyacid reducto-isomerase, 4 of dihydroxyacid dehydratase, 5 of branched chain amino-acid transaminase, 6 of α-isopropylmalate synthetase, 7 of α-isopropylamalate isomerase, and 8 of β-isopropylmalate dehydrogenase.

V. L-ISOLEUCINE, L-LEUCINE AND L-VALINE

A. L-Isoleucine

Biosynthesis of branched amino acids in bacteria is controlled by the following regulatory mechanisms: (i) feedback inhibition of L-threonine dehydratase by isoleucine, (ii) inhibition of α-acetolactate (ALA) synthetase by valine, (iii) multivalent repression of isoleucine-valine biosynthetic enzymes by all four end-products of this sequence, i.e. valine, leucine, isoleucine and pantothenic acid, (iv) feedback inhibition of α-isopropylmalate (IPM) synthetase by leucine, (v) regulation of intracellular levels of threonine, a precursor of isoleucine (Umbarger, 1969). Figure 9 illustrates the regulation. Thus wild-type strains of bacteria do not excrete appreciable amounts of isoleucine in a salts–sugar medium. Microbial production of isoleucine was, therefore, first established by addition of precursors such as α-aminobutyrate(αAB), D-threonine and α-hydroxybutyric acid.

Fermentation methods using the precursors have been established using various bacteria by several groups and have been utilized industrially. These processes were reviewed recently in detail (Shimura, 1972b). D L-α-Bromobutyric acid could also be utilized as a precursor for isoleucine production (Matsushima *et al.*, 1974). In addition to its role as a precursor by de-amination to α-oxobutyrate, αAB stimulates isoleucine production by releasing the feedback inhibition of threonine dehydratase by isoleucine, and acts as a regulatory factor on acetohydroxyacid (AHA) synthetase, stimulating synthesis of acetohydroxybutyrate (isoleucine precursor) and inhibiting that of acetolactate (valine precursor) in *Bacillus subtilis*. D-Threonine induced synthesis of D-threonine dehydratase which is resistant to end-product inhibition by L-threonine in contrast to L-threonine dehydratase in species of *Serratia* and *Pseudomonas*. Hence, D-threonine supplies α-oxobutyrate continuously for isoleucine biosynthesis, avoiding feedback control by isoleucine. In addition, αAB formed from an excess of α-oxobutyrate exerts a stimulating effect on the enzyme systems in these bacteria in favour of isoleucine biosynthesis. In *Serratia marcescens* no. 1, D-threonine derepresses synthesis of D-threonine dehydratase and AHA synthetase. In an αAB-resistant mutant derived from *S. marcescens* no. 1, D-threonine dehydratase and AHA synthetase are derepressed. This strain produced 15–16 g of L-isoleucine per litre in the presence of D-threonine.

Recently a process for L-isoleucine production directly from carbohydrates has been established. Isoleucine hydroxamate (IH) shows false feedback inhibition on L-threonine dehydratase activity (Kisumi *et al.*, 1971b). A mutant of *Serratia marcescens* which is resistant to IH and is to desensitized L-threonine dehydratases, and αAB-resistant mutants which are genetically derepressed for the isoleucine–valine biosynthetic enzyme, produce no isoleucine in a salts–sugar medium in the absence of isoleucine precursors (Kisumi *et al.*, 1971a, c). A double mutant (IHA 818), resistant to both IH and αAB and isolated from the IH-resistant strain, produced 6–7 g of

L-isoleucine per litre in a medium lacking threonine. The double mutant lacked both feedback inhibition and repression of isoleucine biosynthesis. Increasing the αAB resistance led strain GIHVLA 2795 to produce 12 g of L-isoleucine per litre in a glucose medium (Komatsubara *et al.*, 1973).

In *Brevibacterium flavum*, growth inhibition by α-amino-β-hydroxyvaleric acid(AHV), an analogue of threonine, was reversed not only by L-threonine but also partially by L-isoleucine. Threonine-producing mutants, which were isolated as analogue-resistant mutants, produced a small amount of L-isoleucine (Shiio and Nakamori, 1970). Strain ARI-129, which is resistant to AHV and was isolated on a medium supplemented with 2 mg of AHV per ml, produced 11 g of L-isoleucine per litre (Shiio *et al.*, 1973b). Homoserine dehydrogenase from strain ARI-129 was insensitive to feedback inhibition by L-isoleucine and threonine, while no difference was observed in the activity of threonine dehydratase between strains ARI-129 and 2247A, a wild-type parent strain. Another AHV-resistant mutant (ARI-199), which was selected on a plate containing 3 mg of AHV per litre, produced 15.1 g of L-isoleucine per litre. The medium had the following composition: 10% glucose, 5% $(NH_4)_2SO_4$, 0.15% KH_2PO_4, 0.05% $MgSO_4 . 7H_2O$, 2 p.p.m. Fe^{2+}, 2 p.p.m. Mn^{2+}, 200 μg biotin per litre, 400 μg thiamin HCl per litre, 4 mg Mieki per litre and 5% $CaCO_3$. Maximum production of L-isoleucine was obtainable after 66 h. Strain ARI-129 grew slightly better than ARI-199, and maximum production of L-isoleucine was obtained after 44 h. The productivity of strain ARI-129 was more stable than that of ARI-199. The difference between the isoleucine- and L-threonine-producing mutants was attributed to differences in permeability of L-threonine. In isoleucine-producing mutants, intracellular accumulation of threonine overcomes feedback inhibition of threonine dehydratase by isoleucine because inhibition of threonine dehydratase by isoleucine is competitive with respect to L-threonine. An *O*-methylthreonine-resistant mutant (AORI-126), which was derived from ARI-129, produced 14.5 g of L-isoleucine per litre. The specific activity of threonine dehydratase from strain AORI-126 increased about two-fold over that of strains no. 2247A and ARI-129, whereas the degree of inhibition of the enzyme by L-isoleucine was the same (Shiio *et al.*, 1973b).

An ethionine (20 mg per ml)-resistant mutant (no. 168) of *Brev.*

flavum was isolated after mutagenic treatment of strain BB-69, an AHV-resistant mutant which produces L-threonine in a yield of 13% from glucose. Mutant strain no. 168 produced L-isoleucine in a yield of 10% (Ikeda *et al.*, 1974). Another ethionine (15 mg per ml)-resistant mutant (no. 14083) was derived from *Brev. flavum* FAB-3-1, which was resistant to both thialysine, a lysine analogue, and AHV and produced L-threonine in 11% yield. The mutant (no. 14083) produced L-isoleucine in 12% yield. Production of L-isoleucine by the process using acetic acid as the carbon source reached 33.5 g per litre, a 10% yield (Ikeda *et al.*, 1974).

B. L-Leucine

Kisumi *et al.* (1973a) obtained a leucine-producing mutant as a leaky-type isoleucine auxotroph from an αAB-resistant mutant of *Serratia marcescens* (Ar 130-1). The isoleucine auxotroph (no. 149) produced L-leucine by long incubation in a medium lacking isoleucine. Partial revertants from strains no. 149, S-13, S11 and others produced 13 g of L-leucine per litre in 48 h in a medium lacking isoleucine. The medium had the following composition: 2% glucose, 10% dextrin, 1% urea, 0.1% K_2HPO_4, 0.05% $MgSO_4 . 7H_2O$, 2% $CaCO_3$ The original αAB-resistant mutant (Ar 130-1) was derepressed for isoleucine–valine biosynthetic enzymes. Aceto-hydroxyacid synthetase and transaminases of strains S-3 and S-11 were found to be derepressed 10–20 fold compared with the parent strain, no. 149 (Kisumi *et al.*, 1971d). Furthermore, α-isopropyl-malate (α-IPM) synthetase, the first enzyme on the leucine biosynthetic pathway of the αAB-resistant mutant, was also found to be constitutive. Reversion of the isoleucine auxotrophy was accompanied by desensitization of the leucine biosynthetic enzyme. On the other hand, α-IPM synthetase in the wild-type strain is repressed by leucine. L-Threonine dehydratase activity was not detected in leucine-producing revertants by the assay method used. These results indicate that leucine production was due to a lack of both feedback inhibition and repression. In the L-leucine-producing revertant, leucine enzymes catalysed α-oxobutyrate formation from pyruvate via citramalate, citraconate and β-methylmalate (Komatsubara *et al.*, 1974).

Tsuchida *et al.* (1973b) found a L-leucine producer in the 2-thiazolealanine-resistant mutants derived from an isoleucine-methionine double auxotroph (no. 6) of *Brev. lactofermentum* 2256. One of the mutants, no. 128, produced 19 g of L-leucine per litre in a medium having the following composition: 8% glucose, 4%

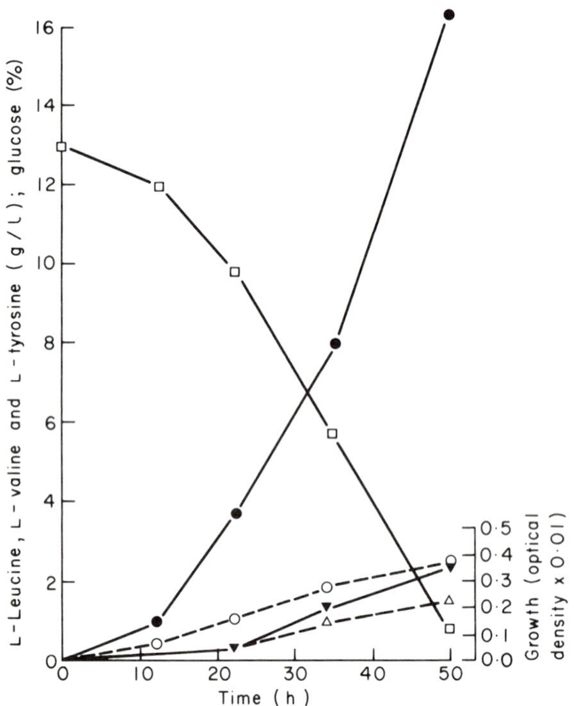

Fig. 10. Time-course of L-leucine production by *Corynebacterium glutamicum* Pα-219. ● indicates production of L-leucine, □ glucose utilization, ○ growth, △ L-valine production, and ▼ L-tyrosine production.

$(NH_4)_2SO_4$, 0.1% KH_2PO_4, 0.04% $MgSO_4 . 7H_2O$, 50 μg biotin per litre, 300 μg thiamin HCl per litre, 2 p.p.m. Fe^{2+}, 2 p.p.m. Mn^{2+}, 40 mg DL-methionine per litre, 20 mg L-isoleucine per litre and 5% $CaCO_3$. In a wild-type strain of *Brev. flavum*, α-IPM synthetase is sensitive to inhibition and repression by L-leucine, and AHA synthetase is weakly sensitive to inhibition by valine, leucine or isoleucine and is strongly repressible by a combination of the three amino acids (Tsuchida and Momose, 1974). On the other hand, in the leucine producer, no. 218, the regulatory property of AHA synthetase was almost the same as that of a wild-type strain while α-IPM synthetase was found to be derepressed three-fold and resistant to feedback inhibition. From these results, the cause of leucine overproduction is attributed to genetic release of feedback inhibition by the mutation leading the 2-thiazolealanine-resistance and derepression of AHA synthetase brought about by the mutation leading the isoleucine auxotrophy (Tsuchida and Momose, 1974).

A L-leucine production process using an auxotrophic mutant of *Corynebacterium glutamicum* has also been developed (Araki *et al.*, 1974c). Mutant no. 190 was found to produce a large amount of L-leucine in the culture medium. The nutritional requirements of the mutant are rather complex but its growth was remarkably stimulated by L-phenylalanine. Acetate (1.5–3.0%) or pyruvate (3%) stimulated L-leucine production. A histidine-requiring auxotrophic derivative, $P\alpha'$-219, produced twice as much L-leucine as the parent strain, i.e. L-leucine production by the strain reached 16 g per litre in a medium containing 12% glucose, 2.5% CH_3COONH_4 and other ingredients. The chemical changes which take place during the fermentation with strain $P\alpha$-219 are shown in Fig. 10. Fermentation was carried out with a five-litre jar fermentor containing three litres of fermentation medium which contained 13.2% glucose, 1% $(NH_4)_2SO_4$, 2% CH_3COONH_4, 1% peptone, 0.7% meat extract, 0.001% $(NH_4)_6[Mo(Mo_6O_{24})].4H_2O$, 0.15% KH_2PO_4, 0.05% K_2HPO_4, 0.05% $MgSO_4.7H_2O$, 0.002% $FeSO_4.7H_2O$, 0.002% $MnSO_4.4H_2O$, 50 μg biotin per litre, 1 mg thiamin HCl per litre and 3% $CaCO_3$. The pH value of the culture was automatically controlled with NH_4OH during fermentation. The amount of L-leucine produced increased remarkably after a long lag, being paralleled with consumption of glucose. By-production of L-valine and L-tyrosine in small amounts was noted. Significant amounts of L-isoleucine were not detected by a biological assay. The mechanism of the high production of L-leucine by this mutant is not known at present.

C. L-Valine

L-Valine production by a fermentation process was reviewed recently (Uemura *et al.*, 1972). Valine-producing bacteria are often found in nature and numerous studies on them have appeared. Processes using auxotrophic or regulatory mutants have also been reported. Addition of certain drugs to the culture medium induces valine production.

1. Valine Production by Wild-Type Bacteria

Most of the wild-type organisms capable of copious valine production belong to the family Enterobacteriaceae, especially to the genera *Aerobacter* and *Escherichia*.

Optimal carbon : nitrogen ratios for valine production seem to be approximately 100 : 7 to 100 : 4. It is noteworthy that heavy metal ions profoundly influence valine production. Ferrous ions seem to be essential for valine production largely because α-acetolactate synthetase (ALSase) and β-hydroxy acid dehydratase are Fe^{2+}-requiring enzymes. Control of oxygen supply to the culture medium is one of the most important factors for valine production. Using a selected wild-type culture of a strain of Enterobacteriaceae, 12–13 g of L-valine per litre, equivalent to 20% conversion of consumed glucose, could be produced.

2. Mechanism of Valine Production in Wild-Type Bacteria

Since the early studies of Udaka and Kinoshita (1959) and Sugisaki (1959), most prototrophic valine-producing bacteria have been shown to belong to the group of bacteria which have two α-acetolactate synthetases (ALSase), namely pH 8 ALSase and pH 6 ALSase. The former enzyme has optimum activity at pH 8 and functions as the initial enzyme in the biosynthesis of L-valine. The second enzyme has optimum activity at pH 6, and performs a biodegradative function leading to acetoin formation, although it can function as a biosynthetic enzyme at lower pH values. In the bacteria described above, pH 6 ALSase, which is valine insensitive, is induced by growing in fermentation medium and functions in overproduction of valine. This was clearly demonstrated in *Paracolobactrum coliforme* (Udaka and Kinoshita, 1959). Production of valine seemed to be further enhanced due to a lack of ALA decarboxylase in this bacterium. In *Aerobacter aerogenes* no. 19–35, studied by Uemura *et al.* (1972), the pH 6 enzyme was formed inductively in the presence of an optimal concentration of inorganic phosphate or 6-thiol-purines. In addition to *P. coliforme* and *A. aerogenes, Aerobacter cloacae* and some *E. coli* strains belong to the group of bacteria described above.

Another group of valine-producing bacteria have only pH 8 ALSase. *Escherichia freundii, Serratia marcescens,* some *E. coli* strains, *Bacillus subtilis* and *Brevibacterium ammoniagenes* belong to this group. In these bacteria, the pH 8 ALSase was insensitive or less sensitive to feedback inhibition by valine.

Under circumstances in which the altered regulatory mechanisms are participating, valine overproduction may be understood as a consequence of the extent to which pyruvate accumulates as a catabolic intermediate, the extent to which ALA is efficiently synthesized from pyruvate as a first precursor leading to valine, and to which an amino donor in the form of glutamic acid is available for a transamination reaction. Each of these processes is influenced by several external factors, e.g. the presence of amino acids, drugs and heavy metal ions, partial pressure of oxygen, pH value, and the quality and quantity of carbon and nitrogen sources. Hence valine production is dependent on environmental changes, often being converted into another mode of fermentation.

Acetohydroxy acid (AHA) synthetase is sensitive to catabolite repression in wild-type *E. coli* B (Coukell and Polglase, 1969a). The synthetase in a streptomycin-dependent mutant of *E. coli* B which excretes L-valine was relatively resistant to catabolite repression (Coukell and Polglase, 1969b). $3',5'$-Cyclic AMP derepresses catabolite repression. Furthermore, this nucleotide is an allosteric inhibitor of the malic enzyme in *E. coli* (Sanwal and Smando, 1969). Since the malic enzyme is significantly inhibited at concentration ranges ($\sim 10^{-4}$ M) of $3',5'$-cyclic AMP which are only found in cells starved of an energy source, it is reasonable to suppose that this inhibition is directed towards prevention of wasteful reactions during starvation. The malic enzyme presumably qualifies as one of these dispensable enzymes. This presumption is supported by the observation (Raunio, 1966) that, under normal conditions of growth, i.e. in the absence of limitation of energy source, *E. coli* secretes large quantities of pyruvate part of which possibly arises by the withdrawal of malate from the TCA cycle by action of the malic enzyme. This diversion of malate to pyruvate may lead to production of valine or alanine.

3. Valine Production by Auxotrophic and Regulatory Mutants

Isoleucine and leucine auxotrophic mutants of *C. glutamicum* produce a large amount of L-valine in culture broths. Some other auxotrophs produce L-valine in smaller amounts. L-Valine production by a isoleucine-requiring auxotroph reached 11 g per litre in a medium containing 7.5% glucose (Nakayama *et al.*, 1961c). Addition of leucine and valine markedly increased valine production in a synthetic medium. Antagonisms among L-isoleucine, L-leucine and L-valine were observed in growth and valine production. It was speculated that L-leucine and L-valine compete with the isoleucine, reversing the inhibition of isoleucine on valine production although permitting protein formation to a certain extent to enable overproduction of valine. Permeation to the cell was considered as a probable step for antagonism by the three amino acids. Multivalent repression detected in *E. coli* and *Salmonella* sp. (Umbarger, 1969) tempts us to study the regulation mechanism in *C. glutamicum*.

A mutant of *Serratia marcescens* resistant to αAB produced a remarkable amount of L-valine (Kisumi *et al.*, 1971a). Overproduction of valine by the mutant was correlated with genetic release from multivalent repression on the formation of isoleucine–valine enzymes, particularly pH 8 ALSase which was sensitive to feedback inhibition by valine to the same extent as the enzyme in the parent strain. In a number of analogue-resistant mutants, the best valine producer, strain no. 9, showed high activity of pH 8 ALSase, being also less sensitive to valine inhibition. A thiazolealanine-resistant mutant of *Brev. lactofermentum* produced 23.1 g of L-valine per litre in a medium containing 8% glucose (Tsuchida *et al.*, 1973a).

VI. L-TRYPTOPHAN AND OTHER AROMATIC AMINO ACIDS

A. L-Tryptophan

Various attempts using precursors, such as indole and anthranilic acid, have been made for microbial production of L-tryptophan. These were reviewed by Terui (1972). With *Hansenula anomala*, L-tryptophan production from anthranilic acid reached 5.7 g per litre (Terui, 1972). By feeding anthranilic acid to a derepressed anthranilic acid-requiring auxotroph of *Bacillus subtilis*, 5.5 g of L-tryptophan per litre were produced from 5 g of anthranilic acid per litre (Arima *et al.*, 1971). *Candida utilis* (synonym. *Torulopsis utilis*) 295-t produced 6.4 g of L-tryptophan per litre from 4.2 g of anthranilic acid per litre in 36 h (Bekers *et al.*, 1971). By feeding indole, a 5-methyl-tryptophan-resistant mutant of *B. subtilis* A.T.C.C. 21336 produced 10.4 g of L-tryptophan per litre in 96 h with a medium containing 7% glucose (Thieman and Pagani, 1972). A strain carrying a *F Try*-episome in addition to the chromosomal *try* operon was obtained by sexduction from a feedback-resistant and derepressed mutant of *E. coli* K12 which is resistant to 5-methyltryptophan (5MT) and 5-fluoro-tryptophan (5FT). This strain produced 5 g of L-tryptophan per litre by feeding indole (1.5 g per litre) and L-serine (7 g per litre) (Sahm and Zähner, 1971).

Tryptophanase, which catalyses synthesis of L-tryptophan by reversal of the α,β-elimination reaction at rates similar to the forward reaction, was utilized for production of L-tryptophan and related compounds such as 5-hydroxytryptophan (Nakazawa *et al.*, 1972). The culture broth of *Proteus rettgeri* (AJ 2770) was used as the enzyme for the reaction. For the synthesis of L-tryptophan, a reaction mixture contained 6.0 g indole in 10 ml of methanol, 8.0 g of sodium pyruvate, 8.0 g ammonium acetate, 0.001 g of pyridoxal phosphate, 0.1 g of Na_2SO_4 and 100 ml of the culture broth in a total volume of 120 ml. After the pH value of the mixture was adjusted to 8.8 with 6 N KOH, it was incubated at 34°C for 48 h. Under these conditions, 7.5 g of L-tryptophan were synthesized. Similarly, 5-hydroxy-L-tryptophan was synthesized from 5-hydroxy-indole, pyruvate and ammonia:

$$Indole + CH_3COCOOH + NH_3 \rightarrow \text{L-Tryptophan} + H_2O$$

The above methods using precursors are not advantageous since the precursor compounds are expensive at present. Recently, direct production of L-tryptophan from carbohydrate by fermentation has been developed to the practical level.

Figure 11 shows the genealogy of L-tryptophan-producing mutants of *Corynebacterium glutamicum*. Mutants producing a large amount of L-tryptophan were derived from a phenylalanine- and tyrosine-requiring double auxotroph of *C. glutamicum* KY 9456 which produced only a trace amount of L-tryptophan and anthranilate (Nakayama *et al.*, 1974). A mutant (4MT-11), which stepwise acquired resistance to 5MT, tryptophan hydroxamate (TrpHx), 6-fluorotryptophan (6FT) and 4-methyltryptophan (4MT), produced L-tryptophan to a concentration of 4.9 g per litre in a cane molasses

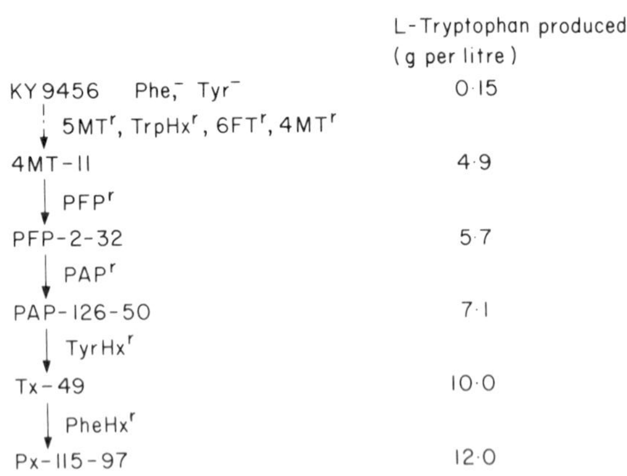

Fig. 11. Genealogy of L-tryptophan-producing mutants of *Corynebacterium glutamicum* and their tryptophan-producing ability. The bacteria were grown in a cane molasses medium containing 10% (w/v) sugar as glucose.

medium containing 10% sugar as glucose. L-Tryptophan production with this mutant was inhibited by L-phenylalanine and L-tyrosine. Accordingly, mutants resistant to phenylalanine and tyrosine analogues such as *p*-fluorophenylalanine (PFP), *p*-aminophenylalanine (PAP), tyrosine hydroxamate (TyrHx) and phenylalanine hydroxamate (PheHx), were derived from this mutant. One of the mutants thus obtained (Px-115-97) produced 12 g of L-tryptophan per litre in the molasses medium. The medium used had the following composition: 10% (as glucose) cane molasses, 0.05% KH_2PO_4, 0.05% K_2HPO_4, 0.025% $MgSO_4$. $7H_2O$, 2% $(NH_4)_2SO_4$, 1% corn-steep liquor, 2% $CaCO_3$ (pH 7.2). Production of L-tryptophan with the mutant was still sensitive to L-phenylalanine and L-tyrosine. Hence, further genetic improvement of the strain may be possible. The regulatory mechanism of aromatic amino-acid biosynthesis in *C. glutamicum*, and the derangement of metabolic controls in L-tryptophan-, L-phenylalanine- and L-tyrosine-producing mutants, were reported by Nakayama *et al.* (1974).

A 5FT-resistant mutant derived from *B. subtilis* produced 4 g of L-tryptophan and L-phenylalanine per litre. A leucine auxotroph derived from the mutant produced 6.15 g of L-tryptophan per litre in a medium containing 300 μg of L-leucine per ml (Shiio *et al.*, 1973a). Suppressor prototrophic revertants from no. 149 *rec⁻* histidine auxotroph of *B. subtilis* produced L-tryptophan in the medium. Elimination of *rec⁻* mutation from the gene using genetic transformation resulted in an increase in tryptophan productivity by 60%. This strain produced 5–6 g of L-tryptophan per litre (Alikhanian, 1972).

Mutants of *Brev. flavum*, resistant to 5MT and derived from the auxotroph for tyrosine and phenylalanine, produced 1.9 g of L-tryptophan per litre (Shiio *et al.*, 1972). A 5FT-resistant mutant of *Brev. flavum* produced 2.1 g of L-tryptophan per litre. The production was increased to 7.0 g per litre using a *m*-fluorophenylalanine (MFP)-resistant mutant derived from the 5MT-resistant (Sugimoto *et al.*, 1974).

B. L-Phenylalanine

L-Phenylalanine production has been much improved using a regulatory mutant after the review of Ōishi (1972). A tyrosine-requiring auxotroph of *C. glutamicum* produced a little phenylalanine (Nakayama *et al.*, 1961a). A prototrophic mutant resistant to *p*-fluorophenylalanine (PFP) produced 5.5 g of L-phenylalanine per litre and trace amounts of L-tryosine in a cane molasses medium containing 10% sugar calculated as glucose (Hagino and Nakayama, 1974). A tyrosine-requiring auxotrophic mutant, resistant to PFP and PAP, produced 9.5 g of L-phenylalanine per litre in the molasses-containing medium. The medium used had the following composition: 10% (as glucose) cane molasses, 2% $(NH_4)_2SO_4$, 0.05% KH_2PO_4, 0.05% K_2HPO_4, 0.025% $MgSO_4 . 7H_2O$, 0.25% NZ-Amine (enzymic digest of casein), 2% $CaCO_3$ (pH 7.2). L-Phenylalanine production in these mutants was inhibited by L-tyrosine and was stimulated by L-tryptophan (Hagino and Nakayama, 1974).

A mutant of *Brev. flavum* resistant to *m*-fluorophenylalanine produced 2.2 g of L-phenylalanine per litre (Sugimoto *et al.*, 1973). A mutant of *B. subtilis* resistant to 5FT produced 6.0 g of L-phenylalanine per litre in addition to 4.0 g of L-tryptophan per litre (Shiio *et al.*, 1973a). A tyrosine-requiring auxotroph of a *Corynebacterium* species, a hydrocarbon-utilizing glutamic acid producer, has also been reported to produce L-phenylalanine (Tokoro *et al.*, 1970).

C. L-Tyrosine and L-DOPA

In 1961, L-tyrosine production by a phenylalanine-requiring auxotroph of *C. glutamicum* at the concentration of 2–3 g per litre was reported. The production was attained in the presence of a low concentration of L-phenylalanine (Nakayama *et al.*, 1961b). The phenylalanine-requiring auxotroph was improved in its tyrosine productivity by endowing certain analogue-resistances (Hagino and Nakayama, 1973). Among the analogues of L-phenylalanine and L-tyrosine tested, PFP and MFP were the most effective inhibitors on growth of a wild-type strain of *C. glutamicum*. Their inhibitory effects were released by L-phenylalanine and slightly by L-tyrosine and L-tryptophan. 3-Aminotyrosine (3AT), PAP, *o*-fluorophenylalanine and β-2-thienylalanine were weak inhibitors. Figure 12 shows the genealogy of the finally selected L-tyrosine producers. A combination of auxotrophy and multiple resistance to analogues of aromatic amino acids was necessary to yield a large amount of L-tyrosine. A phenylalanine-requiring auxotroph (97-Tx-71), which became multiply resistant to the analogues of phenylalanine and tyrosine (PFP, PAP, 3AT and tyrosine hydroxamate (TyrHx)), produced 13.5 g of L-tyrosine per litre in a cane molasses medium containing 10% sugar. This strain is a so-called leaky mutant and has a little L-phenylalanine-synthesizing activity similar to the original strain, KY 10233. In fact, it sometimes excreted a trace amount of L-phenylalanine. The L-phenylalanine pool of this mutant may reach a size enough to inhibit biosynthesis of L-tyrosine owing to deviation of the regulation for L-phenylalanine synthesis, in addition to deviation of regulation of L-tyrosine synthesis of this mutant. Therefore, mutants which are defective in L-phenylalanine biosynthesis more strictly were expected to produce higher amounts of L-tyrosine. Such a mutant may be selected as an L-tyrosine-sensitive strain which grows slowly in the minimal medium supplemented with excess L-tyrosine because L-tyrosine antagonizes L-phenylalanine in entering the cells and inhibits growth of a phenylalanine-requiring auxotroph in proportion to the degree of the requirement for L-phenylalanine. Colonies grown slowly in the presence of tyrosine

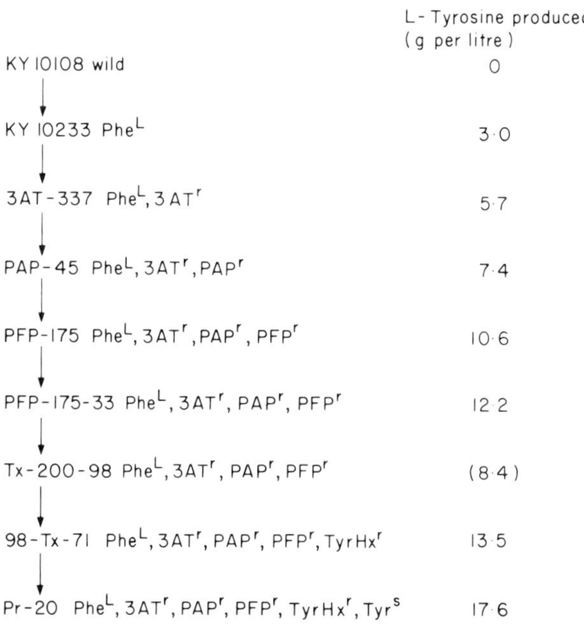

Fig. 12. Genealogy of L-tyrosine-producing mutants of *Corynebacterium glutamicum*. The yield for strain Tx-200-98 is indicated in parentheses since it was obtained when the yield produced by the parent (PFP-175-33) was only 6.2 g tyrosine per litre.

were selected as mutants sensitive to L-tyrosine. Some mutants thus obtained produced larger amounts of L-tyrosine than the parent strain, notably strains Pr-20 and Pr-102 which produced L-tyrosine at a concentration of 17.6 g and 17.3 g per litre, respectively. The medium had the following composition: 10% (as glucose) cane molasses, 2% $(NH_4)_2SO_4$, 0.05% K_2HPO_4, 0.05% KH_2PO_4, 0.025% $MgSO_4 . 7H_2O$, 2% $CaCO_3$ (pH 7.2). An increase in L-tyrosine production was also noted in many auxotrophic mutants derived from a phenylalanine-requiring auxotroph of *C. glutamicum*. Amongst them, LM-96, a phenylalanine requiring and purine-requiring double auxotrophic strain, produced L-tyrosine at a concentration of 15.1 g per litre in a medium containing 20% sucrose (Hagino *et al.*, 1973). A MFP-resistant mutant of *Brev. flavum* produced 1.9 g of L-tyrosine per litre (Sugimoto *et al.*, 1973).

A reversion of the α,β-elimination reaction catalysed by β-tyrosinase was utilized for preparation of L-tyrosine and L-DOPA (Yamada *et al.*, 1972).

$$\text{pyruvic acid} + NH_3 + \text{phenol} \rightarrow \text{L-tyrosine} + H_2O$$

$$\text{pyruvic acid} + NH_3 + \text{pyrocatechol} \rightarrow \text{L-DOPA} + H_2O$$

Cells of *Erwinia herbicola*, prepared by growing at 28°C for 28 h in an appropriate medium, were used as a source of enzyme. A reaction mixture (100 ml) containing 0.5 g sodium pyruvate, 1.0 g phenol or 0.8 g pyrocatechol, 5 g ammonium acetate and cells was incubated. At intervals, sodium pyruvate and phenol or pyrocatechol were added. Under these conditions, 6.05 g of L-tyrosine or 5.85 g L-DOPA were synthesized.

A 3AT-resistant mutant of *Pseudomonas maltophila* (synonym *Ps. melanogenum*) produced 14–15 g of L-DOPA per litre from 26 g of L-tyrosine per litre (68% in molar conversion ratio). An enzymic basis of the high L-DOPA productivity of the improved mutants was found to be due to the increased tyrosinase activity of the mutants (Tanaka *et al.*, 1974).

VII. L-ARGININE, L-ORNITHINE AND L-CITRULLINE

A. L-Arginine

Efficient production of L-arginine was attained using regulatory mutants of *Bacillus subtilis* (Kisumi *et al.*, 1971e), *C. glutamicum* (Nakayama and Yoshida, 1972) and *Brev. flavum* (Kubota *et al.*, 1973). An L-arginine and hydroxamate-resistant mutant of *B. subtilis* produced 4.5 g of L-arginine per litre in shaken culture. This mutant by-produced N^δ-acetylornithine and its ornithine carbamoyl transferase was derepressed strongly. Therefore, a shortage of carbamoylphosphate was deemed to be the cause of by-production of N^δ-acetylornithine. A 6-aza-uracil-resistant mutant derived from the arginine- and hydroxamate-resistant produced 28 g of L-arginine per litre without by-production of N^δ-acetylornithine in a medium containing 8% glucose and 3.5% glutamic acid.

Corynebacterium glutamicum DSS8, isolated as a D-serine-sensitive mutant from an isoleucine-requiring auxotroph (KY 10150), was found to be sensitive to D-arginine and arginine hydroxamate.

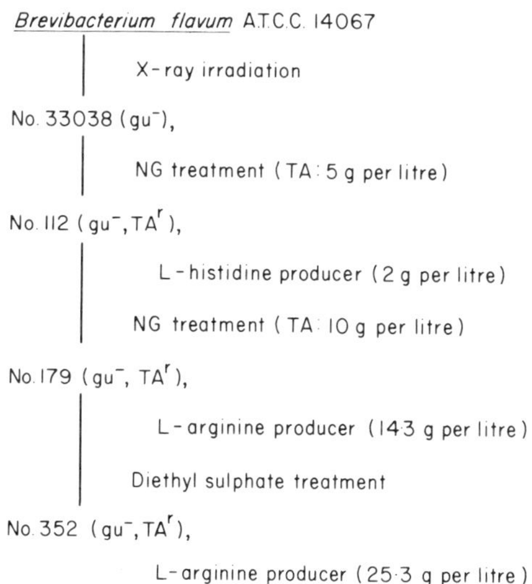

Fig. 13. Genealogy of L-arginine-producing mutants of *Brevibacterium flavum*. TA is the abbreviation for 2-thiazolealanine, and NG for N-methyl-N'-nitro-N-nitrosoguanidine. Reproduced from Kubota *et al.* (1971).

Furthermore, strain DSS8 produced L-arginine in culture medium. Most of the L-arginine analogue-resistant mutants derived from DSS8 produced a large amount of L-arginine. An isoleucine revertant from one of these mutants produced 19.6 g of L-arginine per litre in a medium containing 15% (as sugar) molasses. Strain DSS8 seems to be a mutant with increased permeability to D- and L-arginine.

An efficient L-arginine producer was derived through several mutation and selection steps from strain no. 33038, a guanine-requiring auxotroph of *Brev. flavum* (Fig. 13). Strain no. 352, a finally isolated mutant, produced 25.3 g of L-arginine per litre. Maximum L-arginine production of 28.4 g per litre was found at a concentration of 0.1 mg of guanine per ml which was a suboptimum concentration for growth of the mutant. Increasing the concentration of ammonium sulphate in the medium increased L-arginine production. Maximum production of 29.4 g per litre was attained with the 7% $(NH_4)_2SO_4$. L-Histidine markedly retarded growth of strain no. 352 and also that of the parent strain A.T.C.C. 14067. The time course of L-arginine production is presented in Fig. 14.

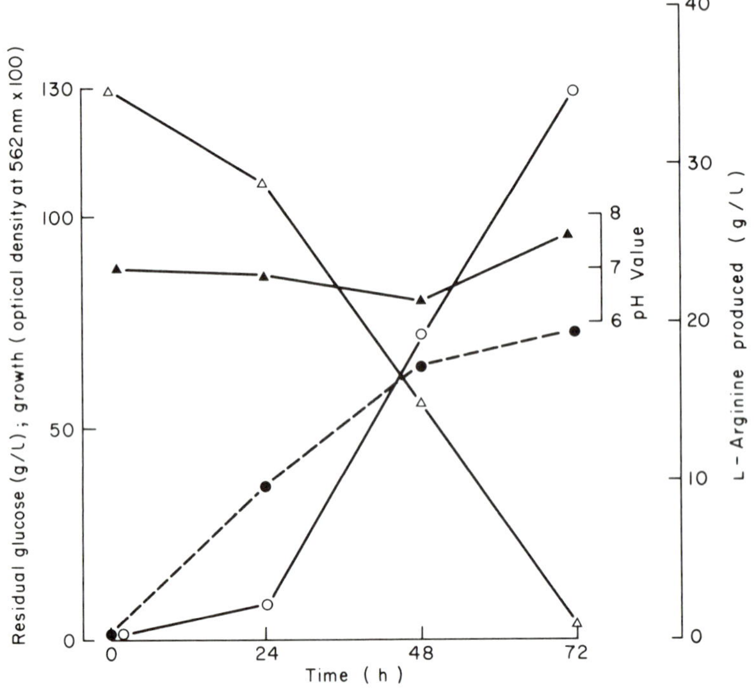

Fig. 14. Time-course of L-arginine production by *Brevibacterium flavum* no. 352 in a medium containing 13% (w/v) glucose. ○ indicates L-arginine production, ● growth, △ residual glucose, and ▲ pH value of the culture. Reproduced from Kubota *et al.* (1971).

B. L-Ornithine

L-Ornithine production from carbohydrate is now known among arginine- or citrulline-requiring auxotrophs of many micro-organisms such as species of *Corynebacterium, Brevibacterium, Arthrobacter, Bacillus, Escherichia* (Udaka, 1972) and *Streptomyces* (Kondo *et al.*, 1970). L-Ornithine production from hydrocarbons by the same type of auxotroph of *C. hydrocarboclastus* and *Arthrobacter paraffineus* is also known. Among these, a *C. glutamicum* mutant is the first reported efficient L-ornithine producer. The molar yield was as high as 36%. The fermentation conditions are similar to those for glutamic acid production except that the medium contained an appropriate concentration of arginine and a large concentration of biotin.

C. L-Citrulline

Arginine-requiring auxotrophs of *B. subtilis* and *C. glutamicum* produced 16.5 g and 10.7 g of L-citrulline per litre in a medium containing 13% and 10% glucose, respectively. An arginine-requiring auxotroph of *Corynebacterium* sp. produced 8 g of L-citrulline per litre in a medium containing 10% (v/v) *n*-paraffin (Udaka, 1972). Production of L-citrulline by an arginine-requiring auxotroph resistant to arginine hydroxamate was not influenced by the concentration of arginine in the medium (Kisumi *et al.*, 1973b). L-Citrulline production reached 26 g per litre using an arginine-requiring auxotroph derived from a mutant resistant to both arginine hydroxamate and 6-aza-uracil.

D. Regulation of Arginine Biosynthesis and Mechanism of Overproduction of Arginine, Ornithine and Citrulline

Arginine is synthesized from glutamic acid via ornithine and citrulline in micro-organisms as shown in Fig. 15. Eight different enzymes participate in the chain of reactions. The first and fifth reactions are different depending on the micro-organism. In bacteria of the Enterobacteriaceae and species of *Bacillus*, ornithine is formed by hydrolytic cleavage of acetylornithine with an acylase. On the other hand, in species of *Pseudomonas*, *Corynebacterium*, *Streptomyces* and in yeast, transacetylase catalyses the transfer of an acetyl residue from acetylornithine to glutamic acid forming ornithine and acetylglutamic acid. In the first group of micro-organisms, endproduct inhibition acts on the first enzyme in the pathway but, in the second group micro-organisms, arginine regulates the activity of the second enzyme, and probably the first enzyme as well. In the second group, if the second enzyme is not regulated, arginine synthesis proceeds without control because acetylglutamic acid would be formed by cyclic utilization of the acetyl residue regardless of the activity of the first enzyme. The end product (arginine) has the role of corepressor in some bacteria, i.e. enzyme formation is repressed only in the presence of arginine. Other bacteria may have a repression mechanism since a much larger amount of enzyme is synthesized after a single mutation. Overproduction of ornithine becomes possible

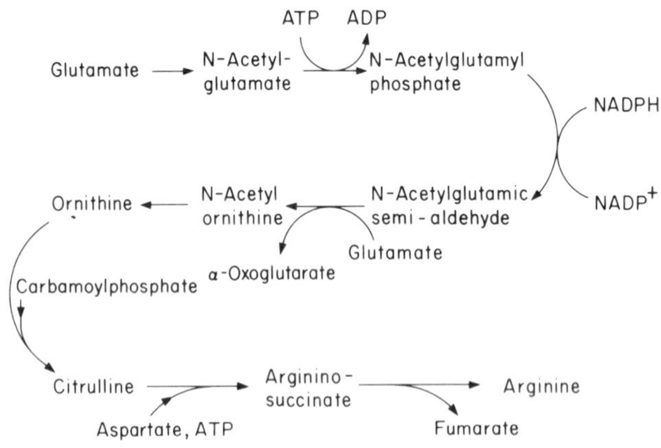

Fig. 15. Biosynthetic pathway for L-arginine.

because of an increase in the amount of enzyme and the weak or non-existent inhibition of enzyme activity by arginine when supplied in limited concentrations to cultures of an arginine-requiring auxotroph. When the concentration of arginine included in the medium is below certain levels, arginine may be utilized exclusively for cell growth, and the concentration of free arginine within the cell would be very low, giving no inhibition of enzyme activity. If arginine is included in large concentration, the second enzyme would be inhibited, halting ornithine formation, and the cell physiology would be shifted back to that of the parent. A similar mechanism also explains citrulline production by arginine auxotrophs. Details of the deviation of regulation in arginine producers are not clarified at present.

VIII. L-HISTIDINE

A. L-Histidine Production

L-Histidine is an essential amino acid to some animals though not for humans. Its effect in the treatment of rheumatoid arthritis has been recently reported (Gerber, 1971). Its deamination product, urocanic acid, has a 'sun-screening' effect (Ženišek *et al.*, 1955) and is used in toiletries. Hence, an efficient enzymic process for urocanic acid production from L-histidine has been developed (Kajiwara *et al.*, 1971; Shibatani *et al.*, 1974).

Before the establishment of L-histidine production by a 'direct fermentation' method, histidinol production by a histidine-requiring auxotrophic mutant and its conversion to L-histidine were first studied. A histidine-requiring auxotroph of *Brev. flavum* produced 9.2 g L-histidinol (as dihydrochloride) per litre of a medium containing 10% glucose (Kubota *et al.*, 1971). L-Alanine, L-phenylalanine, L-tryptophan, propionic acid, adenine and 4-amino-5-imidazolecarboxamide riboside depressed L-histidinol production. L-Histidine production from L-histidinol was carried out by various bacteria including *Brev. flavum* A.T.C.C. 14067 which converted with a molar efficiency of 49%. A histidine-requiring auxotroph of *C. glutamicum* also produced L-histidinol at a concentration of 11 g per litre (Kurihara *et al.*, 1971). For conversion of L-histidinol, *E. coli* (Kurihara *et al.*, 1972) and *Torulopsis* sp. 942 (Nakajima *et al.*, 1972) were selected as the most efficient converters.

A 1,2,4-triazole-3-alanine (TRA)-resistant mutant (KY 10260) and a 2-thiazolealanine (TA)-resistant mutant derived from *C. glutamicum* A.T.C.C. 13761, a wild-type strain, produced several grams of L-histidine per litre in a cane molasses medium (Araki and Nakayama, 1971). The histidine productivity of the TRA-resistant mutant could be improved stepwise by successively endowing resistance to 6-mercaptoguanine, 8-azaguanine, 4-thiouracil and 6-mercaptopurine, and increased resistance to triazolealanine and 5-methyltryptophan (Araki *et al.*, 1974b). Improvement of L-histidine productivity in each step was rather minor but, as a total, a finally selected mutant strain, KY 10522, produced about twice the amount of L-histidine compared with the original L-histidine producer. Among the steps, insertion of 4-thiouracil-resistance resulted in a most significant increase in L-histidine productivity. The rationale of the improvement was an increased supply of 5-phosphoribosyl pyrophosphate and adenine nucleotide for L-histidine biosynthesis by releasing the supposed feedback regulation on their biosynthesis, based on the speculation that the regulatory mechanism of L-histidine biosynthesis and related biosynthesis known in some other micro-organisms (Brenner and Ames, 1971; Gots, 1971; Stadtman, 1966) would be applicable to this bacterium. An improvement by increasing resistance to TRA could be explained in terms of a further release of end-product regulation on the histidine pathway resulting from the additional mutation other than the original one.

Strain KY 10522 produced 15 g of L-histidine per litre equivalent to 10% (w/w) of the initial sugar. The medium used had the following composition: 6% (as glucose) cane molasses, 9% sucrose, 4% $(NH_4)_2SO_4$, 0.2% KH_2PO_4, 0.1% K_2HPO_4, 0.05% $MgSO_4 \cdot 7H_2O$, 0.2% urea, 0.75% meat extract, 1 mg thiamin HCl per litre, 80 μg biotin per litre, 3% $CaCO_3$. Amongst the auxotrophic derivatives of strain KY 10260, a leucine-requiring auxotroph, produced L-histidine at a concentration of 11 g per litre equivalent to 5.8% (w/w) of the initial sugar (Araki *et al.*, 1974a).

L-Histidine-producing mutants were also isolated from *Brev. flavum* (Kamijo *et al.*, 1973; Mihara *et al.*, 1973). A TA-resistant mutant of *Brev. flavum* (no. 2247) produced 4 g of L-histidine per litre. Production was improved by adding the following markers successively; sulpha drug-resistance, resistance to α-amino-β-hydroxyvaleric acid, a threonine analogue, and to 2-aminobenzothiazole. A L-histidine-producing mutant was also found among sulphisoxazole-resistant mutants of *Brev. flavum* no. 2247. The productivity was improved by successive insertion of the following markers: resistance to sulphadiazine, S-(2-amino-ethyl)-L-cysteine, a lysine analogue, and ethionine. The mutant thus derived produced 9.7 g of L-histidine per litre. An isoleucine–valine double auxotroph of *Proteus rettgeri* produced 3.8 g of L-histidine per litre in a medium containing 10% glucose (Nakamura *et al.*, 1973).

B. Mechanism of L-Histidine Production

Genetic and enzymic studies on *Salmonella typhimurium* by Brenner and Ames (1971) have provided much information on the biosynthetic pathway of histidine and control mechanism on the pathway. The first step on the pathway is a condensation of 5 phosphoribosyl pyrophosphate (PRPP) and ATP to form N-1-(5'-phosphoribosyl) adenosine triphosphate (phosphoribosyl-ATP) and pyrophosphate. The reaction is catalysed by phosphoribosyl-ATP pyrophosphorylase, and is subject to feedback inhibition by L-histidine. Certain mutants of *S. typhimurium*, resistant to TA, were found to have a pyrophosphorylase resistant to feedback inhibition. This property offered an explanation for L-histidine excretion by TA-resistant mutants of this bacterium (Sheppard, 1964) and *E. coli* (Moyed and Friedman, 1959; Moyed, 1961). The histidine pathway is also under feedback repression control. Mutants of *S. typhimurium* resistant to TRA have been found to be relieved of the repression control (Brenner and Ames, 1971; Roth *et al.*, 1966).

Phosphoribosyl-ATP pyrophosphorylases in two L-histidine producers of *C. glutamicum*, each selected as a TA-resistant and a TRA-resistant strain, were found to be 100-fold resistant to L-histidine inhibition in comparison with the wild-type enzyme (Araki and Nakayama, 1974). It was also resistant to inhibition by TA, but was still as sensitive as the wild-type enzyme to inhibition by α-methylhistidine. Formation of the pyrophosphorylase in these mutants was not significantly derepressed. However, two-fold derepression was noted with a further improved L-histidine producer, namely strain KY-10522. Phosphoribosyl-ATP pyrophosphorylase of strain KY-10522 was found to be resistant to feedback inhibition like its parent strain. Thus, loss of feedback inhibition and derepression of phosphoribosyl-ATP pyrophosphorylase synthesis offer an explanation for L-histidine production in the above described mutants of *C. glutamicum*.

IX. OTHER AMINO ACIDS

A. Alanine

1. Direct Fermentation

Many bacteria, fungi, yeasts and actinomycetes isolated from natural sources produce alanine in culture (Kitai, 1972). Prominent producers are *Corynebacterium gelatinosum*, *Brevibacterium monoflagellum*, *Brev. alanicum*, *Brev. amylolyticum*, *Brev. pentoso-alanicum*, *Brev. pentoso-aminoacidicum*, *Bacillus coagulans* and a mutant strain (13W) of *Brevibacterium* 22 (Grivina, 1968). Differently from other amino acids, the alanine produced in these cultures is usually the racemic form. *Pseudomonas* sp. no. 483 (Samejima *et al.*, 1960) and *Micrococcus sodonensis* (Perry, 1967) are rare examples of L-alanine producers. D-Alanine is produced by *Corynebacterium fascians* under certain conditions (Yamada *et al.*, 1973).

The most common substrate for alanine production is D-glucose, although other carbohydrates are also used. Some alanine producers which utilize hydrocarbons as substrate have been reported, but the amounts of alanine produced are small. The concentration of the nitrogen source and the degree of aeration are important factors in producing alanine. Manganese ions (*Brev. pentoso-alanicum*) and Zn^{2+} (*Fusarium moniliforme*) stimulate L-alanine production. Pyruvate and ammonium lactate increase alanine production with *Pseudomonas* sp. no. 483. Yields of 40% by a *C. gelatinosum* strain have been reported (Kitai *et al.*, 1961).

In *Pseudomonas* sp. no. 483, alanine is formed from pyruvate by the action of alanine dehydrogenase. Alanine biosynthesis by *Arthrobacter* sp. 19d in a molasses-containing medium also involves reductive amination (Setty and Bhat, 1973). On the other hand, in *C. gelatinosum*, transamination from glutamate to pyruvate was suggested as the mechanism of alanine formation. Alanine racemase catalyses conversion of L-alanine to D-alanine yielding an equilibrium mixture.

2. *Enzymic Conversion of L-Aspartic acid to L-Alanine*

The difficulty in obtaining optically active alanine led to the development of an enzymic process for producing L-alanine from L-aspartic acid (Kitai, 1972). *Pseudomonas dacunhae* and *Xanthomonas oryzae* have been selected for the purpose. With *Ps. dacunhae*, L-aspartic β-decarboxylase activity as high as $3910\,\mu$ litres of carbon dioxide per h per ml of medium was obtained by shaking a culture at $30°C$ in a medium containing ammonium fumarate, sodium fumarate, corn-steep liquor, peptone, and inorganic salts. For the enzymic conversion of L-aspartic acid to L-alanine, the whole culture broth was employed as an enzyme source. Large amounts of L-aspartic acid (as much as 40% of the broth) were converted stoicheiometrically to L-alanine in 72 h at $37°C$. The control of pH value at around 5.0 was essential for achieving a yield of more than 90%, since otherwise decomposition and racemization of the accumulated L-alanine occurred.

B. L-Aspartic Acid

Aspartase activity, known from the beginning of this century, was exploited for production of L-aspartic acid from fumarate. *Escherichia coli, E. freundii* and *Ps. fluorescens* are good sources of aspartase (Kitahara, 1972). *Escherichia coli* is a facultative aerobe, so that cell yield is strongly affected by culture conditions such as the degree of aeration. To avoid glucose repression of aspartase formation, the glucose concentration in the medium should be low though a small amount of glucose sometimes gives good results causing improved growth of the bacterium. Kitahara's procedure is as follows. A large concentration of ammonium fumarate, corresponding to 50 g fumaric acid, was suspended in 100 ml water and subjected to the action of aspartase. Crystals of ammonium fumarate gradually disappeared and were converted to a solution of acid ammonium aspartate. Hydrochloric acid was added to bring the pH value of the solution to 2.8, the iso-electric point of aspartic acid. The solubility of aspartic acid being only 0.6% at this point, the greater part immediately crystallizes out.

$$\begin{array}{ccc}
\begin{array}{l} CHCOONH_4 \\ \parallel \\ CHCOONH_4 \end{array}
\rightleftharpoons
\begin{array}{l} CHCOONH_4 \\ \parallel \\ CHCOONH_4 \end{array}
\rightleftharpoons
\begin{array}{l} CHNH_2COOH \\ | \\ CH_2COONH_4 \end{array}
\xrightarrow[pH\ 2.8]{HCL}
\end{array}$$

solid phase   liquid phase   solubility 60%

              solubility 20%

$$\begin{array}{l} CHNH_2\ COOH \\ | \\ CH_2\ COOH \end{array} + NH_4\ CL$$

Later *Alcaligenes faecalis* and *Ps. ovalis* (Ōtsuka, 1961) were selected as bacteria which produce L-aspartate from maleate. The activity of *A. faecalis* was high when it was grown in an acidic medium due to the permeation of maleate, an inducer of maleate cis-trans isomerase (Takamura *et al.*, 1966a). Malonic acid was found to be a gratuitous inducer (Takamura *et al.*, 1966b).

Continuous production of L-aspartic acid from ammonium fumarate was attained by employing an enzyme column packed with the immobilized aspartase, a preparation of partially purified aspartase from *E. coli* entrapped in a polyacrylamide gel lattice (Chibata *et al.*, 1974). Furthermore, *E. coli* was used in place of the enzyme. When a solution of 1 M ammonium fumarate (pH 8.5) containing 1 mM Mg^{2+} was passed through the immobilized cell column at a flow rate of space velocity 0.8 at 37°C, the highest rate of reaction was attained. L-Aspartic acid was obtained in good yield from the column effluent. The immobilized column was very stable (Tosa *et al.*, 1974).

C. L-Proline

Production of L-proline by fermentation was reviewed by Okumura (1972). Some auxotrophic mutants of coryneform glutamic acid-producing bacteria, represented by *C. glutamicum*, produced large amounts of L-proline. These include isoleucine-, histidine- and ornithine-requiring auxotrophs. According to K. Araki, Y. Takasawa and J. Nakajima (unpublished observations), a certain base auxotroph *C. glutamicum* produced 31 g of L-proline per litre in a medium containing 15% (as glucose) cane molasses. A tyrosine-phenylalanine double auxotroph of *Corynebacterium melassecola* also produced a fairly large amount of L-proline (Kanamitsu, 1971).

Some wild-type strains of *C. glutamicum* also produce significant amounts of L-proline under certain conditions (Nakanishi *et al.*, 1973). Production of L-proline by *Kurthia catenaforma* increased in a serine-requiring auxotrophic mutant (Kato *et al.*, 1968).

One of the characteristic conditions for L-proline production is a high concentration of ammonium ions. An excessive supply of biotin and unusually high concentrations of magnesium ion were also required for proline production by auxotrophic mutants of coryneform glutamic acid-producing bacteria. In the case of *K. catenaforma*, co-existence of L-aspartic acid and a high concentration of potassium ions in the medium markedly stimulated L-proline production. L-Glutamic acid was also effective when some surfactants were added to the medium to increase transport of L-glutamic acid (Kato *et al.*, 1972a, b). The nutrients required for auxotrophic mutants should also be limited to an appropriate concentration which is suboptimal for their growth. Addition of high concentrations of both ammonium and chloride ions to the medium was effective for L-proline production by wild-type strains of *C. glutamicum*. Highly aerobic conditions, a temperature near 30°C, and an initial pH value of 7.0–8.0 were also required for good production of L-proline by *C. glutamicum*. Adding alcohols increased L-proline production by *C. glutamicum* KY 9003 (Nakanishi *et al.*, 1973).

Yoshinaga (1969) explained L-proline production by *Brev. flavum* no. 14–5, an isoleucine-requiring auxotroph, based on an intracellular accumulation of threonine resulting from a blockage of threonine dehydratase activity, availability of a high level of ATP resulting from inhibition of aspartate kinase and of homoserine kinase by the threonine, promotion of glutamate kinase activity by the ATP, insensitivity of one of the two isozymic kinases to proline and the availability of intracellular glutamate under biotin-rich conditions (Okumura, 1972). However, this explanation seems to be deficient because many isoleucine-requiring auxotrophs deficient in threonine dehydratase are unable to produce a large amount of proline. The mechanism of L-proline production is still waiting to be clarified.

REFERENCES

Abe, S., Takayama, K. and Kinoshita, S. (1967). *Journal of General and Applied Microbiology* **13**, 279.

Alikhanian, S. I. (1972). *In* 'Fermentation Technology Today', Proceedings of the IVth International Fermentation Symposium, (G. Terui, ed.), pp. 233–237. Society of Fermentation Technology Japan.

Araki, K. and Nakayama, K. (1971). *Agricultural and Biological Chemistry* **35**, 2081.

Araki, K. and Nakayama, K. (1974). *Agricultural and Biological Chemistry* **38**, 2209.

Araki, K., Kato. F., Arai, Y. and Nakayama, K. (1974a). *Agricultural and Biological Chemistry* **38**, 189.

Araki, K., Shimojo, S. and Nakayama, K. (1974b). *Agricultural and Biological Chemistry* **38**, 837.

Araki, K., Ueda, H. and Saigusa, S. (1974c). *Agricultural and Biological Chemistry* **38**, 565.

Arima, K., Nagami, I. and Yoneda, M. (1971). *Meeting of The Agricultural Chemical Society of Japan, Abstracts*, pg. 53.

Bekers, M., Abolins, T., Viesturs, U., Selga, S., Ramina, L. and Ozolins, S. (1971). *Prikladnaya Biokhimiya i Mikrobiologiya, Moscow* **7**, 103 (*Chemical Abstracts* **74**, 98116t).

Brenner, M. and Ames, B. N. (1971). *In* 'Metabolic Pathways', (H. H. Vogel, ed.), vol. 5, pp. 349–387. Academic Press, New York and London.

Casida, L. E. and Baldwin, N. Y. (1956). U.S. Patent 2,771.396.

Chibata, I., Tosa, T. and Sato, T. (1974). *Applied Microbiology* **27**, 878.

Coukell, M. B. and Polglase, W. J. (1969a). *Biochemical Journal* **111**, 273.

Coukell, M. B. and Polglase, W. J. (1969b). *Biochemical Journal* **111**, 279.

Demain, A. L. and Birnbaum, J. (1968). *Current Topics in Microbiology* **46**, 1.

Fukumura, T. (1974). *23rd Amino Acid and Nucleic acid Symposium, Abstracts* (*Japanese*), pg. 1.

Gerber, D. A. (1971). *Clinical Research* **19**, 442.

Gots, J. S. (1971). *In* 'Metabolic Pathways' (H. H. Vogel, ed.), vol. 5, pp. 225–255. Academic Press, New York and London.

Grivina, P. (1968). *Prikladnaya Biokhimiya i Mikrobiologiya, Moscow* **4**, 122.

Hagino, H. and Nakayama, K. (1973). *Agricultural and Biological Chemistry* **37**, 2013.

Hagino, H. and Nakayama, K. (1974). *Agricultural and Biological Chemistry* **38**, 157.

Hagino. H., Yoshida, H., Kato, F., Arai, Y., Katsumata, R. and Nakayama, K. (1973). *Agricultural and Biological Chemistry* **37**, 2001.

Hirakawa, T. and Watanabe, K. (1974). *Agricultural and Biological Chemistry* **38**, 77.

Hirakawa, T., Tanaka, T. and Watanabe, K. (1973). *Agricultural and Biological Chemistry* **37**, 123.

Huang, H. T. (1961). *Applied Microbiology* **9**, 419.

Ikeda, S., Fujita, I., Yoshinaga, F. and Kobayashi, K. (1974). *Meeting of The Agricultural Chemical Society of Japan, Abstracts*, No. 2A-9.

Kajiwara, M., Yasunaga, M., Naganuma, F., Shibuya, M., Shiro, T. and Okumura, S. (1971). *Meeting of The Agricultural Chemical Society of Japan, Abstracts*, p. 124.

Kamijo, H., Mihara, O. and Kubota, K. (1973). *Meeting of The Agricultural Chemical Society of Japan, Abstracts* No. 1j-2.

Kanamitsu, O. (1971). *Annual Meeting of The Agricultural Chemistry Society of Japan, Abstracts*, p. 156.

Kanzaki, T., Isobe, K., Okazaki, H., Mochizuki, K. and Fukada, H. (1967). *Agricultural and Biological Chemistry* 31, 1307.

Kanzaki, T., Kitano, K., Sumino, Y. and Okazaki, Y. (1972). *Nippon Nōgei-Kagaku Kaishi* (Japanese) 46, 95.

Kanzaki, T., Nakatsu, I., Kitano, K., Sugiyama, Y., Nishio, Y. and Ishikawa, M. (1973). *Agricultural and Biological Chemistry* 37, 1407.

Kase, H. and Nakayama, K. (1972). *Agricultural and Biological Chemistry* 36, 1611.

Kase, H. and Nakayama, K. (1973). *Agricultural and Biological Chemistry* 37, 1643.

Kase, H. and Nakayama. K. (1974a). *Agricultural and Biological Chemistry* 38, 993.

Kase, H. and Nakayama, K. (1974b). *Nippon Nōgei-Kagaku Kaishi* 48, 209.

Kase, H., Tanaka, H. and Nakayama, K. (1971). *Agricultural and Biological Chemistry* 35, 2089.

Kato, J., Fukushima, H., Kisumi, M. and Chibata, I. (1972a). *Applied Microbiology* 23, 699.

Kato, J., Kisumi, M. and Chibata, I. (1972b). *Applied Microbiology* 23, 758.

Kato, J., Horie, S., Komatsubara, S., Kisumi, M. and Chibata, I. (1968). *Applied Microbiology* 16, 1200.

Kikuchi, M. and Nakao, Y. (1973). *Agricultural and Biological Chemistry* 37, 509; 515.

Kikuchi, M., Doi, M., Suzuki, M. and Nakao, Y. (1972). *Agricultural and Biological Chemistry* 36, 1141.

Kinoshita, S. and Tanaka, K. (1972). *In* 'The Microbial Production of Amino Acids', (K. Yamada, S. Kinoshita, T. Tsunoda and K. Aida, eds.), pp. 263–324. Kōdansha Ltd., Tokyo.

Kinoshita, S., Nakayama, K. and Kitada, S. (1958). *Journal of General and Applied Microbiology, Tokyo* 4, 128.

Kinoshita, S., Tanaka, K., Udaka, S. and Akita, S. (1957). *Proceedings of the International Symposium on Enzyme Chemistry* 2, 464.

Kisumi, M., Komatsubara, S. and Chibata, I. (1971a). *Journal of Bacteriology* 106, 493.

Kisumi, M., Komatsubara, S., Sugiyama, M. and Chibata, I. (1971b). *Journal of Bacteriology* 107, 741.

Kisumi, M., Komatsubara, S., Sugiyama, M. and Chibata, I. (1971c). *Journal of General Microbiology, Tokyo* 69, 291.

Kisumi, M., Komatsubara, S. and Chibata, I. (1971d). *Journal of Bacteriology* 107, 824.

Kisumi, M., Kato, J., Sugiyama, M. and Chibata, I. (1971c). *Applied Microbiology* 22, 987.

Kisumi, M., Komatsubara, S. and Chibata, I. (1973a). *Journal of Biochemistry, Tokyo* 73, 107.

Kisumi, M., Kato, J., Takagi, T. and Chibata, I. (1973b). *Meeting of The Agricultural Chemical Society of Japan, Abstracts*, p. 114.

Kitahara, K. (1972). *In* 'The Microbial Production of Amino Acids', (K. Yamada, S. Kinoshita, T. Tsunoda and K. Aida, eds.), pp. 533–537. Kōdansha Ltd., Tokyo.

Kitai, A. (1972). *In* 'The Microbial Production of Amino Acids', (K. Yamada, S. Kinoshita, T. Tsunoda and K. Aida, eds.), pp. 325–337. Kōdansha Ltd., Tokyo.

Kitai, A., Tone, H., Sasaki, H., Miyachi, N. and Yamanoi, A. (1961). *Nippon Nōgei-Kagaku Kaishi*, (Japanese) **35**, 862.

Kobayashi, K., Ikeda, S., Takahashi, K., Hirose, Y. and Shiro, T. (1971). *Agricultural and Biological Chemistry* **35**, 1241.

Komatsubara, S., Kisumi, M. and Chibata, I. (1973). *Meeting of Society of Fermentation Technology Japan, Abstracts*, pg. 98.

Komatsubara, S., Kisumi, M. and Chibata, I. (1974). *23rd Amino Acid and Nucleic Acid Symposium Abstracts*, pg. 14.

Kōmura, I., Komagata, K. and Mitsugi, K. (1973). *Journal of General and Applied Microbiology, Tokyo* **19**, 161.

Kondo, E., Mitsugi, T. and Hanei, F. (1970). *Amino Acid and Nucleic Acid* No. 22, 151.

Kubota, K., Shiro, T. and Okumura, S. (1971). *Journal of General and Applied Microbiology, Tokyo* **17**, 1.

Kubota, K., Kageyama, K., Maeyashiki, I., Yamada, U. and Okumura, S. (1972). *Journal of General and Applied Microbiology, Tokyo* **18**, 365.

Kubota, K., Onda, T., Kamijo, H., Yoshinaga, F. and Okumura, S. (1973). *Journal of General and Applied Microbiology, Tokyo* **19**, 339.

Kurihara, S., Araki, K. and Takazawa, M. (1971). Japanese Patent 46-28822.

Kurihara, S., Araki, K. and Takazawa, M. (1972). Japanese Patent 47-4506.

Matsushima, H., Murata, K. and Mase, Y. (1974). *Hakkō Kōgaku Zasshi* **52**, 20.

Mihara, O., Kamijo, H. and Kubota, O. (1973). *Meeting of The Agricultural Chemical Society of Japan, Abstracts*, No. 1J-3.

Moyed, H. S. (1961). *Journal of Biological Chemistry* **226**, 2261.

Moyed, H. S. and Friedman, M. (1959). *Science, New York* **129**, 968.

Nakajima, J., Araki, K. and Morinaga, K. (1972). U.S. Patent 3,676,301.

Nakamori, S. and Shiio, I. (1972). *Agricultural and Biological Chemistry* **36**, 1209.

Nakamura, T. and Aida, K. (1970), *Meeting of The Agricultural Chemical Society of Japan, Abstracts*, pg. 9.

Nakamura, T., Ebihara, M. and Shirai, T. (1973). Japanese Patent-Kōkai 48-92588.

Nakanishi, T. (1972). *Meeting of Society of Fermentation Technology Japan, Abstracts*, p. 110.

Nakanishi, T., Yokote, Y. and Taketsugu, Y. (1973). *Hakkō Kōgaku Zasshi* **51**, 742.

Nakao, Y., Kikuchi, M., Suzuki, M. and Doi, M. (1970). *Agricultural and Biological Chemistry* **34**, 1875.

Nakao, Y., Kikuchi, M., Suzuki, M. and Doi, M. (1972). *Agricultural and Biological Chemistry* **36**, 490.

Nakao, Y., Kanamaru, T., Kikuchi, M. and Yamatodani, S. (1973). *Agricultural and Biological Chemistry* **37**, 2399.

Nakayama, K. (1972). *In* 'Fermentation Technology Today', Proceedings of the IVth International Fermentation Symposium, (G. Terui, ed.), pp. 433–438. Society of Fermentation Technology, Japan.

Nakayama, K. (1972). *In* 'The Microbial Production of Amino Acids', (K. Yamada, S. Kinoshita, T. Tsunoda and K. Aida, eds.), pp. 369–397. Kōdansha, Ltd., Tokyo.

Nakayama, K. and Yoshida, H. (1972). *Agricultural and Biological Chemistry* **36**, 1675.

Nakayama, K. and Araki, K. (1973). U.S. Patent 3,708,395.

Nakayama, K., Sato, Z. and Kinoshita, S. (1961a). *Nippon Nōgei-Kagaku Kaishi* (Japanese) **35**, 142.

Nakayama, K., Sato, Z. and Kinoshita, S. (1961b). *Nippon Nōgei-Kagaku Kaishi* (Japanese) **35**, 146.

Nakayama, K., Kitada, S. and Kinoshita, S. (1961c). *Journal of General and Applied Microbiology, Tokyo* **7**, 52.

Nakayama, K., Tanaka, H., Hagino, H. and Kinoshita, S. (1966). *Agricultural and Biological Chemistry* **30**, 611.

Nakayama, K., Araki, K., Hagino, H., Kase, H. and Yoshida, H. (1974). *In* 'Genetics of Industrial Microorganisms, Second International Symposium', Abstracts, pg. 47. Academic Press, London.

Nakazawa, H., Enei, H., Okumura, S., Yoshida, H. and Yamada, H. (1972). *Federation of European Biochemical Societies Letters*, **25**, 43.

Nara, T. (1972). *In* 'The Microbial Production of Amino Acids', (K. Yamada, S. Kinoshita, T. Tsunoda and K. Aida, eds.), pp. 417–434. Kōdansha Ltd., Tokyo.

Ōishi, K. (1972). *In* 'The Microbial Production of Amino Acids', (K. Yamada, S. Kinoshita, T. Tsunoda and K. Aida, eds.), pp. 435–452. Kōdansha Ltd., Tokyo.

Oki, T., Sayama, Y., Nishimura, Y. and Ozaki, A. (1968). *Agricultural and Biological Chemistry* **32**, 119.

Okumura, S. (1972). *In* 'The Microbial Production of Amino Acids', (K. Yamada, S. Kinoshita, T. Tsunoda and K. Aida, eds.), pp. 473–490. Kōdansha Ltd., Tokyo.

Ōtsuka, K. (1961). *Agricultural and Biological Chemistry* **25**, 726.

Perry, J. J. (1967). *Journal of Bacteriology* **94**, 1249.

Raunio, R. (1966). *Acta Chemica Scandinavica* **20**, 11.

Roth, J. R., Anton, D. N. and Hartman, P. E. (1966). *Journal of Molecular Biology* **22**, 305.

Sahm, H. and Zähner, H. (1971). *Archiv für Microbiologie* **76**, 223.

Samejima, H., Nara, T., Fujita, C. and Kinoshita, S. (1960). *Nippon Nogei-Kagaku Kaishi* (Japanese) **34**, 832.

Sano, K. and Shiio, I. (1970). *Journal of General and Applied Microbiology, Tokyo* **16**, 373.

Sanwal, B. D. and Smando, R. (1969). *Biochemical and Biophysical Research Communications* **35**, 486.

Sato, E., Kawabata, Y. and Fukumura, T. (1974). *23rd Amino Acid and Nucleic Acid Symposium, Abstracts*, pg. 2.

Seto, T. A. (1962). U.S. Patent 3,056,729.

Setty, T. M. R. and Bhat, J. V. (1973). *Current Science* **42**, 484.

Sheppard, D. E. (1964). *Genetics* **50**, 611.

Shibatani, T., Ninimura, N., Abe, R., Kakimoto, T. and Chibata, I. (1974). *Applied Microbiology* **27**, 688.

Shibukawa, M., Ōsawa, T., Nobukuni, T. and Yamamoto, S. (1964a). *Nippon Nōgei-Kagaku Kaishi* **38**, 285.

Shibukawa, M., Komatsu, K. Ōsawa, T. and Yamamoto, S. (1964b). *Nippon Nōgei-Kagaku Kaishi* **38**, 291.

Shigeto, M. (1962). *Nippon Nōgei-Kagaku Kaishi* **36**, 809.

Shiio, I., Ōtsuka, S. and Katsuya, N. (1963). *Journal of Biochemistry, Tokyo* **53**, 333.

Shiio, I. and Miyajima, R. (1969). *Journal of Biochemistry, Tokyo* **65**, 849.

Shiio, I. and Nakamori, S. (1970). *Agricultural and Biological Chemistry* **34**, 448.

Shiio, I., Sato, H. and Nakazawa, H. (1972). *Agricultural and Biological Chemistry* **36**, 2315.

Shiio, I., Ishii, K. and Yokozeki, K. (1973a). *Agricultural and Biological Chemistry* **37**, 1991.

Shiio, I., Sasaki, A., Nakamori, S. and Sano, K. (1973b). *Agricultural and Biological Chemistry* **37**, 2053.

Shimura, K. (1972a). *In* 'The Microbial Production of Amino Acids', (K. Yamada, S. Kinoshita, T. Tsunoda and K. Aida, eds.), pp. 453–472. Kōdansha Ltd., Tokyo.

Shimura, K. (1972b). *In* 'The Microbial Production of Amino Acids', (K. Yamada, S. Kinoshita, T. Tsunoda and K. Aida, eds.), pp. 491–513. Kōdansha Ltd., Tokyo.

Sommerson, N. and Phillips, T. (1962). U.S. Patent 3,080,297.

Stadtman, E. R. (1966). *Advances in Enzymology* **28**, 41.

Sugimoto, S., Nakagawa, M., Tsunoda, T. and Shiio, I. (1973). *Agricultural and Biological Chemistry* **37**, 2327.

Sugimoto, S., Nakagawa, M. and Shiio, I. (1974). *Meeting of The Agricultural Chemical Society of Japan, Abstracts*, No. 2A-16.

Sugisaki, Z. (1959). *Journal of General and Applied Microbiology, Tokyo* **5**, 138.

Sugiyama, Y., Kitano, K. and Kanzaki, T. (1973). *Agricultural and Biological Chemistry* **37**, 1837.

Suzuki, T., Yamaguchi, K. and Tanaka, K. (1971). *Agricultural and Biological Chemistry* **35**, 2135.

Takamura, Y., Kitamura, I., Iikura, M., Kōno, K. and Ozaki, A. (1966a). *Agricultural and Biological Chemistry* **30**, 338.

Takamura, Y., Kitamura, I., Iikura, M., Kōno, K. and Ozaki, A. (1966b). *Agricultural and Biological Chemistry* **30**, 345.

Takayama, K., Abe, S. and Kinoshita, S. (1969). *Amino Acid and Nucleic Acid*, No. 19, 121.

Takinami, K., Okada, H. and Tsunoda, T. (1963). *Agricultural and Biological Chemistry* **27**, 858.

Takinami, K., Okada, H. and Tsunoda, T. (1964). *Agricultural and Biological Chemistry* **28**, 114.

Tanaka, K., Iwasaki, T. and Kinoshita, S. (1960). *Nippon Nōgei-Kagaku Kaishi* **34**, 593.

Tanaka, K., Suzuki, T. and Okumura, S. (1971). *5th World Petroleum Congress, Proceedings* No. 5, pp. 165–170, Applied Science Publication Ltd., London.

Tanaka, Y., Yoshida, H. and Nakayama, K. (1974). *Agricultural and Biological Chemistry* **38**, 633.

Terui, G., (1972). *In* 'The Microbial Production of Amino Acids', (K. Yamada, S. Kinoshita, T. Tsunoda and K. Aida, eds.), pp. 515–531. Kōdansha, Ltd., Tokyo.

Thieman, J. E. and Pagani, H. (1972). U.S. Patent 3,700,558.

Tokoro, Y., Ōshima, K., Okii, M., Yamaguchi, K., Tanaka, K. and Kinoshita, S. (1970). *Amino Acid and Nucleic Acid*, No. 21, 49.

Tosa, T., Sato, T., Mori, T. and Chibata, I. (1974). *Applied Microbiology* **27**, 886.

Tsuchida, T. and Momose, H. (1974). *Meeting of The Agricultural Chemical Society of Japan, Abstracts*, No. 2A-7.

Tsuchida, T., Yoshinaga, F., Kubota, K. and Momose, H. (1973a). *Meeting of The Agricultural Chemical Society of Japan, Abstracts*, No. 1J-4.

Tsuchida, T., Yoshinaga, F., Kubota, K. and Momose, H. (1973b). *Meeting of The Agricultural Chemical Society of Japan, Abstracts*, No. 1J-7.

Tsukada, Y. and Sugimori, T. (1971). *Agricultural and Biological Chemistry* **35**, 1.

Udagawa, K., Abe, S. and Kinoshita, S. (1962). *Hakkō Kōgaku Zasshi* **40**, 614.

Udaka, S. (1972). *In* 'The Microbial Production of Amino Acids', (K. Yamada, S. Kinoshita, T. Tsunoda and K. Aida, eds.), pp. 399–415. Kōdansha Ltd., Tokyo.

Udaka, S. and Kinoshita, S. (1959). *Journal of General and Applied Microbiology, Tokyo* **5**, 139.

Uemura, T., Sugisaki, Z. and Takamura, T. (1972). *In* 'The Microbial Production of Amino Acids', (K. Yamada, S. Kinoshita, T. Tsunoda and K. Aida, eds.), pp. 339–368. Kōdansha Ltd., Tokyo.

Umbarger, H. E. (1969). *Current Topics in Cellular Regulation* **1**, 57.

Umbarger, H. E. and Brown, B. (1958). *Journal of Biological Chemistry* **233**, 415.

Wada, H. (1967). Japanese Patent 42-17728.

Woodruff, H. B. and Jackson, M. (1970). U.S. Patent 3,524,797.

Yamada, H., Kumagai, H., Enei, H., Matsui, H. and Okumura, S. (1972). *In* 'Fermentation Technology Today', Proceedings of the IVth International Fermentation Symposium, (G. Terui, ed.), pp. 445–454. Society of Fermentation Technology Japan.

Yamada, K. and Komagata, K. (1972). *Journal of General and Applied Microbiology, Tokyo* **18**, 417.

Yamada, S., Maeshima, H., Wada, H. and Chibata, I. (1973). *Applied Microbiology* **25**, 636.

Yates, A. B. and Pardee, A. B. (1957). *Journal of Biological Chemistry* **277**, 677.

Yoshida, H. and Nakayama, K. (1974). *Nippon Nōgei-Kagaku Kaishi* **48**, 201.

Yoshinaga, F. (1969). *Amino Acid and Nucleic Acid* No. 19, 25.

Ženišek, A., Král, J. A. and Hais, I. M. (1955). *Biochimica et Biophysica Acta* **18**, 589.

7. Lipids and Fatty Acids

COLIN RATLEDGE

Department of Biochemistry, The University of Hull, Hull, HU6 7RX, England

I. INTRODUCTION AND HISTORICAL NOTE

There is, at the present time, no industrial production of any microbial lipid. This chapter, therefore, assesses the potential of micro-organisms as a source of these materials and, because of their economic importance, will deal primarily with triglycerides (otherwise known as triacylglycerols) and fatty acids.

Fatty acids, whether from micro-organisms or more conventional sources such as plants and animals, occur in an esterified form which is usually a triglyceride. With plants and animals, these triglycerides usually constitute at least 95% of the material extracted as an 'oil' or 'fat' (these terms are used according to the fluidity of the material at ambient temperature). 'Lipid', however, is the more correct term to use for a *fatty* material as it embraces, besides the triglycerides, such materials as waxes, sterols, sterol esters, polar lipids of a wide variety of types, hydrocarbons such as squalene, di- and monoglycerides and even free fatty acids themselves. However, in this review, the term 'lipid' will not be used except where there is a need for clarity; instead 'fat' or 'oil' will be used in keeping with previous reviews on this subject to describe the fatty materials in general of micro-organisms.

Depending upon the ultimate use, one may be interested in either the intact triglyceride or in the fatty acids derived from the acylglycerols. In the latter case, these are widely used in the chemical industry for the manufacture of soaps, shampoos and other toiletries and cosmetics, lubricating greases and oils, polishes, waxes, alkyd and epoxy resins, printing inks and putty. Triglycerides, on the other hand, may be incorporated, with or without modification, into edible oil products, such as cooking oils and fats, margarine and salad oils, or may be used in the production of a wide range of materials ranging from paints and varnishes to candles and linoleum (Swern, 1964; Tooley, 1971). It follows, therefore, that if only a source of fatty acids is required, one could look at the total fatty-acyl composition of a micro-organism whereas, if the tri-glycerides were desired, these may have to be separated from unwanted lipids including phospholipids and sterols.

Fat production by micro-organisms was first considered as a commercial proposition by Germany during the First World War (1914–1918) although industrial production was not realized (Lindner, 1922). Lindner's work, with *Endomyces* (*Endomycopsis*) *vernalis*, however, was extended independently by two German industrial groups just prior to and then during the Second World War (1939–1945); Zellstoff-Fabrik Mannheim Waldhof used species of *Torula* (*Candida*) in submerged culture, and achieved an organism containing 30% of its weight as fat and 20% as protein (Schmidt, 1947), whereas Henkel and Co. of Dusseldorf used a species of *Fusarium*, also in submerged culture, which attained up to 50% of its weight as fat (Damm, 1943). Neither of these processes was commercially viable once hostilities had ended and normal international commerce was resumed. Accounts of these early ventures and of the relevant researches are given by Hesse (1949), Woodbine (1959) and to a lesser extent by Prescott and Dunn (1959).

Since 1959 there has not been any attempt to survey in detail the feasibility of producing fats microbiologically save for a review by Watanabe (1967a, b), which being in Japanese has only a limited readership, and an earlier review by the present author which was concerned solely with a discussion of alkanes as a possible feedstock for fat production (Ratledge, 1970). The purpose of this chapter is to describe the advances which have occurred since 1959, and readers requiring information prior to this date are asked to consult Woodbine's authoritative and detailed review (Woodbine, 1959). However, before discussing recent developments, a brief analysis of the World situation in the oil and fat commodity market is necessary, for if microbial oils and fats are ever to be produced commercially it is against this market and its values which they will have to compete.

II. ECONOMICS OF WORLD TRADE IN OILS AND FATS

Useful data on the trends in commodity markets, statistics of trade and production as well as World prices may be derived by consultation of: *The Monthly Bulletin of Agricultural Statistics* (published by F.A.O. of the U.N., Rome); *Foreign Agriculture* (a biweekly publication of the U.S. Department of Agriculture); *Overseas Trade Statistics* (published monthly by the U.K. Board of Trade and Industry); *The Oil, Paint and Drug Recorder—The Chemical Marketing Newspaper* (a weekly American newspaper giving prices of most industrial chemicals) and *The Public Ledger* (a daily British newspaper detailing prices and trends of a range of market commodities).

About 70% of the World's production of oils and fats is from plant sources of which soybeans, groundnuts, cotton-, sunflower- and palm-seeds predominate (Table 1). Total production showed little real expansion in the last decade but, in the two years 1973 and 1974, production suddenly increased by 3 million tons to reach new all-time high levels of 48.0 million tons. This expansion continued into 1975 but dropped by a small margin in 1976 mainly due to the adverse dry summer of that year. This increase in production has been mainly due to the expansion in soybean production not only in the U.S.A. but in Brazil and other South American countries. Although

                                              C. RATLEDGE

the expectations are that soybean production will continue to expand in the next ten years (Holz, 1974), 1976 saw a 9% drop in the crop from the 1975 figures.

The main producing countries of seed oils are U.S.A. (soybeans, linseed and cottonseed), West Africa and India (groundnuts), China

Table 1

World production of oils and fats (1973 to 1976) and current prices. From *Monthly Bulletin of Agricultural Statistics* and the *Commodity Review and Outlook*, both published by Food and Agriculture Organization of the United Nations

	1973	1974	1975	1976	Current U.K. prices[a]
		(millions metric tons)			(£/metric ton)
Total plant oils	30.1	33.0	36.4[b]	35.1[b]	
Major individual commodities					
Soybean oil	7.8	9.6	10.9	10.0	307
Groundnut oil	3.1	3.3	4.0	4.1	527
Cottonseed oil	3.0	2.9	2.4	2.6	372
Sunflowerseed oil	3.6	4.5	3.9	4.5	440
Rapeseed oil	2.5	2.5	3.2	2.4	355
Palm oil	2.6	2.9	3.1	3.3	402
Coconut oil	2.5	2.2	2.7	2.9	301
Olive oil	1.6	1.6	1.9	1.6	1250
Linseed oil	0.8	0.8	0.9	0.8	300
Palm kernel oil	0.7	0.7	0.7	0.8	292
Fish and marine mammal oil	0.9	1.1	1.2	1.2	240
Animal fats (lard and tallow)	8.4	8.8	8.8[b]	8.8[b]	335[c]
TOTAL (all sources)[d]	44.59	48.00	51.5[b]	50.4[b]	

[a] 17th September 1977; prices for crude oil, i.e. not de-odourized or hardened.
[b] Estimated.
[c] Price quoted for lard; tallow is somewhat cheaper.
[d] Includes butter but not other dairy fat products.

(groundnuts and cottonseed,) Malaysia (palm and palm kernel oil), U.S.S.R. (cotton- and sunflowerseed) and Canada (linseed). Brazil must also be included in this list as its current (1976) production of soybeans is now 11 million tons being almost one-third of the massive U.S. output. Although some attempt is currently being made to grow soybeans in Europe (Bulgaria, Romania, Yugoslavia and also in France), the only general oilseed crop of any value grown in Europe are those of rape and olive oil. The advent of zero-erucic acid-containing rapeseed has led to a better acceptance of this crop and because of its tolerance of more temperate climates is now being widely cultivated in the United Kingdom. Our present agriculture programme anticipates that by 1978 production of rapeseed oil will have reached about 250,000 tons (Pritchard, 1974). Europe also provides 80% of the

Table 2

Prices for selected oils and fats from 1969 to 1977 (U.K. imported quoted price: pounds sterling/metric ton; source *The Public Ledger*)

Year and month		Linseed oil	Groundnut oil	Soybean oil	Palm oil	Lard	Fish oil
1969		101	140	90	71	92	68
1970		96	160	136	110	115	113
1971		81	185	138	109	109	97
1972		84	173	108	89	102	80
1973	Jan.	115	194	111	92	102	118
	Oct.	318	229	185	181	190	*unq*
1974	Jan.	466	355	323	214	252	*unq*
	Oct.	532	515	*unq*	362	300	*unq*
1975	Jan.	477	470	380	300	312	250
	Oct.	400	408	270	200	285	175
1976	Jan.	375	360	205	185	240	175
	April	390	400	235	220	250	200
	Aug.	450	420	270	250	280	240
	Oct.	425	540	335	325	350	260
1977	Jan.	412	560	325	350	340	260
	April	388	520	428	400	380	330

unq indicates unquoted.

World's output of olive oil coming principally from Italy, Spain and Greece. This, however, is a highly specialized commodity and is not used other than for cooking or as a salad oil.

The prices of all oils and fats, including those from animals and fish, increased dramatically during 1973 as did all other basic commodities. Stabilization of prices has only partially been achieved since that time. Price trends of some oils and fats over the last four years are shown in Table 2 and current prices are given in Table 1. The most dramatic increases in price occurred with linseed oil with lard and sunflowerseed oil not far behind. This has led to changes in consumption:

industry has modified its products where possible in order to take advantage of the more abundant and cheaper soybean oil. Although production of soybeans is continuing to expand so to is the production of palm kernel oil. Currently Malaya accounts for 8% of the World trade in oils through palm oil and palm kernel oil. By 1980, output of palm oil from Malaya is expected to reach 1.6 million tons, this being over twice the 1972 production level. Copra expansion in the Philippines has also recently increased; a 15% improvement was recorded in 1976 alone over the previous year's production.

Although one may predict a continual expansion in the output of seed oils, one must nevertheless remember that crop yields can be severely hit by unfavourable weather or changes in climate. Thus export of groundnuts from Nigeria and other West African countries was halted in 1972/73 because of severe drought; Australia, after a brief attempt to become an oilseed exporter, was finally obliged to withdraw from World trading in 1974 because of poor crop yields. The sunflowerseed crop failed badly in the U.S.S.R. in 1976: a shortfall of 6 million tons was experienced and, as already mentioned, the same dry summer reduced the U.S. soybean harvest by over 7 million tons. Even the production of fish oil is subject to the vagaries of the climate as witness the change in the Humboldt current off the coasts of Peru and Chile resulting in a dramatic decline in fishmeal and fish oil production in 1972 and 1973.

Demand for oils, however, is increasing faster than production and is likely to remain so at least until the end of this Century. Prices of oils and fats will, therefore, remain under pressure for many years and any shortfalls in production, for any reason, can only exacerbate future prices. Prices may rise faster than expected if more of the oil seed crops are crushed and extracted in their country of origin. European nations, which rely heavily on importing uncrushed oilseeds for their main supply of oils and fats, would be most vulnerable to the developing countries deciding that they wanted a greater share in the profits derived from their products.

The range of prices of the various oils and fats is, of course, a reflection of their respective demands and to the applications to which they can be put. Where a specific and very desirable property is required, the price is likely to be high. Such is the case with linseed oil which is used extensively in the paint and varnish industry or with olive oil for which no substitute is acceptable.

Table 3

Fatty-acyl composition of some major commercial oils and fats

	Fatty-acyl composition (percent w/w),										
Commodity	10:0 and under	12:0	14:0	16:0	16:1	18:0	18:1	18:2	18:3	20:0	20:1+
Soybean oil	—	—	0.5	11	—	4.5	22	53	8	0.5	0.5
Cottonseed oil	—	—	1	25	1	3	18	51	0.5	0.5	—
Sunflowerseed oil	—	—	0.5	6.5	—	3.5	23	64.5	0.5	1	0.5
Groundnut oil											
(Indian)	—	—	—	10.5	—	3	50	30	—	5	1.5
Palm oil	—	—	1	43	0.5	5	40	10	—	0.5	—
Palm kernel oil	6.5	47	16	9	—	2.5	18	1	—	—	—
Coconut oil	13	47	19	8.5	—	3	8	1.5	—	—	—
Linseed oil	—	—	—	7	—	14	18	14	47	—	—
Rapeseed oil											
(European)	—	—	—	4	0.5	1	14	15	9	1	55.5[a]
Animal lard	—	0.5	3.5	28.5	4	20	39	3	0.5	0.5	—
Fish oil											
(herring)	—	—	7	18	10	2	10	2	0.5	0.5	50[b]

[a] This acid is mainly erucic acid ($C_{22:1}$).
[b] This is a mixture of di-, tri- and tetra-enoic C_{20} and C_{22} acids.

The uses of an oil or fat depend upon its outward characteristics which, in turn, are brought about by the fatty-acyl compositions of the triglycerides. The fatty acids of some of the more commercially important oils and fats are given in Table 3. The compositions of most naturally occurring fats and oils are given by Hilditch and Williams (1964).

A successful process for production of a microbial fat would seek to manufacture a material which was acceptable to industrial users as a substitute for one of the existing materials. The properties, and therefore the fatty-acyl composition, of the microbial oil should be as close as possible to one of these. The most desirable oils to aim for from microbial sources are obviously those which will command the highest prices and, as can be seen from Table 1 (pg. 266) and 3 (pg. 269), oils containing a high proportion of polyunsaturated acids ($C_{18:2}$ and $C_{18:3}$) would be most in demand. Not apparent from Tables 1 and 3 are the high prices of certain individual fatty acids. Caprylic acid (C_8) and capric acid (C_{10}) both cost over £1,000/ton and even a mixture of the two sells at £900/ton. Prices for acids from C_{12} to C_{18} drop to between £500 to £625/ton though rise again for the longer chain acids, erucic acid ($C_{22:1}$) and behinic acid ($C_{22:0}$), up to £950/ton. Any microbial fat containing very high concentrations of either very short or long chain fatty acids would therefore be of considerable commercial interest. For some of the more expensive commodities, such as cocoa butter or olive oil (costing £3,000 and £1,300/ton), no substitute, no matter how much like the original, would be acceptable as legislation does not permit substitutions to be made without there having to be a renaming of the end product. Thus 'chocolate' can only be so called if it contains not more than 10% cocoa-butter substitute; 'olive oil' is the oil from olives and, if obtained from elsewhere, cannot be given the same name. Hence a cocoa butter or olive-oil substitute would probably command only about half of the price of the original material. Substitutions are, however, quite permissible where a particular oil is not specified as in margarines, cooking oils and detergents. Here, therefore, a microbial oil or fat would be expected to be on an acceptable equal price basis as its plant or animal counterpart.

III. CHOICE OF ORGANISM

The criteria to be used in the selection of a micro-organism for oil or fat production are similar to those which must be satisfied for any industrial fermentation process. Some of the more self-evident points

Table 4

Desirable properties of fat or fatty acids for production from micro-organisms

1. PRODUCT

General
High content within the micro-organism, at about 40% (w/w).
High commercial value.
Acceptable as a substitute for an existing plant or animal oil or fat.
Extracted cell residue saleable as animal fodder.

Edible oil or fat
Non-toxic.
Readily digested.
No unusual taste which persists after de-odourization.
Should consist mainly (>90%) of triglycerides.
Low sterol (<1%) content.

Non-edible fatty acids
No materials present likely to interfere with their intended use.
All fatty acids to be usable, preferably without fractionation.

Desirable properties of fat or fatty acids for production from micro-organisms

2. PROCESS

Preferably a single-stage continuous process.
High productivity output from the fermenter.
Constant composition of fat and fatty acids in cells.
Fat easily extracted from harvested cells.
Efficient utilization of carbon source.
Ability to use a variety of cheap substrates without loss of quantity or quality of fat.
Organism or growth medium able to withstand chance bacterial contamination.

are listed in Table 4. Clearly, fat of the highest quality is needed in greatest yield in the shortest possible time. Undesirable materials should be absent or at as low a concentration as possible in keeping with the applications to which the microbial oil is to be put. For an edible oil, the content of total sterols should be as low as possible, the usual content of sterols in most vegetable and animal oils and fats being below 0.5% (Swern, 1964).

Table 5

Yeasts and moulds examined since 1959 as potential fat producers or having been reported as having a minimum fat content of about 30%

Organism	Substrate	Fat content (% dry wt)	Fat co-efficient (g fat produced/ 100 g substrate utilized)	Lipid/fatty acid analysis given	Reference
YEASTS					
Candida guilliermondii	*n*-alkanes	30	—	no	Patent (1972)
Candida intermedia	*n*-alkanes	20	—	no	Patent (1969)
Candida tropicalis	*n*-alkanes	32	—	no	Patent (1969)
Candida sp. no. 107	glucose	42	22.5	yes	Thorpe and Ratledge (1972); Gill (1973)
Candida sp. no. 107	*n*-alkanes	15–37	25	yes	Ratledge (1968a); Thorpe and Ratledge (1972)
Crytococcus terricolus	glucose	55–65	21	no	Pedersen (1962a)
Hansenula anomala	glucose	17	—	no	Hopton and Woodbine (1960)
Hansenula ciferrii	molasses	22	—	no	Hopton and Woodbine (1960)
Hansenula saturnus	molasses	20	—	no	Hopton and Woodbine (1960)
Hansenula saturnus	glucose	28	8	no	Fahmy *et al.* (1962)
Lipomyces lipofer	glucose	38	—	yes	McElroy and Stewart (1967)
Lipomyces lipofer	peat moss hydrolysate	48	—	yes	Zalashko *et al.* (1972, 1973)
Lipomyces sp.	glucose	67	20	no	Watanabe (1974a)
	xylose	48	17	no	Watanabe (1974a)
	various wastes and molasses	66	up to 24	no	Watanabe (1974b)

Lipomyces starkeyi	lactose	31	10	no	Cullimore and Woodbine (1961)
Lipomyces starkeyi	glucose	31–38	9–15	yes	Suzuki and Hasegawa (1974a, b)
Rhodotorula gracilis	molasses	40	44 (short time only)	yes	Allen *et al.* (1964)
Rhodotorula gracilis	glucose	64	—	no	Enebo and Iwamoto (1966)
Rhodotorula gracilis	sugar-cane syrup	67	21	no	Protiva and Negrete (1968)
Rhodotorula gracilis	glucose	64	15 overall 44 (short time only	yes	Kessell (1968)
Rhodotorula gracilis	ethanol	62	15	yes	
	synthetic ethanol	60	14	yes	Krumphanzl *et al.* (1973)
	glucose	66	17	yes	
	alkanes	32	—	yes	

MOULDS

Aspergillus fischeri	sucrose	32–53	12–20	no	Gad and Hassan (1962)
Aspergillus fumigatus	maltose and other sources	20	—	no	Osman *et al.* (1969)
Aspergillus nidulans	glucose	27	9	no	Garrido and Walker (1959)
Aspergillus nidulans	glucose	15	7	no	Naguib and Saddik (1973)
Aspergillus ochraceus	sucrose	48	13	yes	Singh and Sood (1972) Sood and Singh (1973)
Aspergillus terreus	sucrose	51–57	10–13	yes	Singh and Sood (1972, 1973)
Aspergillus terreus	starch	18–24	6	no	Naguib and Yassa (1973a, b, 1974)
Aspergillus ustus	lactose	36	12.7	no	Wix and Woodbine (1959a, b)
Chaetomium globosum	glucose	54	—	yes	Mumma *et al.* (1970, 1971)
Cladosporium fulvum	sucrose	22–14	7	no	Singh and Sood (1972)
Cladosporium herbarum	sucrose	20–29	7–11	no	Singh and Sood (1972)
Gibberella fujikuroi (*Fusarium moniliforme*)	glucose	45	7.8	no	Borrow *et al.* (1961)

Table 5—*continued*

Organism	Substrate	Fat content (% dry wt)	Fat co-efficient (g fat/ 100 g substrate utilized)	Lipid/fatty acid analysis given	Reference
Malbranchea pulchella	glucose	27	—	yes	Mumma *et al.* (1970)
Mortierella vinacea	acetate	28	—	no	
	glucose	66	18	no	Chesters and Peberdy (1965)
	maltose	34	—	no	
Mucor miehei	glucose	24	—	yes	Sumner and Morgan (1969)
Mucor pusillus	glucose	26	—	yes	Sumner and Morgan (1969)
Myrothecium sp.	not given	30	—	yes	Pavlovica *et al.* (1972)
Penicillium funiculosum	*n*-alkanes	22	—	yes	Ratledge (1968b)
Penicillium gladioli	sucrose	32	5.7	no	Singh and Sood (1972)
Penicillium javanicum	glucose	39	9	no	Garrido and Walker (1959)
Penicillium lilacinum	date extract	23	—	no	Naguib *et al.* (1973a)
Penicillium lilacinum	sucrose	35	25	no	Duncan (1973)
Penicillium soppi	*n*-alkanes sucrose	11–25	—	yes	Ratledge (1968b)
Penicillium soppi	molasses	19	—	no	Naguib *et al.* (1973b)
Penicillium spinulosum	molasses and sucrose	25–64	6–16	no	Khan and Walker (1961) Garrido and Walker (1959)
Pythium irregulare	glucose	30–42	—	yes	Bhatia *et al.* (1972)
Pythium ultimum	glucose	48	—	yes	Bowman and Mumma (1967)
Rhizopus arrhizus	glucose + maltose	20	—	yes	Shaw (1966a)
Rhizopus sp.	glucose	27	—	yes	Sumner and Morgan (1969)
Stilbella thermophila	glucose	38	—	yes	Mumma *et al.* (1970, 1971)

From these criteria, bacteria and algae can probably be eliminated from serious consideration as, with the former, low contents of lipid tend to prevail or else the lipids may be toxic. With algae, growth is not only slow but is restricted to locations where there is an abundance of cheap land, warm water and sunshine. The amount of lipid in algae is unlikely to exceed 10% in practice (Hudson and Karis, 1974; Fogg and Collyer, 1955; Iwata, 1964; Priestley, 1976) although Spoehr and Milner (1949) demonstrated that, under artificial laboratory conditions, *Chorella pyrenoidosa* could contain up to 70% of its dry weight as lipid. This latter work was reviewed by Milner (1951).

Organisms which can be, and have been, used for fat production are yeasts and moulds. Each group has its own particular attributes thus making selection of the 'best' organism far from easy. A good deal of work has been carried out with many yeasts and moulds, particularly prior to 1960, and a detailed summary of fat yields by micro-organisms up to this time was given by Woodbine (1959). Table 5 lists those organisms which have been examined specifically for fat accumulation or have shown fat contents of over about 30% since this time. Not included in this table are organisms which would be difficult to grow in the laboratory, such as the plant-rust fungi shown by Tulloch and Ledingham (1962) to contain up to 37% lipid.

Of the various yeasts examined, the most promising are:

a. *Rhodotorula gracilis*, syn. *Rh. glutinis*, which has been extensively examined by Lundin and his coworkers (Enebo *et al.*, 1946; Lundin, 1950). Lipid contents up to 74%, with a fat co-efficient (i.e. g lipid/100 g substrate used) of 21 have been reported (Blinc and Hočevar, 1953; see also Bunker, 1963).

b. *Lipomyces lipofer* and *L. starkeyi*. Although Starkey (1946), in his original report of the isolation and characteristics of *L. starkeyi*, assessed its lipid content as 65% of the cell dry weight, more recent reports have indicated that lower concentrations of fat are likely to be attained (Suzuki and Hasegawa, 1974a, b; see Table 5). *Lipomyces lipofer*, although producing less lipid than *Rhodotorula gracilis*, has been examined by workers in the U.S.S.R. because of its ability to utilize a large number of very cheap raw materials, such as hydrolysed peat, and thus may have a useful advantage (Zalashko *et al.*, 1972, 1973; Obraztsova and Andreevskaya, 1973).

c. *Endomycopsis vernalis* syn. *Trichosporon pullulans* (not listed in

Table 6

Fatty-acyl compositions of lipids from fat-producing micro-organisms or related strains grown on carbohydrates

Organism	Substrate	Relative proportion of fatty acids (%, w/w)												Reference
		12:0	14:0	14:1	16:0	16:1	17:0 and 17:1	18:0	18:1	18:2	18:3	20:0 and 22:0	20+ un-saturated	
YEASTS														
Candida sp. no. 107 (batch culture)	glucose	—	1	—	22	2	—	8	31	26	—	3	7	Ratledge (1968a)
Candida sp. no. 107 (continuous culture)	glucose	—	1	—	37	1	—	14	36	8	—	—	4	Gill *et al.* (1977)
Hansenula anomola (triglyceride)	glucose	—	tr	—	20	2	1	2	49	26	—	—	—	Thorpe and Ratledge (1972)
Hansenula anomala	glucose	—	—	—	12	10	—	—	28	24	19	—	—	Bracco and Muller (1969)
Hansenula anomala	glucose	tr	3	tr	35	2	2	36	3	—	3	—	—	Maurice and Baraud (1967)
Lipomyces lipofer	glucose	—	tr	—	17	4	—	10	48	16	3	—	—	McElroy and Stewart (1967)
Lipomyces lipofer (triglyceride)	glucose	—	tr	—	12	3	—	6	77	3	tr	tr	—	Jack (1965)
Lipomyces lipofer (triglyceride)	glucose	—	2	—	16	7	—	3	62	9	1	tr	—	Haley and Jack (1974)
Lipomyces starkeyi (triglyceride)	glucose	—	—	—	40	6	—	5	44	4	—	—	—	Suzuki and Hasegawa (1974a)
Rhodotorula gracilis	glucose	tr	1	—	24	2	1	11	45	12	3	2	—	Enebo and Iwamoto (1966)
Rhodotorula gracilis	glucose	—	1	—	20	2	4	1	42	21	8	—	—	Kessell (1968)
Rhodotorula gracilis	glucose	—	1	—	31	—	—	9	53	1	5	—	—	Krumphanzl *et al.* (1973)

Organism	Substrate													Reference
Rhodotorula gracilis	ethanol	—	1	—	35	—	—	11	46	1	6	—	—	Krumphanzl *et al.* (1973)
Rhodotorula graminis	starch and glucose	1	4	1	32	tr	—	3	37	10	5	—	3	Hartman *et al.* (1959)
Rhodotorula glutinis	glucose	—	1	—	12	2	1	7	50	21	6	—	—	Thorpe and Ratledge (1972)
MOULDS														
Aspergillus nidulans	glucose	—	1	—	21	1	—	16	40	17	tr	1	2	Shimi *et al.* (1959)
Aspergillus niger	glucose	1	2	—	22	3	—	5	7	46	11	1	1	Salmonowicz and Niewiadomski (1965)
Aspergillus niger	lactose	3	5	—	50	2	2	10	11	14	2	—	—	Salmonowicz and Niewiadomski (1965)
Aspergillus ochraceus	sucrose	—	tr	—	38	—	—	tr	15	45	2	—	—	Sood and Singh (1973)
Aspergillus terreus	sucrose	tr	2	—	23	tr	—	tr	14	40	21	—	—	Singh and Sood (1973)
Chaetomium globosum	glucose	—	—	—	58	3	4	8	27	—	—	—	—	Mumma *et al.* (1970)
Fusarium moniliforme	glucose	—	1	1	14	—	—	11	30	42	1	—	—	Shaw (1965)
Malbranchea pulchella	glucose	—	—	—	11	—	11	27	51	—	—	—	—	Mumma *et al.* (1970)
Mucor globosus	glucose	2	8	—	26	8	7	26	8	16	—	—	—	Mumma *et al.* (1970)
Mucor ramannianus	glucose	—	2	—	19	3	—	4	28	14	31[a]	—	—	Sumner and Morgan (1969)
Penicillium chrysogenum	sucrose	tr	tr	—	18	1	tr	9	11	53	—	6	tr	Divikaran and Modak (1968)
Penicillium chrysogenum	glucose	—	—	—	13	1	—	5	14	61	7	—	—	Mumma *et al.* (1971)
Penicillium chrysogenum	glucose	tr	tr	—	24	3	tr	9	5	48	—	5	4	Divikaran and Modak (1968)
Penicillium lilacinum	glucose	—	tr	—	16	3	—	2	40	13	—	—	1	Shimi *et al.* (1959)
Penicillium soppi	glucose	—	1	—	20	2	1	5	9	50	13	tr	—	Salmonowicz and Niewiadomski (1965)
Penicillium soppi	lactose	2	2	—	41	1	tr	8	12	32	3	tr	—	Salmonowicz and Niewiadomski (1965)
Penicillium spinulosum	glucose	—	tr	—	18	4	—	12	43	21	tr	1	tr	Shimi *et al.* (1959)
Pythium irregulare	glucose	3	7	2	15	13	—	3	30	4	2[a]	13	8	Bhatia *et al.* (1972)
Pythium ultimum	glucose	—	8	—	23	9	—	7	22	15	2[a]	5	11	Bowman and Mumma (1967)
Rhizopus arrhizus	glucose	—	—	—	21	4	—	9	42	17	8[a]	—	—	Shaw (1966a)
Rhizopus sp.	glucose	—	1	—	21	2	—	5	30	29	12[a]	—	—	Sumner and Morgan (1969)
Stilbella thermophila	glucose	—	2	—	43	2	14	25	14	—	—	—	—	Mumma *et al.* (1970)

[a] γ-Linolenic acid (octadeca-6,9,12-trienoic acid).

Table 5) was originally studied by Lindner (1922) (see pg. 272) and, although it can attain 57% cell matter as lipid (Heide, 1939), it is uneconomical in terms of sugar utilization.

d. *Candida* sp. no. 107, although not producing the very high fat/lipid concentrations of other yeasts, is able to utilize a range of carbon compounds from *n*-alkanes (Ratledge, 1968a) to lactose and ethanol.

e. *Cryptococcus terricolus*, syn. *C. albidus* var. *albidus* is probably the most exceptional of all organisms, yeasts as well as moulds, for fat production. It produces a high concentration of fat (55 to 68%) irrespective of the growth rate and growth conditions (Pedersen, 1961, 1962a, b) and thus is completely different from all other organisms which must have an extended time of culture to ensure maximum fat formation (see Section IV, pg. 279). It has a very high efficiency of glucose utilization (Pedersen, 1961). It is somewhat surprising that no work beyond that of Pedersen, who isolated this yeast (Pedersen, 1958), has been carried out.

Of the moulds examined for fat production, the highest producers are: *Mortierella vinacea*, 66% lipid (Chesters and Peberdy, 1965); *Mucor circinelloides*, 65% lipid (Bernhauer and Rauch, 1948a, b; Bernhauer *et al.*, 1948), *Aspergillus terreus*, 57% lipid (Singh and Sood, 1972) and *Penicillium lilacinum*, 56% lipid (Philip, 1957). The last organism has also been accredited with a fat co-efficient of 24.95 which is the highest efficiency of conversion of carbohydrate to fat yet recorded for any micro-organism (Duncan, 1973). This, however, should be treated with caution until verified independently, as fat co-efficients for moulds rarely exceed 15 (see Woodbine, 1959) although, with yeasts, values over 20 are often attained (see Table 5, pg. 272).

It should be emphasized that the fat contents given for the various organisms listed in Table 5 (pg. 272) may not be the maxima the organism are capable of producing, for, as will be discussed in Section V.C (pg. 288), the fat content of an organism can vary considerably according to the growth conditions. Furthermore, the quantity of fat in a micro-organism, although important, is not the sole consideration in the selection of an organism for fat production. The quality of the fat (i.e. the nature of its fatty-acyl residues and its triglyceride content), the effeciency of its formation from the substrate used and the rate at which it can be produced are probably equally, if not more, important factors. These points would thus militate against moulds being superior to yeasts as they are slower growing than yeasts and do not, for the most part, have the same efficiency of fat production as yeasts. However, they can produce oils containing higher concentrations of polyunsaturated fatty acids than yeasts (see Table 6) and this makes them still worthy of serious consideration.

IV. CHOICE OF PROCESS

A. Course of Lipid Formation

Although lipids are synthesized throughout growth, accumulation of fat in the most organisms does not begin until a nutrient, other than the carbon source, has become exhausted from the medium. Thus, a fat-storage phase occurs at the end of growth and this is illustrated in Fig. 1 with respect to *Rhodotorula gracilis* as described by Kessell (1968). Similar observations for this organism had been previously made by Enebo *et al.* (1946), Lundin (1950), Tornqvist and Lundin (1951) and Rehaček and Beran (1964) amongst others. There are many similar reports for other yeasts as well as for moulds having to be incubated for an extended period of growth to allow maximum fat accumulation to occur (for example, see Garrido and Walker, 1959).

The nutrient that usually is allowed to become exhausted from the medium is that containing nitrogen, and there are numerous publications describing the most conducive C : N ratios to use to give maximum fat production (see Woodbine, 1959, and Table 5 for other references). With some organisms, fat synthesis can be stimulated by an exhaustion from the medium of a nutrient other than that containing nitrogen. Depletion of phosphate, sulphate and iron have been shown to induce fat formation in *Rhodotorula gracilis* (Nielsen and Nilsson, 1953; Maas-Forster, 1955; Neilsen and Rojowski, 1950). However, not all organisms will accumulate fat if a nutrient other than the nitrogenous one becomes exhausted. Fahmy *et al.* (1962) found that, with *Hansenula saturnus* when phosphate alone was made the limiting nutrient, there was no build-up of fat. Thus, the optimum balance of nutrients for maximum fat formation must be determined for each individual organism.

An exception to the necessity for there to be a fat-storage phase following growth has been noted by Pedersen (1961, 1962a, b) with *Cryptococcus terricolus* and to a lesser extent by Shaw (1966a) for *Rhizopus arrhizus*. Both organisms produced fat throughout the growth phase, and at all rates of growth, so that at all stages of harvesting the cells contained near to their maximum amount of fat (about 65% and 20%, respectively).

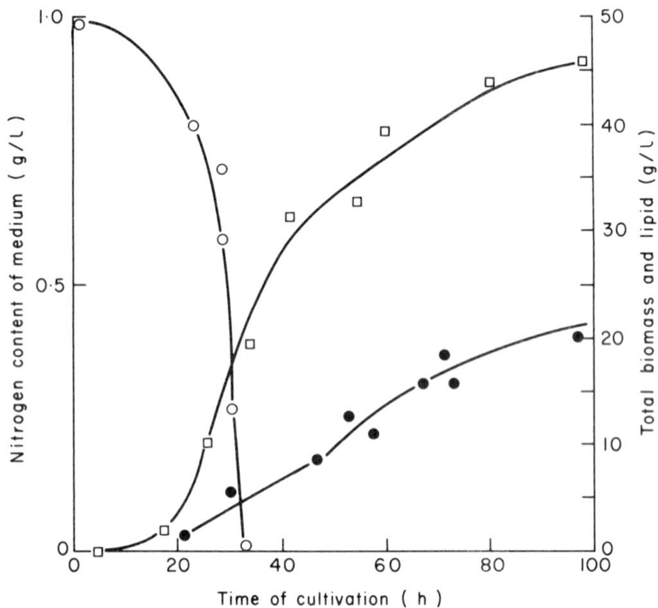

Fig. 1. Relationships between nitrogen content of medium (○) and growth (□) and lipid content (●) of *Rhodotorula gracilis*. Cultures were grown in 5-litre baffled fermenters in a medium formulated by Lundin (1950) containing glucose (100 g/l) and ammonium sulphate (0.6 g/l) and held at pH 6.0. From Kessell (1968) with kind permission of the author.

It should be noted, however, that the growth rates of both these organisms were slow. With the former organism there seems to be a metabolic inability to synthesize amino acids quickly enough and the organism thus grows with a permanent metabolic imbalance (Pedersen, 1962a). Attempts to achieve more rapid growth of this yeast in the author's laboratory have failed, indicating that it probably has little commercial potential although it does suggest that other yeasts might be discovered which would still have this property but which had a higher growth rate.

A means of increasing the lipid content of yeasts (*Saccharomyces cerevisiae*) has been suggested through the work of Challinor and Daniels (1955) who found that inositol deficiency led to the yeast having a fragile cell envelope enabling its fat content to be easily extracted. Other workers discovered that inositol deficiency led to substantial relative increases in the total lipid of the cell, though levels ultimately reached only 13% of the biomass (Lewin, 1965; Shafai and Lewin, 1968; Paltauf and Johnson, 1970; Hayaishi *et al.*, 1976). The applicability of this method does not seem to have been extended to recognized fat-accumulating micro-organisms to see if their fat contents could be raised still further. However, inositol deficiency may be difficult to achieve in large-scale fermentations where crude substrates (e.g. molasses) would be employed. Furthermore, deprivation of this vitamin, if overdone, could lead to a decrease in the yeasts's growth rate and in its ultimate yield. The main point perhaps to be concluded from these studies is that the lipid content of a micro-organism can be raised by appropriate manipulation, and even those yeasts which are not normally considered as high-fat producers (see Rattray *et al.*, 1975) might be exploitable if they were containing high-value fatty acids.

B. Fermentation Systems

If an organism is to be used for fat production which requires a fat-storage stage following the initial phase of growth, the most obvious choice of process would be a batch-culture method. Indeed, almost all work carried out so far on fat formation in micro-organisms has used either static, shake-flask or stirred-fermenter methods of cultivation as batch processes. Although it is obviously unrealistic to consider static cultures as means of commercial production of fat, nevertheless the use of static cultures has been advocated, particularly with moulds, as there have been various reports that agitation leads to an appreciable lowering of the fat content (Fink *et al.*, 1937; Witter and Stotz, 1946; Woodbine *et al.*, 1951; Naguib *et al.*, 1973b). These results, however, are probably explained by the fact that, in aerated or shaken cultures, the moulds grow initially more quickly than in static cultures, the pH value drops rapidly, and fat synthesis in the later stages of growth is therefore impaired due to the unfavourably low pH value. Cultivation of the organisms in submerged culture under properly controlled conditions, i.e. with the pH value and the rate of aeration controlled, should lead to a more rapid attainment of a high fat-production rate. Unfortunately, apart from the work of Borrow *et al.* (1961), there has been little work published on the formation of fat in moulds grown under such carefully controlled conditions. Clearly, such a study could be most rewarding. The time of cultivation in a batch fermentation process would seem, however, to be at least three to four days even with fast-growing yeasts under well-controlled conditions (Kessell, 1968).

Application of continuous-culture methods has recently been examined in this field. Krumphanzl *et al.* (1973), in a study of *Rhodotorula gracilis* growing in a medium containing glucose or ethanol, concluded, not surprisingly that, as the fermentation could be separated into two phases, the first cell propagation and the second fat accumulation, the culture could be adapted into a two-stage continuous process. They did not, however, establish such a system but indicated that it would probably work from their studies with a semicontinuous method of cultivation. The use of continuous culture as a means of accumulating fat in a species of *Candida* was, however, achieved in the author's laboratory by Gill (1973) (see also Gill *et al.*, 1977). Fat accounting for about 40% cell dry weight was attained in a single-stage process even though a fat-storage phase was needed. The amount of lipid produced was dependent upon the dilution rate (see Fig. 2) with the highest concentration being produced at a dilution rate of about 35% of the maximum dilution rate. Thus, even in a single-stage system, there is still time for fat accumulation to occur and the medium needed to induce fat formation is no different from that needed in batch culture (see Ratledge, 1968a). When the same yeast was grown in a two-stage continuous-culture system there was no improvement in either the rate of lipid formation or in the amount which could be induced to accumulate (Hall and Ratledge, 1977). Although this system offered some advantages for studying the effect of various growth conditions on lipid formation, in practical terms it had no advantage over the simpler one-stage continuous process.

In a recent patent (Patent, 1972) *Candida tropicalis, C. intermedia* and *C. lipolytica* were grown in one- and two-stage continuous units on *n*-alkanes as carbon source. In the two-stage process, all three yeasts produced about 22% lipid but, surprisingly, this was bettered by *C. tropicalis* when grown in the single-stage process by giving up to 32% lipid. The dilution rates were not, however, specified for either system.

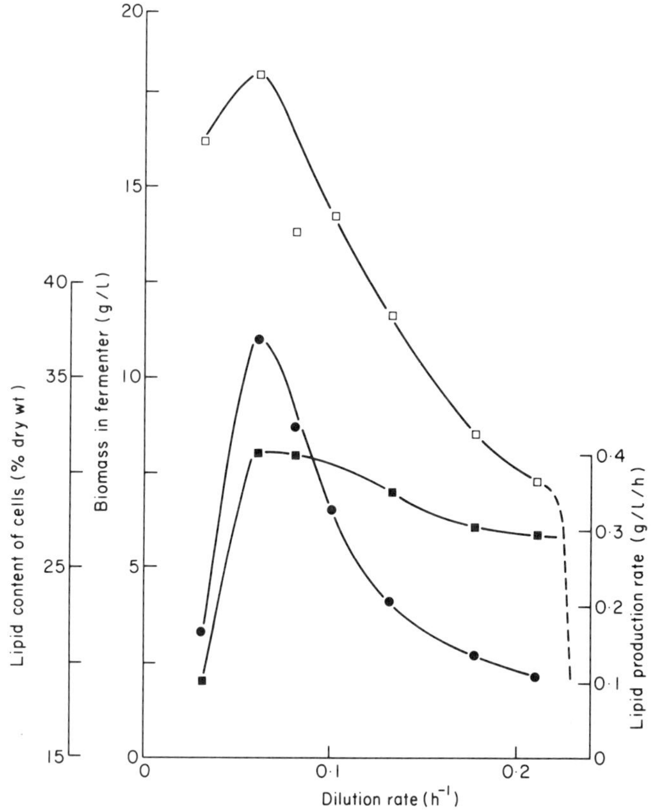

Fig. 2. Fat production in *Candida* sp. no. 107 growing in single-stage continuous culture with glucose as the sole source of carbon and with nitrogen as the limiting nutrient. The medium was that of Ratledge (1968b) at pH 5.5 containing glucose (30 g/l) and ammonium chloride (1.5 g/l). Each sample was taken after the culture had been held at a steady state for at least 48 h. At a dilution rate of $0.06 \, h^{-1}$, all glucose was consumed giving an economic coefficient of 54 and a fat coefficient of 22.5. ● indicates content of cells, ■ lipid production rate, and □ total biomass in the fermenter. From Gill *et al.* (1977).

It is apparent, therefore, that future developments in fat production should centre around exploitation of the continuous method of microbial cultivation. It will be noted from the details given in Fig. 2 that the fat coefficient (g fat/100 g substrate used) and also the economic coefficient (g biomass/100 g substrate used) were slightly higher than most reports given for batch cultivation (Table 5, pg. 272). A more economic utilization of the substrate is therefore occurring during continuous cultivation than during batch culture.

V. NATURE OF LIPID PRODUCED

A. Analysis of Total Lipid

Analysis of lipid constituents has now reached a highly refined state, and a complete description of all of the lipid components produced by micro-organisms has been carried out in a large number of cases. Unfortunately, however, only a few of the potentially useful fat-producing organisms have been completely examined. For general reviews on microbial lipids, those by Brennan *et al.* (1974) and Weete (1974) on lipids of fungi, and those by Hunter and Rose (1971) and Rattray *et al.* (1975) on the lipids of yeasts should be consulted. The monograph by Gurr and James (1975) can be recommended for a comprehensive and very readable account of lipid biochemistry in general.

With respect to fat-producing yeasts, the lipids of *Candida* sp. no. 107 (Thorpe and Ratledge, 1972; Gill *et al.*, 1977), *Cryptococcus terricolus* (Pedersen, 1962c), *Lipomyces lipofer* (McElroy and Stewart, 1967; Jack, 1965, 1966; Haley and Jack, 1974), *L. starkeyi* (Suzuki and Hasegawa, 1974; Uzuka *et al.*, 1975), *Hansenula anomala* (Thorpe and Ratledge, 1972), *Rhodotorula glutinis* (Kates and Baxter, 1962; Thorpe and Ratledge, 1972) have been examined in some detail. In all cases, except the last, the triglycerides represented at least 70%, and in most cases were over 80%, of the total lipid extracted; that from *C. terricolus* indeed contained over 90% as triglycerides. The remaining material is a mixture of sterols (usually ergosterol and sterol esters) with polar lipids usually at 5 to 8%. Some glycolipids may also occur (Gill *et al.*, 1976) and diol esters have also been recognized in *Lipomyces starkeyi* (Bergelson *et al.*, 1966), but these are usually minor components.

Yeasts grown on *n*-alkanes in general appear to contain higher concentrations of phospholipids than when grown on glucose, possibly to maintain the integrity of the membrane which becomes convoluted and invaginated (Thorpe and Ratledge, 1972). The amount of phospholipid formed, however, can be modulated, at least in *Candida tropicalis*, *C. lipolytica* and *C. guilliermondii*, by growth temperature, pH value of the medium and choice of nitrogen source (Greshnykh *et al.*, 1968; Dyatlovitskaya *et al.*, 1968).

With fat-producing moulds, there is even less information available than with yeasts. The data which are available from other species are, however, somewhat confusing and may not be too much of a guide. Merdinger and Devine (1965) found with *Derbaryomyces hansenii*, which produced only 11% of its cell dry matter as lipid, that 67% of this was neutral lipid but triglycerides were only about 20% of this fraction. The majority of the remainder was a mixture of free fatty acids and naturally occurring hydrocarbons. Jack (1964), however, noted that, with *Glomerella cingulata*, triglycerides accounted for more than 95% of its neutral lipid although Safe and Duncan (1974) and Weete *et al.* (1970) found only 22% triglycerides in the total lipid extracted, respectively, from *Mucor rouxii* and *Rhizopus arrhizus*. Osman *et al.* (1969) found that, in *Aspergillus fumigatus*, non-saponifiable lipids consistently comprised at least 50% of the total lipid under a variety of growth conditions whereas Bhatia *et al.* (1972), in a more detailed study of the lipids of fungi, found that, in seven moulds examined, the triglycerides rarely exceeded 50% of the total lipid extracted. Mumma *et al.* (1971) also examined a number of moulds with a range of lipid contents and found that the neutral-lipid content of the total lipid could range from as low as 33% with *Humicula insolens* to about 80% with *Malbranchea pulchella* and *Mucor globosus*. The lipids of *Penicillium chrysogenum* contained 56% neutral fat. Thus, there seems to be a bewildering array of values for the triglyceride content of the fat or oil from a mould. Each organism of potential interest in this field will have to be examined individually as no generalizations can be drawn from the available information.

The nature of the extra lipid which is accumulated in a fat-producing organism has not been extensively examined. Bowman and Mumma (1967) found with *Pythium ultimum* that there was little change in the content of free fatty acids and triglyceride during the fat-storage phase of growth; indeed the percentage of triglycerides in the total lipid tended to decrease. Suzuki and Hasegawa (1974a) also found that the proportion of triglyceride in low-fat cells of

Lipomyces starkeyi was higher than that in high-fat cells. Gill *et al.* (1977), too, found that the proportions of triglyceride in *Candida* sp. no. 107 did not show much variation as the lipid content of the cells varied. Thus, on the basis of these few results, it would seem that fat accumulation is not solely biosynthesis of triglycerides but represents an all-round increase in all of the various lipid components of the cell.

The nature of the triglycerides, i.e. the fatty-acyl distribution on the glycerol moiety, has been examined in a few micro-organisms though this has been investigated extensively for plant and animal triglycerides. Harries and Ratledge (1969) were the first to carry out such an examination with a microbial triglyceride, and their work with *Candida* sp. no. 107 was subsequently extended by Thorpe and Ratledge (1972) who found that, in the same yeast as well as in *Candida tropicalis*, *Candida lipolytica*, *Hansenula anomala* and *Rhodotorula graminis*, each grown on glucose, a saturated acid was rarely introduced into the 2-position of the glycerol. This type of triglyceride is typical of plant triglycerides but, in animal fats, a saturated acid residue, usually palmitic acid or stearic acid, is often present at C-2. Suzuki and Hasegawa (1974a) with *Lipolyces starkeyi* and Haley and Jack (1974) with *Lipomyces lipofer* have made similar observations. The commonest triglyceride species are of the type 1-oleyl (or palmityl)-2-oleyl-3-palmityl (or oleyl)-glycerol. In yeasts, such as *Candida tropicalis* and *Candida* sp. no. 107 grown on alkanes, myristic acid and pentadecanoic acid, but not palmitic acid, are present in the 2-position of glycerol, and this considerably widens the range of triglyceride species which can be recovered from micro-organisms (Thorpe and Ratledge, 1972).

The toxicity or the presence of toxic factors in microbial fats and oils has not been examined extensively by modern methods of evaluation. However, earlier trials of feeding experimental animals with microbial fat (see Woodbine, 1959) and a detailed examination of yeast fat for the presence of potentially deleterious ω-1 and ω-2 methyl-branched fatty acids (see Kessell, 1968; Harries and Ratledge, 1969) have not shown any reason why microbial fats could not be used in human foods. Stringent testing will obviously be needed in all cases before such an event could occur.

B. Fatty Acids of Fat-Accumulating Micro-Organisms

A great deal of information is now known about microbial fatty acids, and detailed lists of fatty acids of a wide range of yeasts, moulds and bacteria are provided in the reviews of Shaw (1966b), Erwin (1973), O'Leary (1973) and Weete (1974). The reviews of Hunter and Rose (1971), Erwin (1973) and Whitworth and Ratledge (1974) give information regarding fatty-acid biosynthesis in yeasts and moulds. For information regarding the control of fatty-acid biosynthesis, the references quoted in Gill and Ratledge (1973) and Whitworth and Ratledge (1975) may be found useful.

Table 6 lists the fatty-acid compositions of various fat-producing yeasts and moulds and related species grown on carbohydrate materials as principal carbon sources. It should be appreciated that many micro-organisms containing higher levels of the more expensive fatty acids could probably be induced into a fat-accumulating phase and thus may be more commercially attractive than the ones listed here. For example, Tyrrell (1967) found that several species of *Entomophthora* contained very high relative proportions of C_{10} to C_{14} fatty acids. In some cases up to 55% of the total fatty acids were composed of capric, lauric and myristic acids. More recently, Tyrrell and Weatherston (1976) reported that some species of *Conidiobolus* contained up to 50% myristic acid whilst other species had up to 20% as lauric acid. Safe and Duncan (1974) succeeded in producing 80% of the total fatty acids of *Mucor rouxii* as capric, lauric and myristic acids during anaerobic growth and over 60% as these acids under aerobic conditions, thus showing that there is no biochemical reason why such acids cannot be accumulated in relatively high proportions. Gordon *et al.* (1971) similarly found that anaerobic growth led to increases in the short-chain fatty acids in *Mucor genevensis*. Whether these or related organisms could be exploited for production of specific types of oils would clearly depend upon the ability to find or produce oleaginous strains which still retained the original fatty acid spectrum.

Organisms which accumulate large amounts of polyunsaturated acids (linoleic acid, $C_{18:2}$ or linolenic acids, $C_{18:3}$) are more prevalent than the shorter chain fatty-acid producers. The best organisms in this respect are the moulds rather than the yeasts. In particular, species of *Mucor* seem to be the highest producers of linolenic acid;

Mucor genevensis, under aerobic conditions, produced 33% linolenic acid in its total fatty acids (Gordon *et al.*, 1971) and *Mucor ramannianus* produced 31% (Sumner and Morgan, 1969). These observations are of particular interest as, in both cases, the acid was the γ-isomer, i.e. it was the dietary essential fatty acid rather than the commoner α-linolenic acid which is found in plants and other micro-organisms. This ability to produce γ-linolenic acid is however restricted to members of the Phycomycetaceae (Shaw, 1965, 1966a, b; Erwin, 1973) but, if these organisms, particularly members of the order of the Mucorales, which produce only linoleic and γ-linolenic acids as their unsaturated fatty acids, could be induced into a high-fat state, they may be worthy of further attention because of the potential value of γ-linolenic acid in the human diet.

High concentrations of linoleic acid seem to occur most prominently amongst species of aspergilli and penicillia (Table 6, pg. 276). As high fat-yielding representatives of species in both of these genera are known (Table 5, pg. 272), further work with these to maximize the content of the polyunsaturated acids may again be worthwhile.

C. Changes in Fatty-Acid Composition

a. *Influence of carbon source.* The greatest changes which can be brought about in the fatty-acyl compositions of micro-organisms are obtained by using *n*-alkanes as carbon source instead of carbohydrates. Changes in fatty-acid composition in going from one carbohydrate source to another (see Salmonowicz and Niewiadomski, 1965; Divakaran and Modak, 1968) or from glucose to such substrates as ethanol or acetic acid (see Krumphanzl *et al.*, 1973; Cerniglia and Perry, 1974), are much less dramatic. In general, those micro-organisms (bacteria, yeasts or moulds) which are capable of growth on *n*-alkanes oxidize the alkane to the corresponding fatty acid and then incorporate it, directly or after some modification (elongation or desaturation), into the lipids of the cell. In this way, myristic acid comprised 45% of the total fatty acids of the triglycerides from *Candida* sp. no. 107 following growth on tetradecane (Thorpe and Ratledge, 1972). Instances of a high accumulation of lauric acid, even after growth on *n*-dodecane as the sole carbon source, are still quite rare however. Alterations in the fatty-acid

composition of a variety of bacteria and yeasts after growth on
n-alkanes have been reviewed by Ratledge (1970) and by Bird and
Molton (1973).

There have only been a few attempts to cultivate fat-accumulating
organisms on n-alkanes. Moulds seem unsuitable (Ratledge, 1968b,
1970) but yeasts such as *Candida* sp. no. 107 and *Rhodotorula
gracilis* have been cultivated with some success (Ratledge, 1968a;
Pelechova *et al.*, 1971; Krumphanzl *et al.*, 1973). Pure n-alkanes
are, of course, too expensive to use in an industrial process and
mixtures (i.e. fractions or 'cuts') of n-alkanes must be employed.
The range of microbial fatty acids synthesized by using such a
mixture can be extensive (see Ratledge, 1968a; Pelechova *et al.*,
1971; Thorpe and Ratledge, 1973; Hornei *et al.*, 1972; Volfova and
Pecka, 1973; Alentyeva *et al.*, 1974) and may, therefore, be unsuitable for direct industrial use without some fractionation after isolation.
Yields of fatty acids representing 35% of the cell dry weight,
corresponding to a conversion of n-alkanes of 25%, have been
obtained with *Candida* sp. no. 107 and *Rhodotorula gracilis* growing
on alkane mixtures (Ratledge, 1968a; Pelechova *et al.*, 1971).

b. *Influence of environmental conditions.* Kates (1966), Hunter
and Rose (1971), Erwin (1973) and Whitworth and Ratledge (1974)
have reviewed the alterations which can be brought about in the
fatty-acyl composition of micro-organisms by changes in growth
conditions. Many variations in media composition, including carbon
and nitrogen sources as well as other nutrients, have been imposed by
workers using fat-accumulating micro-organisms (see Woodbine,
1959; Wix and Woodbine, 1959a, b; Fahmy *et al.*, 1962; Gad and
Hassan, 1962; Allen *et al.*, 1964; Rehaček and Beran, 1964; Bhatia *et
al.*, 1972; Naquib and Saddik, 1973). No generalizations can, however, be drawn. Each organism studied has its own particular nutrient
requirements for induction of maximum formation of fat (besides a
high carbon-to-nitrogen ratio in the nutrients provided). Unfortunately, very few of these investigations, an exception being that of
Bhatia *et al.* (1972), have included an analysis of the lipid and/or
fatty acids produced.

Investigations of the changes in the lipid content of *Rhodotorula
gracilis* grown at various temperatures and at various pH values were
first carried out by Steinberg and Ordal (1954) and later, in greater

detail involving analysis of the various fatty-acid compositions, by Enebo and Iwamoto (1966) and Kessell (1968). All of these experiments were carried out in batch culture, but only Kessell (1968) consistently used a fermenter with abilities to control all of the relevant parameters during growth, and it is this type of approach which is essential to interpret the changes brought about by alterations in growth conditions. The study by Kessell (1968), however, revealed a serious obstacle for anyone attempting to examine such changes in fatty-acyl composition, and this was the degree to which the fatty-acyl composition changed during the course of a single experimental run. For example, the proportion of oleic acid (usually the most abundant single acid residue) in the fatty acids extracted from the yeast at six successive times of sampling between 25 and 71 h cultivation was 21%, 38%, 54%, 41%, 59% and 46%. Other acids, principally linoleic and palmitic acids, varied likewise. Thus, it was extremely difficult to judge what, if any, effect a change in the pH value of the medium was having on the fatty acid composition unless one coud be absolutely certain that samples were taken from cultures at equivalent ages. 'Age of culture' is not, of course, the same as 'time of culture' because of idiosyncratic variations between experimental run. Similar variations have been found for the fatty acids of *Lipomyces starkeyi* (Suzuki and Hasegawa, 1974a), *Penicillium atrovenetum* (Van Etten and Gottlieb, 1965) and *Candida utilis* (Dawson and Craig, 1966; McMurrough and Rose, 1971, 1973) during growth, although the latter two organisms are not recognized as fat-accumulating species.

To overcome these variations, continuous-culture techniques must be used so that cells are generated under steady-state conditions. Here, the fatty-acyl composition does not vary by more than 1 or 2% (i.e. experimental error) provided all of the environmental conditions including the growth rate are kept constant. Unfortunately, however, all studies carried out with the lipids of organisms grown in continuous culture have, except for the most recent ones of Gill *et al.* (1977) and Hall and Ratledge (1977), been confined to non-fat accumulating organisms, and extrapolations of these results to fat-accumulating species may not always be possible. Dawson and Craig (1966) were the first workers to show that, with a yeast (*Candida utilis*), different growth rates produced different relative proportions of fatty acids. Gill *et al.* (1977), working with *Candida* sp. no. 107, found that the greatest variations in fatty-acyl composition at different growth rates occurred with carbon-limited cultures. At the slowest growth rates, up to 48% linolenic acid was found in the total acids extracted. In nitrogen-limited cultures, wherein fat accumulation was highest, the variations were mainly confined to reciprocal changes in the proportions of palmitic and oleic acids.

Effects of growth temperature on fatty-acyl composition are not usually predictable for any particular organism. Brown and Rose (1969), working with *Candida utilis*, showed that acids became more unsaturated with a decrease in the temperature even at the same growth rates, but Hunter and Rose (1972) found that, in *Saccharomyces cerevisiae*, the relative proportions of unsaturated acids were unchanged as the growth temperature was lowered from 30° to 15°C. Changes in the proportion of one unsaturated acid are often compensated by a change in another unsaturated acid. For example, an increase in the content of linolenic acid is often accompanied by a decrease in oleic acid and/or palmitoleic acid contents, and this has been noted in continuous as well as batch cultivation of various yeasts and moulds (Kates and Baxter, 1962; Salmonowicz and Niewiadomski, 1965; Enebo and Iwamoto, 1966; Summer *et al.*, 1969; and Zalashko *et al.*, 1972). It is, therefore, often difficult to detect if any real change in the physical characteristics of the acids has been occurring, particularly when my previous comment on reproducibility is borne in mind. For a fuller discussion of the effects of temperature on the physiology of micro-organisms, see Farrell and Rose, (1967) and also, more briefly, Hunter and Rose (1971, 1972), Thorpe and Ratledge (1973), and Rattray *et al.* (1975).

In *Candida utilis*, Brown and Rose (1969) found that a depletion of oxygen decreased the degree of fatty-acyl unsaturation more markedly than did changes in growth temperature, this being directly attributed to the necessity for oxygen in the desaturation of fatty acids (e.g. in going from stearic acid to oleic acid). The major changes in composition were brought about at very low dissolved oxygen tensions, and increasing the dissolved oxygen tension from 5 mm Hg to 75 mm Hg had virtually no effect, an observation which was corroborated by Babij *et al.* (1969) with the same organism, and by Brown and Johnson (1971) working with *Saccharomyces cerevisiae*.

An increased supply of glucose has been found to cause an increased accumulation of fatty acids in *Candida utilis* (Babij *et al.*,

1961) the major increase being in the oleic-acid content (Johnson *et al.*, 1972). In *Lipomyces starkeyi*, Suzuki and Hasegawa (1974b), however, found only moderate changes in the proportions of palmitic, oleic and linoleic acids when the cultures were transferred from a glucose-deficient state to a glucose-sufficient one. But Gill *et al.* (1976) found, with *Candida* sp. no. 107, that increasing the glucose concentration increased the proportion of oleic acid in the total acids from 26% to 47%; this was at the expense of a decline in linoleic acid concentration, from 42% to 11%.

Clearly, however, these studies can give only an indication of the extent to which the fatty-acyl composition of a micro-organism can be modulated by control of its growth conditions. Again, each organism of potential value must be examined through a range of environmental conditions to ascertain the most appropriate parameters to use to ensure production of maximum fat of the highest quality.

VI. PRODUCTION OF OTHER LIPIDS

The lipids of micro-organisms are most readily thought of as sources of plant-like triglycerides and fatty acids. However, there are a number of other fatty acids which can be derived from microbial sources. Short-chain dicarboxylic fatty acids, such as adipic and azelic acids, are relatively expensive commodities (about £0.5–£0.8/kg) and have been produced by growth of *Candida tropicalis* and *C. guilliermondii* on *n*-alkanes with odd and even numbers of carbon atoms (Okuhara *et al.*, 1971; Krauel *et al.*, 1973). Yields, however, were very low in both cases, namely about 4% (w/w) of alkane used. Longer chain dicarboxylic acids, such as hexadeca-1,16-dioic acid, have been reported to be produced in a 60% yield from hexadecane by a mutant of *Candida cloacae* (Uchio and Shiio, 1972a, b, c) which may seem attractive for commercial exploitation. Accumulation of sebacic acid (deca-1,10-dioic acid), up to 5.4 g/l, has also been found after growth of *Torulopsis* sp. on decane (Ogata *et al.*, 1973) and this could well be worth further attention as sebacic acid currently demands a price in excess of £1.7/kg. Hydroxy fatty acids (ω and ω-1) also seem to be readily produced especially following growth of yeasts on alkanes (Tulloch *et al.*, 1962; Jones and Howe, 1968). These acids are probably intermediates in the synthesis of the dicarboxylic acids. Uses for these acids are very limited, although Howe and Jones (1969) interesterified some of these hydroxy acids into high molecular-weight polymers but, unfortunately, the resulting material showed no unusual properties which warranted further commercial interest. The use of specially selected mutants can be an obvious way in which the yield of a product can be dramatically increased but, with the exception of the work with *Candida cloacae* already described, there has been little successful exploitation of this approach with microbial production of lipids and fatty acids (see Illarionov, 1972; Jenkins *et al.*, 1972; Macham and Heydeman, 1974).

Although waxes, such as cetyl palmitate, have been produced in some abundance by certain bacteria growing on hexadecane (see Klug and Markovetz, 1971), this type of wax is not in great commercial demand. Cetyl palmitate is the principal constituent of spermaceti which, when available, is used mainly as a cheap ointment base in the pharmaceutical industry. Synthetic cetyl palmitate, however, can be easily prepared and used in place of the dwindling supplies of spermaceti. For waxes to be more valuable commodities, such as a substitute for beeswax which can cost up to £2.8/kg, the chain lengths of both the alcohol and acid components need to be at least C_{26}. Such an elaboration seems unlikely to be attained, however, by micro-organisms.

Sterols occur in yeasts and moulds and, although they are expensive (about £2/kg), there has been no development towards their deliberate microbial production. Instead ergosterol, the commonest microbial sterol, is obtained from spent brewer's yeast wherein the concentration is unlikely to be more than 1% (w/w). Obviously, this method is cheaper than cultivating an organism specifically for extraction of ergosterol, as there are few examples of the sterol content of a micro-organism exceeding 10% of the cell dry weight (Dulaney *et al.*, 1954; Osman *et al.*, 1969).

Of the other fat-soluble vitamins and hormones, only β-carotene, as a colourant or dietary supplement for vitamin A, has been considered for commercial production (see Chapter 8, pg. 317) using mated strains of *Blakeslea trispora* (Isler *et al.*, 1970). The other fat-soluble vitamins and hormones are more easily obtained from plant, fish or mammalian sources, although some of them certainly could be produced, but at a price, from micro-organisms.

VII. ECONOMICS OF A PROCESS FOR MICROBIAL PRODUCTION OF FATS

Any process for the microbiological production of fats is unlikely to be markedly different from current processes used in the production of single-cell protein. The only additional step needed will be a fat-extraction. This is unlikely to give much difficulty on a large scale, as such methods are already in use to remove unwanted fat and unchanged hydrocarbons from yeast grown for protein on gas-oil and other petroleum fractions. Solvent extraction methods, which have been well described for small-scale fat extraction (Ratledge and Saxton, 1968; Hunter and Rose, 1971; Suzuki *et al.*, 1973), may be unsuitable for large-scale work. However, as the lipid within fat-accumulating cells often appears in the form of discrete droplets, extrusion methods similar to those currently used for obtaining oil from plant seeds may be applicable.

The production of fatty acids by micro-organisms is not a very efficient process. The percentage conversions of a carbohydrate, such as glucose or sucrose, to fat are rarely above 21–22%, so that to produce one kg of fat about 4.5 kg of substrate are needed. To the cost of substrate, we must, of course, add all of the fermentation costs plus those for extraction and for any processing that may be needed. If we take as an example molasses, which is a cheap form of readily available carbohydrate, this currently costs about £50/ton but has only a 50% content of fermentable materials. Thus, for $4\frac{1}{2}$ tons of fermentable carbohydrate, the basic cost would be about £450. Fermentation costs for converting the molasses into fat would probably be between £100 to £150/ton of micro-organism produced, judging from current costs for large-scale biomass production (Ratledge, 1975). If the micro-organism selected for this process had a fat content of 65%, fermentation costs would rise to at least £150 to £230/ton of fat produced. The overall costs would therefore be between £550 and £680/ton of fat. Extraction of the fat is not likely to be a costly process and would probably be more than offset by sale of the fat-extracted cell residue as cattle fodder. If a 15% margin for profit was required, the selling price of the fat could not be less than £630 and may be as high as £780/ton. These prices, although only rough estimates, indicate that if microbial oils and fats are to be produced the cost of the substrate must be substantially reduced. For a process to be economical, i.e. producing an oil for about £400–£450/ton, one can calculate that the substrate must not cost more than £30 to £35/ton. Of course, if a much more expensive product can be produced, such as one containing high levels of a particularly desirable fatty acid, then the cost of substrate will become of less importance.

Alkanes as a substrate, although having the attraction of 'tailor-making' the desired fatty acids, are likely to be just as expensive as molasses. Conversions of alkanes to fatty acids may be between 25 and 30% but, with the cost of *n*-alkanes at about £130/ton, this would still mean a cost of at least £430 for the necessary quantity of substrate to produce one ton of fat. As fermentation costs are likely to be as high as with molasses as feedstock, the use of alkanes is unlikely to be a realistic alternative.

Synthetic ethanol has been shown by Krumphanzl *et al.* (1973) to be a suitable substrate for fat formation by *Rhodotorula gracilis*, and this substrate can also be used by numerous other yeasts including most species of *Candida*. The cost of ethanol at about £0.08/kg is probably too high, however, to make it a practicable alternative to molasses.

Cheap substrates suitable for growth of fat-producing micro-organisms have been examined by several groups of workers. Lactose, as a byproduct from the recovery of whey protein using large-scale ultrafiltration, can be used by some fat-producing yeasts including *Lipomyces starkeyi* and moulds (Wix and Woodbine, 1959a, b; Cullimore and Woodbine, 1961). Starch is also a cheap material which can be used effectively by some yeasts and moulds (Hartman *et al.*, 1959; Naguib and Yassa, 1973a, b, 1974). More unusual materials examined recently have included a date extract available in Iraq (Naguib *et al.*, 1973a, c) and peat moss, pretreated with sulphuric acid, in the U.S.S.R. (Zalashko *et al.*, 1972, 1973; Obraztsova and Andreevskaya, 1973). This last material consists mainly of hexose and pentose monosaccharides (Bogdanovskaya *et al.*, 1973).

Two abundant and extremely cheap substrates which have not yet been found suitable for fat accumulation are methane and methanol. Their respective costs are about £0.04–£0.05/kg and £0.28–£0.33/kg. Whether this failure to convert these substrates into fats is only a temporary one because of the novelty of finding organisms which will readily utilize such substrates, or is because of a biochemical inability of the organisms concerned to use C_1 compounds and simultaneously store fat, is not yet known.

Whether any of these cheap raw materials could be useful for commercial exploitation cannot be stated because of the paucity of work carried out either with the substrates themselves or into the processes of fat accumulation. It is singularly surprising that very few of the established fat-producing micro-organisms have been examined in any degree of detail using modern methods of culture in controlled environments even using a substrate such as glucose. When this type of work has been accomplished, then perhaps we shall be in a position to decide if the microbiological production of lipids and fatty acids is, indeed, worthy of further industrial consideration.

REFERENCES

Alentyeva, E. S., Garbalinsky. V. A., Korobova, L. T. and Sokolva, Y. I. (1974). *Prikladnaya Biokhemia i Mikrobiologiya* **10**, 196.

Allen, L. A., Barnard, N. H., Fleming, M. and Hollis, B. (1964). *Journal of Applied Bacteriology* **27**, 27.

Babij, T., Moss, F. J. and Ralph, B. J. (1969). *Biotechnology and Bioengineering* **11**, 593.

Bergelson, L. D., Vaver, V. A., Prokozova, N. V., Ushakov, A. N. and Popkova, G. A. (1966). *Biochimica et Biophysica Acta* **116**, 511.

Bernhauer, K. and Rauch, J. (1948a). *Biochemische Zeitschrift* **319**, 77.

Bernhauer, K. and Rauch, J. (1948b). *Biochemische Zeitschrift* **319**, 102.

Bernhauer, K., Niethammer, A. and Rauch, J. (1948). *Biochemische Zeitschrift* **319**, 94.

Bhatia, I. S., Raheja, R. K. and Chahal, D. S. (1972). *Journal of the Science of Food and Agriculture* **23**, 1197.

Bird, C. W. and Molton, P. (1972). *In* 'Topics in Lipid Chemistry', (F. Gunstone, ed.), vol. 3, pg. 125. Paul Elek (Scientific Books) Ltd., London.

Blinc, M. and Hočevar, B. (1953). *Monatshefte fur Chemie* **84**, 1127.

Bogdanovskaya, A. N., Evdokimova, G. A., Gurinovich, E. S., Raitsina, G. I. and Kostyukevitch, L. I. (1973). *In* 'Mikroorganizmy—Produtsenty Biologitcheskoie Activnoie Veshchestvo' (S. A. Samtsevitch, ed.), pg. 77, 'Nauka i Tekhnika': Minsk, U.S.S.R. (Quoted from *Chemical Abstracts* (1974) **80**, 131655.)

Borrow, A., Jefferys, E. G., Kessell, R. H. J., Lloyd, E. C., Lloyd, P. B. and Nixon, I. S. (1961). *Canadian Journal of Microbiology* **7**, 227.

Bowman, R. D. and Mumma, R. O. (1967). *Biochimica et Biophysica Acta* **114**, 501.

Bracco, U. and Muller, H. R. (1969). *Revue Français des corps gras* **16**, 573.

Brennan, P. J., Griffin, P. F. S., Lösel, D. M. and Tyrell, D. (1974). *Progress in the Chemistry of Fats and Other Lipids* **14**, 49.

Brown, C. M. and Johnson, B. (1971). *Antonie van Leeuwenhoek* **37**, 477.

Brown, C. M. and Rose, A. H. (1969). *Journal of Bacteriology* **99**, 371.

Bunker, H. J. (1963). *In* 'Biochemistry of Industrial Micro-organisms', (C. Rainbow and A. H. Rose, eds.), pg. 34. Academic Press, London.

Cerniglia, C. E. and Perry, J. J. (1974). *Journal of Bacteriology* **118**, 844.

Challinor, S. W. and Daniels, N. W. R. (1955) *Nature, London* **176**, 1267.

Chesters, C. G. C. and Peberdy, J. F. (1965). *Journal of General Microbiology* **41**, 127.

Cullimore, D. R. and Woodbine, M. (1961). *Nature, London* **190**, 1022.

Damm, H. (1943). *Chemische Zeitschrift* **67**, 47.

Dawson, P. S. S. and Craig, B. M. (1966). *Canadian Journal of Microbiology* **12**, 775.

Divikaran, P. and Modak, M. J. (1968). *Experientia* **24**, 1102.

Dulaney, E. L., Stapley, E. O. and Simpf, K. (1954). *Applied Microbiology* **2**, 371.

Duncan, B. (1973). *Mycologia* **65**, 211.

Dyatlovitskaya, E. V., Greshnykh, K. P. and Bergelson, L. D. (1968). *Biokhimiya*, **33**, 83.

Enebo, L., Anderson, L. G. and Lundin, H. (1946). *Archives of Biochemistry* **11**, 383.

Enebo, L. and Iwamoto, H. (1966). *Acta Chemica Scandinavica* **20**, 439.

Erwin, J. (1973). *In* 'Lipids and Biomembranes of Eukaryotic Microorganisms', (J. Erwin, ed.), pg. 41. Academic Press, London.

Fahmy, T. K., Hopton, J. W. and Woodbine, M. (1962). *Journal of Applied Bacteriology* **25**, 202.

Farrell, J. and Rose, A. H. (1967). *In* 'Thermobiology', (A. H. Rose, ed.), pg. 147. Academic Press, London.

Fink, H., Haenseler, F. and Schmidt, M. (1937). *Zeitschrift fur Spiritusindustrie* **60**, 74.

Fogg, G. E. and Collyer, D. M. (1955). *Journal of Experimental Botany* **6**, 256.

Gad, A. M. and Hassan, M. M. (1962). *Egyptian Pharmaceutical Bulletin* **44**, 93.

Garrido, J. M. and Walker, T. K. (1959). *Journal of Applied Bacteriology* **21**, 291.

Gill, C. O. (1973). Ph.D. Thesis: University of Hull.

Gill, C. O., Hall, M. J. and Ratledge, C. (1977). *Applied and Environmental Microbiology* **33**, 23.

Gill, C. O. and Ratledge, C. (1973). *Journal of General Microbiology* **78**, 337.

Gordon, P. A., Stewart, P. R. and Clarke-Walker, G. D. (1971). *Journal of Bacteriology* **107**, 114.

Greshnykh, K. P., Grigorian, A. N., Dikanskaja, E. M., Diatlovitskaya, E. V. and Bergelson, L. D. (1968). *Mikrobiologiya* **37**, 251.

Gurr, M. and James, A. T. (1975). *In* 'Lipid biochemistry: An Introduction', (2nd edition). Chapman & Hall Ltd., London.

Haley, J. E. and Jack, R. C. M. (1974). *Lipids* **9**, 679.

Hall, M. J. and Ratledge, C. (1977). *Applied and Environmental Microbiology* **33**, 577.

Harries, P. C. and Ratledge, C. (1969). *Chemistry and Industry* 582.

Hartman, L., Hawke, J. C., Shorland, F. B. and di Menna, M. E. (1959). *Archives of Biochemistry and Biophysics* **81**, 346.

Hayaishi, E., Hasegawa, R. and Tomita, T. (1976). *Journal of Biological Chemistry* **251**, 5759.

Heide, S. (1939). *Archiv für Mikrobiologie* **10**, 135.

Hesse, A. (1949). *Advances in Enzymology* **9**, 653.

Hilditch, T. P. and Williams, P. N. (1964). 'The Chemical Constitution of Natural Fats', (4th ed.). Chapman & Hall, London.

Holz, A. E. (1974). *Foreign Agriculture* **12** (no. 33). pp. 4 and 20.

Hopton, J. W. and Woodbine, M. (1960). *Journal of Applied Bacteriology* **23**, 283.

Hornei, S., Köhler, M. and Weide, H. (1972). *Zeitschrift für Allgemeine Mikrobiologie* **12**, 19.

Howe, R. and Jones, D. F. (1969). *Chemistry and Industry* pg. 1181.

Hudson, B. J. F. and Karis, I. G. (1974). *Journal of the Science of Food and Agriculture* **25**, 759.

Hunter, K. and Rose, A. H. (1971). *In* 'The Yeasts', (A. H. Rose and J. S. Harrison, eds.), vol. 2, pg. 211. Academic Press, London.

Hunter, K. and Rose, A. H. (1972). *Biochimica et Biophysica Acta* **260**, 639.

Illarionov, E. F. (1972). *Mikrobiologiya* **41**, 126.

Isler, O., Solms, U. and Wursch, J. (1970). *In* 'Fat-Soluble Vitamins', (R. A. Morton, ed.), pg. 99. Pergamon Press, London.

Iwata, I. (1964). *Agricultural and Biological Chemistry* **28**, 610.

Jack, R. C. M. (1964). *Contributions from the Boyce Thompson Institute* **22**, 311.

Jack, R. C. M. (1965). *Journal of the American Oil Chemists' Society* **42**, 1051.

Jack, R. C. M. (1966). *Journal of Bacteriology* **91**, 2101.

Jenkins, P. G., Raboin, D. and Moran, F. (1972). *Journal of General Microbiology* **72**, 395.

Johnson, B., Nelson, S. J. and Brown, C. M. (1972). *Antonie van Leeuwenhoek* **38**, 129.

Jones, D. F. and Howe, R. (1968). *Journal of the Chemical Society* 2801.

Kates, M. and Baxter, R. M. (1962). *Canadian Journal of Biochemistry and Physiology* **40**, 1213.

Kates, M. (1966). *Annual Review of Microbiology* **20**, 13.

Kessell, R. H. J. (1968). *Journal of Applied Bacteriology*, **31**, 220.

Khan, A. W. and Walker, T. K. (1961). *Canadian Journal of Microbiology* **7**, 895.

Klug, M. J. and Markovetz, A. J. (1971). *Advances in Microbial Physiology* **5**, 1.

Krauel, H., Kunze, R. and Weide, H. (1973). *Zeitschrift für Allgemeine Mikrobiologie* **13**, 55.

Krumphanzl, V., Gregr, V., Pelechova, J. and Uher, J. (1973). *In* 'Advances in Microbial Engineering', (Part I), (B. Sikyta, A. Prokop and M. Novak, eds.), pg. 245. John Wiley & Sons, New York.

Lewin, L. M. (1965). *Journal of General Microbiology* **41**, 215.

Lindner, P. (1922). *Angewandte Chemie* **35**, 110.

Lundin, H. (1950). *Journal of the Institute of Brewing* **56**, 25.

Maas-Forster, M. (1955). *Archiv für Mikrobiologie* **22**, 115.

McElroy, F. A. and Stewart, H. B. (1967). *Canadian Journal of Biochemistry* **45**, 171.

Macham, L. P. and Heydeman, M. T. (1974). *Journal of General Microbiology* **85**, 77.

McMurrough, I. and Rose, A. H. (1971). *Journal of Bacteriology* **107**, 753.

McMurrough, I. and Rose, A. H. (1973). *Journal of Bacteriology* **114**, 451.

Maurice, A. and Baraud, J. (1967). *Revue Français des corp gras* **14**, 713.

Merdinger, E. and Devine, E. M. (1965). *Journal of Bacteriology* **89**, 1488.

Milner, H. W. (1948). *Journal of Biological Chemistry* **76**, 813.

Milner, H. W. (1951). *Journal of the American Oil Chemists' Society* **28**, 120.

Mumma, R. O., Fergus, C. L. and Sekura, R. D. (1970). *Lipids* **5**, 100.

Mumma, R. O., Sekura, R. D. and Fergus, C. L. (1971). *Lipids* **6**, 584.

Naguib, K. and Saddik, K. (1973). *Zentralblatt für Bakteriologie, Parasitenkunde, Infektionskrankheiten und Hygiene (Abteilung II)* **128**, 445.

Naguib, K. and Yassa, E. S. (1973a). *Mycopathologia et Mycologia Applicata* **51**, 163.

Naguib, K. and Yassa, E. S. (1973b). *Zentralblatt für Bakteriologie, Parasitenkunde, Infektionskrankheiten und Hygiene (Abteilung II)* **128**, 491.

Naguib, K. and Yassa, E. S. (1974). *Mycopathologia et Mycologia Applicata* **52**, 177.

Naguib, K., Al-Sohaily, I. A. and Al-Sultan, A. S. (1973a). *Chemie Mikrobiologie Technologie der Lebensmittel* **2**, 7.

Naguib, K., Al-Sohaily, I. A. and Al-Sultan, A. S. (1973b). *Mycopathologia et Mycologia Applicata* **49**, 217.

Naguib, K., Al-Sohaily, I. A. and Al-Sultan, A. S. (1973c). *Journal of the Science of Food and Agriculture* **24**, 97.

Nielsen, N. and Nilsson, N. G. (1953). *Acta Chemica Scandinavica* **7**, 984.

Nielsen, N. and Rojowski, P. (1950). *Acta Chemica Scandinavica* **4**, 1309.

Obraztsova, N. V. and Andreevskaya, V. D. (1973). *In* 'Ispolzovanie Mikroorganizmov Ikh Metabolizm narostenie Khoz', (S. A. Samtsevitch, ed.), pg. 13. 'Nauka i Tekhnika', Minsk, U.S.S.R. (Quoted from *Chemical Abstracts* (1974) **81** 48508.)

Ogata, K., Kaneyuki, H., Kato, N., Tani, Y. and Yamada, H. (1973). *Journal of Fermentation Technology* **51**, 227.

Okuhara, M., Kubochi, Y. and Harada, T. (1971). *Agricultural and Biological Chemistry* **35**, 1376.

O'Leary, W. M. (1973). *In* 'Handbook of Microbiology', (A. I. Laskin and H. A. Lechevalier, eds.), vol. 2, pg. 275. C.R.C. Press, Cleveland, Ohio.

Osman, H. G., Mostafa, M. A. and El-Refai, A. H. (1969). *Journal of Chemistry of the United Arab Republic* **12**, 185.

Paltauf, F. and Johnson, J. M. (1970). *Biochimica et Biophysica Acta* **218**, 424.

Patent (1969). Vsesojuzny Nauchno-Issledovatelsky Institut Biosinteza Belkovkh Veschestv, U.K. Patent 1,173,484.

Patent (1972). Vsesojuzny Nauchno-Issledovatelsky Institut Biosinteza Belkovkh Veschester. D. L. Patent 88,294.

Pavlovica, D., Jakobsons, J. and Roze, I. (1972). *In* 'Mikroorganizmy—Produtsenty Biologitcheskoie Aktivnoie Veshchestovo', (M. Beker, ed.), pg. 117. 'Zinatne': Riga, U.S.S.R. (Quoted from *Chemical Abstracts* (1973) **78**, 2725).

Pedersen, T. A. (1958). *Comptes rendus des travaux du Laboratoire Carlsberg* **31**, 93.

Pedersen, T. A. (1961). *Acta Chemica Scandinavica* **15**, 651.

Pedersen, T. A. (1962a). *Acta Chemica Scandinavica* **16**, 359.

Pedersen, T. A. (1962b). *Acta Chemica Scandinavica* **16**, 374.

Pedersen, T. A. (1962c). *Acta Chemica Scandinavica* **16**, 1015.

Pelechova, J. V., Krumphanzl, V., Uher, J. and Dyr, J. (1971). *Folia Microbiologica* **16**, 103.

Philip, S. E. (1957). Ph.D. Thesis: Manchester University.

Prescott, S. C. and Dunn, C. G. (1959). *In* 'Industrial Microbiology', (3rd ed.). McGraw-Hill Book Co. Inc., New York.

Priestley, G. (1976). *In* 'Food from Waste', (G. G. Birch, K. J. Parker and J. T. Worgan), pg. 114. Applied Science Publishers, London.

Pritchard, J. L. R. (1974). *Chemistry and Industry*, pg. 697.

Protiva, J. and Negrete, M. (1968). *In* 'Members Annual Conference for the Association of Sugar Technologists, Cuba', (no. 38), pg. 938. (Quoted from *Chemical Abstracts* (1973) **78**, 86252).

Ratledge, C. (1968a). *Biotechnology and Bioengineering* **10**, 511.

Ratledge, C. (1968b). *Journal of Applied Bacteriology* **31**, 232.

Ratledge, C. (1970). *Chemistry and Industry* 843.

Ratledge, C. (1975). *Chemistry and Industry* 918.

Ratledge, C. and Saxton, R. K. (1968). *Analytical Biochemistry* **26**, 288.

Rattray, J. B. M., Schibeci, A. and Kidby, D. K. (1975). *Bacteriological Reviews* **39**, 197.

Rehaček, J. and Beran, K. (1964). *Folia Microbiologica* **9**, 214.

Safe, S. and Duncan, J. (1974). *Lipids* **9**, 285.

Salmonowicz, J. and Niewiadomski, H. (1965). *Revue Français des corps gras* **12**, 309.

Schmidt, E. (1947). *Angewandte Chemie* **59**, 16.

Shafai, T. and Lewin, L. M. (1968). *Biochimica et Biophysics Acta* **152**, 787.

Shaw, R. (1965). *Biochimica et Biophysica Acta* **98**, 230.

Shaw, R. (1966a) *Comparative Biochemistry and Physiology* **18**, 325.

Shaw, R. (1966b). *Advances in Lipid Research* **4**, 107.

Shimi, I. R., Singh, J. and Walker, T. K. (1959). *Biochemical Journal* **72**, 184.

Singh, J. and Sood, M. G. (1972). *Journal of the Science of Food and Agriculture* **23**, 1113.

Singh, J. and Sood, M. G. (1973). *Journal of the American Oil Chemists' Society* **50**, 485.

Sood, M. G. and Singh, J. (1973). *Journal of the Science of Food and Agriculture* **24**, 1171.

Spoehr, H. A. and Milner, H. W. (1949). *Plant Physiology* **24**, 120.

Starkey, R. L. (1946). *Journal of Bacteriology* **51**, 33.

Steinberg, M. P. and Ordal, J. Z. (1954). *Agricultural and Food Chemistry* **2**, 873.

Sumner, J. L. and Morgan, E. D. (1969). *Journal of General Microbiology* **59**, 215.

Sumner, J. L., Morgan, E. D. and Evans, H. C. (1969). *Canadian Journal of Microbiology* **15**, 515.

Suzuki, T. and Hasegawa, K. (1974a). *Agricultural and Biological Chemistry* **38**, 1371.

Suzuki, T. and Hasegawa, K. (1974b). *Agricultural and Biological Chemistry* **38**, 1485.

Suzuki, T., Takigawa, A. and Hasegawa, K. (1973). *Agricultural and Biological Chemistry* **37**, 2653.

Swern, D. (1964). Ed. 'Bailey's Industrial Oil and Fat Products'. 1107. Interscience Publishers, New York.

Thorpe, R. F. and Ratledge, C. (1972). *Journal of General Microbiology* **72**, 151.

Thorpe, R. F. and Ratledge, C. (1973). *Journal of General Microbiology* **78**, 203.

Tooley, P. (1971). 'Fats, Oils and Waxes'. 182. John Murray, London.

Tornqvist, E. and Lundin, H. (1951). *International Sugar Journal* 53, 123.

Tulloch, A. P. and Ledingham, G. A. (1962). *Canadian Journal of Microbiology* 8, 379.

Tulloch, A. P., Spencer, J. F. T. and Gorin, P. A. J. (1962). *Canadian Journal of Chemistry* 40, 1326.

Tyrrell, D. (1967). *Canadian Journal of Microbiology* 13, 755.

Tyrrell, D. and Weatherston, J. (1976). *Canadian Journal of Microbiology* 22, 1058.

Uchio, R. and Shiio, I. (1972a). *Agricultural and Biological Chemistry* 36, 426.

Uchio, R. and Shiio, I. (1972b). *Agricultural and Biological Chemistry,* 36, 1169.

Uchio, R. and Shiio, I. (1972c). *Agricultural and Biological Chemistry* 36, 1389.

Uzuka, Y., Kanamori, T., Koga, T., Tanaka, K. and Naganuma, T. (1975). *Journal of General and Applied Microbiology, Tokyo* 21, 157.

Van Etten, J. L. and Gottlieb, D. (1965). *Journal of Bacteriology* 89, 409.

Volfova, O. and Pecka, K. (1973). *Folia Microbiologica* 18, 286.

Watanabe, D. (1967a). *Hakko Kyokai-shi* 25, 366.

Watanabe, D. (1967b). *Hakko Kyokai-shi* 25, 414.

Watanabe, D. (1974a). *Hakko Kyokai-shi* 32, 55.

Watanabe, D. (1974b). *Hakko Kyokai-shi* 32, 62.

Weete, J. D. (1974). 'Fungal Lipid Biochemistry', 393 pp. Plenum Press, New York.

Weete, J. D., Weber, D. J. and Laseter, J. L. (1970). *Journal of Bacteriology* 103, 536.

Whitworth, D. A. and Ratledge, C. (1974). *Process Biochemistry* 9 (no. 11), 14.

Whitworth, D. A. and Ratledge, C. (1975). *Journal of General Microbiology* 88, 275, 90, 183.

Witter, F. R. and Stotz, E. (1946). *Archives of Biochemistry* 9, 331.

Wix, P. and Woodbine, M. (1959a). *Journal of Applied Bacteriology* 22, 14.

Wix, P. and Woodbine, M. (1959b). *Journal of Applied Bacteriology* 22, 175.

Woodbine, M. (1959). *Progress in Industrial Microbiology* 1, 179.

Woodbine, M., Gregory, M. and Walker, T. K. (1951). *Journal of Experimental Botany* 2, 204.

Zalashko, M. V., Andreyevskaya, V. D. and Obraztsova, N. V. (1972). *Prikladnaya Biokhimiya i Mikrobiologiya* 8, 891.

Zalashko, M. V., Obraztsova, N. V. and Andreyevskaya, V. D. (1973). *In* 'Mikro-organizmy—Produtsenty Biologitcheskoie Activnoie Veshchestvo', (S. A. Samtsevitch, ed.), pg. 18 'Nauka i Tekhnika', Minsk, U.S.S.R. (Quoted from *Chemical Abstracts* (1974) 80, 119205).

Note added in proof

Park (1974) has isolated new strains of *Rhodotorula glutinis* producing up to 59% fat, indicating that searches for other fat-accumulating strains may be worthwhile. Algae as means of producing fat have again received attention (Materassi *et al.*, 1977). Growth requirements for *Lipomyces starkeyi* have been investigated with a view of optimizing fat production in batch culture (Korenaga *et al.*, 1976). Oxygen demand of *Rhodotorula* and *Candida* species during fat accumulation in continuous culture is only about 40% of that at maximum growth rate (Ratledge and Hall, 1977). Thus, in a commercial process, yeast producing fat could be maintained at a very high cell density without becoming oxygen limited. A high productivity from a fermenter (cf. Table 4, p. 271) would then be achieved.

Substrates for growth continue to be investigated (Ratledge, 1977). The peat-utilization programme in the U.S.S.R. has acted as a stimulus for one group of workers. Species of

Candida, Sporobolomyces and *Lipomyces* are the best utilizers (Evdokimova *et al.*, 1974; Zalashko *et al.*, 1975, 1976). *Rhodotorula* species have been grown on whey (Atamanyuk and Vakar, 1976) and on domestic waste water (Fustier and Simard, 1976) and on molasses (Atamanyuk and Vakar, 1975).

A novel method for rapidly determining the fat content of dry yeast has been patented (Patent, 1977). Methods for extracting lipid in the laboratory continue to be improved (Sorbus and Holmlund, 1976; Kaneko and Itoh, 1976; Fischer and Kating, 1976) though, on a commercial scale, pressing seems to be a simple and satisfactory method (Patent, 1976).

Micro-organisms which contain high levels of fat can also be used, after the fat has been extracted, as a source of protein for animal feeding (Uzuka, 1977). Direct uses of microbial fats' have begun to be patented (Patent, 1975, 1976). In the first, Czech workers have incorporated the whole biomass of *Rhodotorula gracilis* into cosmetic preparations such as skin masks, face creams and face powder. In the other patent, fat extracted from various yeasts has been used as a cocoa butter substitute (now £3,000/ton).

References

Atamanyuk, D. I. and Vakar, L. I. (1975). *In* 'Lipidy Gribov' (A. I. Garkavenko, ed.) p. 98 'Shtiintsa': Kishinev, U.S.S.R.

Atamanyuk, D. I. and Vakar, L. I. (1976). *Izvestia Akademiia Nauk Moldavskoj SSR, Serjia Biologicheskokh i Khimija Nauk* **4**, 48.

Evdokimova, G. A., *et al.* (1974). *Prikladnaya Biochimiya i Mikrobiologiya* **10**, 780.

Fischer, H. and Kating, H. (1976). *Planta Medica* **29**, 160.

Fustier, P. and Simard, R. E. (1976). *Canadian Institute of Food Science and Technology.* **9**, 182.

Kaneko, H. and Itoh, T. (1976). *Lipids* **11**, 821 and 837.

Korenaga, H., *et al.* (1976). *Nippon Nogei Kagaku Kaishi* **50**, 9.

Materassi, R., *et al.* (1977). *Rivista Italiana delle Sostanze Grasse* **54**, 109.

Park, S. O. (1974). *Hanguk Nonghwa Hakhoe Chi* **17**, 93.

Patent, (1976). (Fuji Oil Co. Ltd.) Netherlands Patent 76, 04177.

Patent, (1977). (Akhinyan, R. M., Erzinkyan, L. A. and Petrosyan, L. G.) U.S.S.R. Patent 558, 936.

Ratledge, C. (1977). *Annual Reports on Fermentation Processes* **1** (in press).

Ratledge, C. and Hall, M. J. (1977). *Applied and Environmental Microbiology* **34**, 230.

Sorbus, M. T. and Holmlund, C. E. (1976) *Lipids* **11**, 341.

Uzuka, Y. (1977) *Kagaku To Seibutsu* **15**, 229.

Zalashko, M. V. and Andreyevskaya, V. D. (1975). *Vestsi Akademiia Navuk Belaruskai SSR Seryia Biialagichmykh navuk Izvestia* **4**, 49.

Zalashko, M. V., *et al.* (1976). *Prikladnaya Biokhimiya i Mikrobiologiya* **12**, 131.

8. Vitamins

D. PERLMAN

School of Pharmacy, University of Wisconsin, Madison, Wisconsin, U.S.A.

I. INTRODUCTION

The usefulness of yeasts and other micro-organisms as sources of the B-group vitamins has been known for more than 50 years, and it was only a 'short step' to consider the use of micro-organisms for production of vitamins. Although there are numerous reports in the scientific and patent literature indicating that micro-organisms can be used to produce large amounts of B-group vitamins including thiamin, biotin, folic acid, pantothenic acid and pyridoxine, only vitamin B_{12} and riboflavin have been produced on a commercial scale by microbial processes. In fact, all of the vitamin B_{12} available commercially is produced by fermentation processes since chemical synthesis is not economic and isolation from liver and other sources is more costly than using fermentations. Fermentation processes for riboflavin compete very favourably with those utilizing chemical synthesis (in 1977) and are an important source of the feed-grade material used to supplement poultry and animal feeds.

The successes achieved with the vitamin B_{12} and riboflavin fermentations have encouraged microbiologists and biochemists to search for microbial processes for production of other vitamins. A process for production of β-carotene, based on the fungus *Blakeslea trispora*, has been studied in several laboratories and at one time was thought to be competitive with the chemical synthesis. This 'near miss' as far as commercialization is concerned makes the fermentation worthy of consideration in this chapter along with the vitamin B_{12} and riboflavin fermentations.

II. MICROBIAL PROCESS FOR VITAMIN B_{12} PRODUCTION

A. Historical Evolution

Vitamin B_{12} is found in practically all animal tissues and products though amounts present cover a wide range. It originates either from commensal organisms within the animal's own digestive tract or from ingestion of animal food, where in turn the vitamin arose in one of two ways. Strictly non-carnivorous, non-ruminant animals must presumably acquire much of their vitamin B_{12} by absorption of what is synthesized within the gut, for they have no obvious food source of the vitamin. Human beings are wholly dependent on dietary vitamin since they appear unable to utilize any of the intestinally synthesized vitamin; it is either not formed or not released from the cells of synthesizing organisms in regions of the gut from which absorption occurs.

It seems probable that the only primary source of vitamin B_{12} in nature is the metabolic activity of micro-organisms; there is no convincing evidence for its elaboration in tissues of higher plants or animals. It is synthesized by a wide range of bacteria and streptomycetes, though not to any extent by yeasts and fungi (Mervyn and Smith, 1964). Liver, containing approximately 1 p.p.m. of cobamides, is entirely uneconomical as a source material for commercial production. Sewage sludge, and especially activated sludge, contain a useful amount of vitamin B_{12} and have been investigated on a semi-commercial scale. A serious disadvantage is its content of B_{12} analogues that so far can only be separated from vitamin B_{12} (cyanocobalamin) by ion-exchange chromatography (Vogelmann and Wagner, 1974).

The only commercial production is by fermentation using either species of *Propionibacterium* or *Pseudomonas*, and among the companies producing the vitamin in 1977 were: Farmitalia S.p.A. (Milano, Italia); Glaxo Laboratories, Ltd. (Greenford, England); Merck and Company, Inc. (Rahway, New Jersey); Rhone Poulenc (Paris, France; Roussel U.C.L.A.F. (Romainville, France); and G. Richter Pharma-

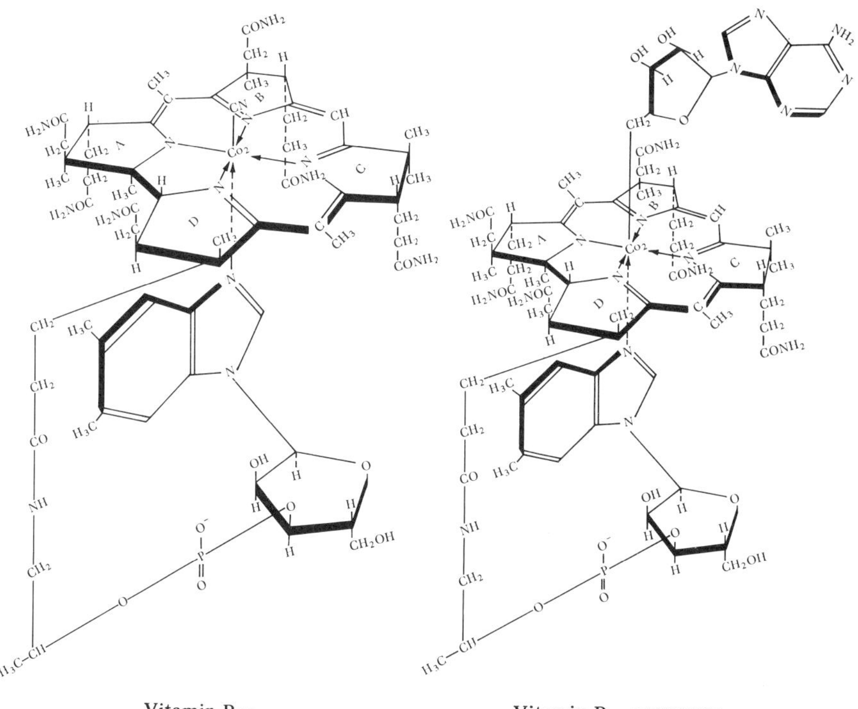

Fig. 1. Structures of vitamin B_{12} and B_{12} coenzme.

ceutical Company (Budapest, Hungary) (Perlman, 1977). Approximately 10,000 kg of vitamin B_{12} is produced annually. For a number of years (Mervyn and Smith, 1964), vitamin B_{12} was recovered from the 'spent liquors' from streptomycete antibiotic fermentations producing streptomycin, chlortetracycline or neomycin. As the demand for vitamin B_{12} increased, these 'dual purpose' fermentations could not supply sufficient amounts of the vitamin, and separate fermentations were developed for the vitamin and for antibiotics.

It is important to select microbial species which make the 5,6-dimethyl-α-benzimidazolylcobamide exclusively; several manufacturers have been led astray by organisms that give high yields of related cobamides including pseudovitamin B_{12} (adeninylcobamide; Mervyn and Smith, 1964). The 'natural form' of the vitamin is *Barker's coenzyme* where a deoxyadenosyl residue replaces the cyano group found in the commercial vitamin (see Fig. 1 for structures).

B. Current Processes

While over 100 fermentation processes have been described for production of vitamin B_{12}, only a half dozen have apparently been used on a commercial scale. These include recovery of vitamin B_{12} as a byproduct of the antibiotic fermentations already mentioned, a process based on a *Bacillus megaterium* fermentation, a process using *Streptomyces olivaceus* and another using an unidentified streptomycete, processes based on the use of strains of *Propionibacterium freudenreichii* and *P. shermanii*, and a process based on *Pseudomonas denitrificans*. Some of the characteristics of the various processes are summarized in Table 1. The processes using the *Propionibacterium* species and *Pseudomonas denitrificans* are the most productive, and are used commercially with broth titres said to be greater than 50 mg/liter (Vogelmann and Wagner, 1974).

As indicated in Table 1, a number of media have been used for growth of these organisms. Some of the formulations have contained a variety of materials of natural origin, including soybean oil meal (a byproduct of the manufacture of soybean oil), fishery wastes and fish meal, yeast preparations, meat extract, animal stick liquor (a

Table 1

Processes for microbial production of vitamin B_{12}

Micro-organism	Ingredients of medium	Yield of vitamin B_{12} (mg/litre)	Comments	Reference
Bacillus megaterium	Beet molasses; ammonium phosphate; cobalt salt; inorganic salts	0.45	Aerated fermentation (18 h)	Lewis *et al.* (1949)
Propionibacterium freudenreichii	Corn-steep liquor; glucose; cobalt salt; maintained at pH 7 with NH_4OH	19	Three days anaerobic and three days aerobic	Speedie and Hull (1960)
Propionibacterium freudenreichii	Corn-steep liquor (or auto-lysed *Penicillium* mycelium); glucose; cobalt salt; maintained at pH 7 with NH_4OH	8	Continuous two-stage fermentation; 33 h retention time	Riley *et al.* (1961)
Propionibacterium shermanii	Corn-steep liquor; glucose; cobalt salt; maintained at pH 7 with NH_4OH	23	Three days anaerobic and four days aerobic	Speedle and Hull (1960)
Streptomyces spp.	Soybean meal; glucose; cobalt salt; K_2HPO_4	5.7	Six-day aerated fermentation	Pagano and Green-span (1954)
Streptomyces olivaceus	Soybean meal; glucose; distiller's solubles; cobalt salt; inorganic salts	3.3	Six-day aerated fermentation	Hall *et al.* (1953)
Micromonospora spp.	Soybean meal; glucose; $CaCO_3$; cobalt salt	11.5	Seven-day aerated fermentation	Wagman *et al.* (1969)
Pseudomonas denitrificans	Sucrose; betaine; glutamic acid; cobalt salt; 5,6-dimethyl-benzimidazole; salts	15	Two-day aerated fermentation	Demain *et al.* (1968)
Butyribacterium rettgeri	Corn-steep liquor; cobalt salt; glucose; pH 7 with NH_4OH	5	Four-day anaerobic fermentation	Perlman and Semar (1963)

While the cultures are presumed to produce the coenzyme form of vitamin B_{12}(5,6-dimethyl-α-benzimidazolylcobamide-5'-deoxyadenosine), the vitamin is usually isolated in the cyanide form.

byproduct of recovery of packing-house wastes), corn-steep liquor (a byproduct of the manufacture of corn starch), casein or casein hydrolysates, and residues from various other fermentation processes, e.g. distiller's solubles, brewer's yeast and a mould mycelium. In some fermentation processes, the vitamin B_{12} yields are correlated with cellular yields, and inclusion in the media of these animal and plant extracts and fractions has increased cell production. Continuous addition of carbohydrates or fermentable vegetable or animal oils (also useful as antifoams in aerated fermentations) has promoted high yields of cells, as has maintenance of the fermentation at a constant pH value, e.g. pH 7.0, by addition of ammonium hydroxide which also is metabolized. The observation that cobalt is part of the cobamide molecule led to the use of media containing this element, and marked increases in production of vitamin B_{12} activity (as measured by microbial assays) were found when cobalt salts were added to media (Hendlin and Ruger, 1950; Wood and Hendlin, 1952).

The Speedie and Hull (1960) process has been one of the more productive of those described in the literature. The *Propionibacterium* cultures are grown anaerobically for two to four days, and then aerated vigorously for three to four days. Large quantities of cobinamide (presumably cobinamide-5′-deoxyadenosine) are formed during the anaerobic growth period, and this compound is converted to 5,6-dimethyl-α-benzimidazolylcobamide-5′-deoxyadenosine during the aerobic phase (Riley *et al.*, 1961). Only traces of other cobamides are formed in this process. A two-stage continuous fermentation based on this process has been described (Riley *et al.*, 1961) with the first stage being anaerobic and the second stage aerobic. As the propionibacteria synthesize 5,6-dimethylbenzimidazole during the aerobic phase, an alternative is to add this compound (prepared by chemical synthesis) during the anaerobic phase of the sequence and thus shorten the processing time, since the aerobic portion can be eliminated.

A completely anaerobic process using *Butyribacterium rettgeri* has been described (Perlman and Semar, 1963) but has apparently not been exploited on a commercial basis.

The *Pseudomonas denitrificans* process, in which betaine and 5,6-dimethylbenzimidazole are included in the medium (Demain *et al.*, 1968; Daniels, 1970; Kaplan and Birnbaum, 1973), has been one of the most productive processes. The function of betaine is still unclear; it probably affects the cell surface of the bacteria and makes them more permeable to the substrates and aids excretion of waste products.

C. Recovery of Vitamin B_{12} from Fermentation Media

Practically all of the cobamides formed in the fermentation are retained in the cells, and the first step is the separation of cells from the fermentation medium. Large, high-speed centrifuges are used to concentrate the propionibacteria to a cream (Sudarsky and Fisher, 1957). The bacteria can then be drum dried, giving a product containing 385 mg of B_{12} activity per kg and suitable for animal or poultry feed supplements. The vitamin B_{12} activity is released from the cells by acid, heating, cyanide or other treatments (McCormack *et al.*, 1954). Addition of cyanide solutions (either directly to the cells or to the filtrate obtained after treating the cells) decomposes the coenzyme form of the vitamin and results in the formation of cyanocobalamin.

The cyanocobalamin thus formed is isolated from the solution by any one of a variety of procedures. Adsorption on and elution from the ion-exchange resin IRC-50 have been reported to give high initial purification (Bungay *et al.*, 1960; Shive, 1953). More recently, use of Amberlite XAD-2, a neutral non-polar resin, has resulted in quantitative adsorption of cyanocobalamin and other corrinoids from solutions, and effective separation of these materials from the fermented media (Vogelmann and Wagner, 1974). A selective separation of corrinoids is achieved by elution with a gradient of water–*t*-butanol. Further purification can be achieved by partition between phenolic solvents (such as phenol-cresol) and water, and the vitamin is finally crystallized from aqueous acetone solutions.

D. Biosynthetic Studies

Investigations into the biosynthesis of the vitamin B_{12} molecule have followed two main pathways, namely formation of the porphyrin-like macroring, and incorporation of the nucleotide. The similarity of the ring structure to the porphyrins suggested common precursors in biosynthesis, and glycine, a precursor of δ-aminolaevulinic acid, and aminolaevulinic acid itself were incorporated in the macroring of vitamin B_{12} by growing micro-organisms. These studies showed that two molecules of δ-aminolaevulinic acid condense to form porphobilinogen, which is the basic unit of the macroring. The point at which biosynthesis of the vitamin B_{12} ring and porpyrins diverge is not yet entirely settled. Biosynthesis of the vitamin may proceed from porphobilinogen to uroporphyrinogen III, which can then undergo methylation and re-arrangement, with the introduction of cobalt, to give the B_{12} chromophore. The source of the 6-methyl groups of the corrin nucleus has been demonstrated to be methionine and neither betaine nor choline is utilized as a methyl-group donor. The regular arrangement of the extra methyl groups, as opposed to the irregular order of the amide substituents, suggests that methylation occurs after formation of the macrocycle. The remaining methyl group in ring C arises by decarboxylation of the acetic acid side-chain originating from δ-aminolaevulinic acid. Some of these origins are outlined in the drawing in Fig. 2.

When labelled threonine is added to growing cultures, the labelled atoms were located in the aminopropanol moiety of cobinamide indicating formation by decarboxylation of threonine, or a closely related derivative. It is not known how or at what stage cobalt is introduced into the corrinoid structure.

If one assumes that cobinamide is an obligatory intermediate, the next steps merely entail net incorporation of the N-α-glycosidic nucleoside 3′-phosphate. Experiments with whole cells established that dimethylbenzimidazole is incorporated into the vitamin as the 5′-nucleotide. The nucleotide ribose becomes esterified at C-3 with the phosphate of GDP-cobinamide, and the product is vitamin B_{12} 5′-phosphate. This phosphate is removed in the last step of the biosynthetic sequence by what appears to be a specific phosphatase. See Friedmann and Cagen (1970) for an extensive review of the mechanisms.

In the presence of dimethylbenzimidazole, cobamides are always formed by growing cells. A wide variety of vitamin B_{12} analogues have been prepared by addition to the growing cultures of *Propionibacterium* spp. of other bases (see Mervyn and Smith, 1964, for a catalogue of the compounds which have been tested). The term

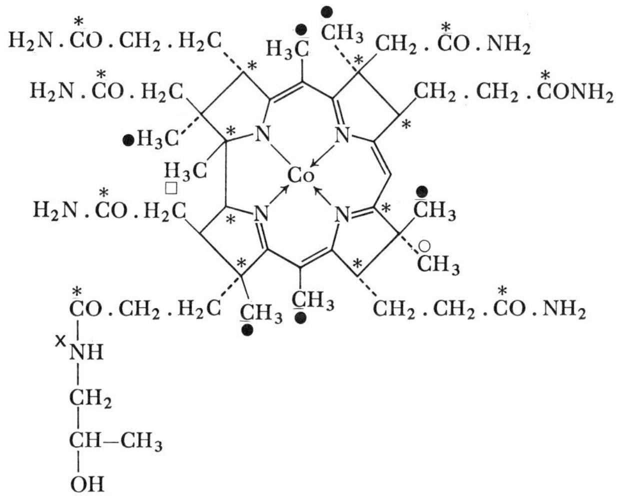

* Carbon atoms originating in δ-aminolaevulinic acid
● Methyl groups originating from methionine
○ Methyl group originating by decarboxylation of the acetic acid side-chain arising from δ-aminolaevulinic acid
x Nitrogen arising from L-threonine
□ Methyl group produced by deamination

Fig. 2. Biosynthetic origins of atoms in the corrin ring of vitamin B_{12}.

'directed biosynthesis' has been applied to those cases where a particular corrinoid has been made by addition of the corresponding base to the growing or resting organism. From a practical viewpoint, the only base that is of interest is 5,6-dimethylbenzimidazole, since the other corrinoids have significantly less bio-activity (Mervyn and Smith, 1964). Analytical methods for detecting the related corrinoids are summarized by Perlman (1971), and methods of preparation of some of the intermediates in the biosynthesis are described by Renz (1971).

III. MICROBIAL PROCESSES FOR RIBOFLAVIN PRODUCTION

A. Historical Evolution

Microbiologically produced riboflavin, or lactoflavin as it is also known, has long been available in yeast and related preparations in association with many other vitamins, particularly those of the B-complex. It is a unique vitamin in that it can be totally synthesized to a very high concentration rather rapidly by certain micro-organisms. Fermentations have been reported in which the riboflavin content of the fermented medium amounted to 7 mg per ml and higher. The organisms involved in such processes are ascomycetes, namely, *Eremothecium ashbyii* and *Ashbya gossypii*.

Table 2

Micro-organisms producing considerable amounts of riboflavin and the effects of iron on biosynthesis of the vitamin

Micro-organism	Riboflavin yield (mg/litre)	Optimum iron concn. (mg/litre)	Reference
Clostridium acetobutylicum	97	1 to 3	Meade *et al.* (1947)
Mycobacterium smegmatis	58	not critical	Mayer and Rodbart (1946)
Mycocandida riboflavina	200	not critical	McClary (1951)
Candida flareri	567	0.04 to 0.06	Levine *et al.* (1949)
Eremothecium ashbyii	2,480	not critical	Moss and Klein (1949)
Ashbya gossypii	6,420	not critical	Szczesniak *et al.* (1973)

Riboflavin is synthesized to some degree by many micro-organisms, including bacteria, yeasts, and moulds. Other micro-organisms require riboflavin for growth and may, therefore, be used for assaying the vitamin. The organisms which have been found to produce sufficient riboflavin to be, or to have been, of interest from a commercial standpoint are relatively few. Some are listed in Table 2 where the yields are listed along with the sensitivity of the micro-organism to iron.

Commercial fermentation processes for production of riboflavin or riboflavin concentrates are relatively recent, having been developed in the past 40 years. Aside from food yeasts, the first organism employed primarily for riboflavin production was *Clostridium aceto-butylicum* which, when grown in grain mashes or on whey, yielded dried residues containing 4–5 mg riboflavin per gram. These fermentations were succeeded in about 1940 by a process using *Eremo-thecium ashbyii* with yields of about 2 mg per ml, and in 1946 processes using *Ashbya gossypii* were started. In 1974, manufacturers using the microbiological process included Grain Processing Corporation (Muscatine, Iowa, U.S.A.), Merck and Company (Rahway, New Jersey, U.S.A.) and Charles Pfizer and Company (New York, New York, U.S.A.) (Perlman, 1974). Today, only Merck and Company are operating a fermentation process for riboflavin (Perlman, 1977).

B. Current Processes

Development of the current processes used for riboflavin production by *Ashbya gossypii* N.R.R.L. Y-1056 focused on three aspects: (1) preparation of the culture medium; (2) selection of mutant cultures and optimization of inoculum preparation; and (3) optimization of the fermentation conditions, e.g. incubation temperature, aeration level and fermentor design. Of this group, the most important advances have resulted from the improvement of culture media and selection of high riboflavin-producing mutant cultures.

Initial studies (Tanner *et al.*, 1949) showed that good growth occurred in a medium containing glucose, corn-steep liquor, and animal stick liquor, tankage or meat scraps. It was also shown that a sterilization time of less than 30 minutes, a low concentration of 'young' inoculum (e.g. 2%, v/v) and efficient aeration were important, with maximum yields of 700 mg/litre (10 days) when the medium was autoclaved for 15 to 30 minutes (as compared to 325 mg/litre when it was autoclaved for 90 minutes). Pilot-plant reproduction of these conditions resulted in yields of 500 to 850 mg/litre.

Further improvements resulted when enzymically degraded collagen and lipids were used as energy sources together with accessory factors present in corn-steep liquor, distiller's solubles and brewer's yeast, were included in the medium with yields of the order of 4200 mg/litre (Malzahn *et al.*, 1959). The lipid used as an energy source was proposed earlier for the *Eremothecium ashbyii* fermentation (Phelps, 1949; Rudert, 1945) and also worked well for the *Ashbya gossypii* fermentation. Typical data on stimulation of riboflavin production as a result of including in the media various peptones are summarized in Table 3(a) while the effect of supplementation with glycine is shown in Table 3(b).

Following earlier studies, Pfeifer *et al.* (1950) suggested that some improvement in productivity was possible as a result of culture selection. Pridham and Raper (1952) reported on methods they used to obtain mutants yielding 200 to 300 mg riboflavin/litre more than the parent. Since only 700 mutants were tested, the chance of obtaining an improved mutant was slim. In addition, they suggested several methods of handling the cultures including: (1) incorporating

Table 3

Effect of supplementing growth medium with peptones and glycine on
riboflavin production by *Ashbya gossypii*

(a) PEPTONES (from Malzahn *et al.*, 1959)

Nature of supplement	Riboflavin yield (mg/litre)
Peptic hydrolysate of animal tissue	1,520
Equal amounts of a peptic hydrolysate of animal tissue and pancreatic digest of casein	1.280
Pancreatic digest of lactalbumin	1,000
Pancreatic digests of casein	340
Pancreatic digest of gelatin	3,620
Papaic digest of soybean meal	673

(b) GLYCINE (from Malzahn *et al.*, 1963)

Glycine added[a] (g/litre)	Riboflavin yield (mg/litre)
0	3,280
1	3,640
2	3,980
3	4,200

[a] The basal medium contained (w/v): corn-steep liquor solid, 2.25%; Wilson's peptone W-809, 3.5%; soybean oil, 4.5%.

a reducing agent, e.g. sodium dithionate, into the stock culture media to decrease the production of low-yielding mutants; (2) re-isolating superior substrains frequently; and (3) preserving proven strains by immediate lyophilization or storage under mineral oil. Note that in 1977 workers would preserve the cultures in liquid-nitrogen storage. Malzahn *et al.* (1963) recommended a procedure in which the cultures were transferred at four-day intervals on malt–yeast extract–glucose–agar slants. Cultures were frozen after four-days growth to maintain their flavinogenic capacity. They also noted that productivity was improved in mutants obtained by treatment of the cells with ultraviolet radiation, uranyl nitrate, nitrogen mustard, or ethyleneamine, with initial selection based on increased pigmentation of colonies when grown on the malt–yeast extract–agar medium. One isolate produced 3,620 mg/litre when soybean oil was the energy source, compared with 1,420 mg/litre for the parent grown in the same medium.

The optimal fermentation conditions include aeration at about one-third volume of air per volume of liquid per minute, and agitation with three impellers at a rate of 1.0 h.p. per 1,000 litres of medium. When foaming became excessive, it was controlled by initial addition of emulsified silicone antifoam, and later by soybean oil (which also acted as a nutrient). Sterilization of the medium was accomplished by heating at 121°C for three hours (which improved productivity), and the incubation temperature was maintained at 28°C for the seven-day incubation period (Malzahn *et al.*, 1963).

C. Recovery of Riboflavin from Fermentation Media

Upon completion of the fermentation, the solids may be dried to a crude product for animal-feed supplementation, or processed to a United States Pharmacopoeia-grade product. In either case, the pH value is adjusted to 4.5. For the feed-grade product, the broth is concentrated to about 30% solids and dried on double-drum driers.

When a crystalline product is required, the broth is heated for one hour at 121°C to solubilize the riboflavin. Insoluble matter is removed by centrifugation, and riboflavin recovered by converting the riboflavin to the less soluble form (Michaelis *et al.*, 1936). Both chemical (Hines, 1945a) and microbiological (Hines, 1945b) methods of conversion have been used. The precipitated riboflavin may then be dissolved in water or polar solvents (Dale, 1947) or an alkaline solution (Morehouse, 1958), oxidized by aeration, and recovered by crystallization from the aqueous or polar solvent solution or by acidification of the alkaline solution. The crystals obtained from the alkaline solution may be type A or B (Dale, 1952), but heating the mixture to boiling results in conversion to the more desirable type A (Dale, 1957).

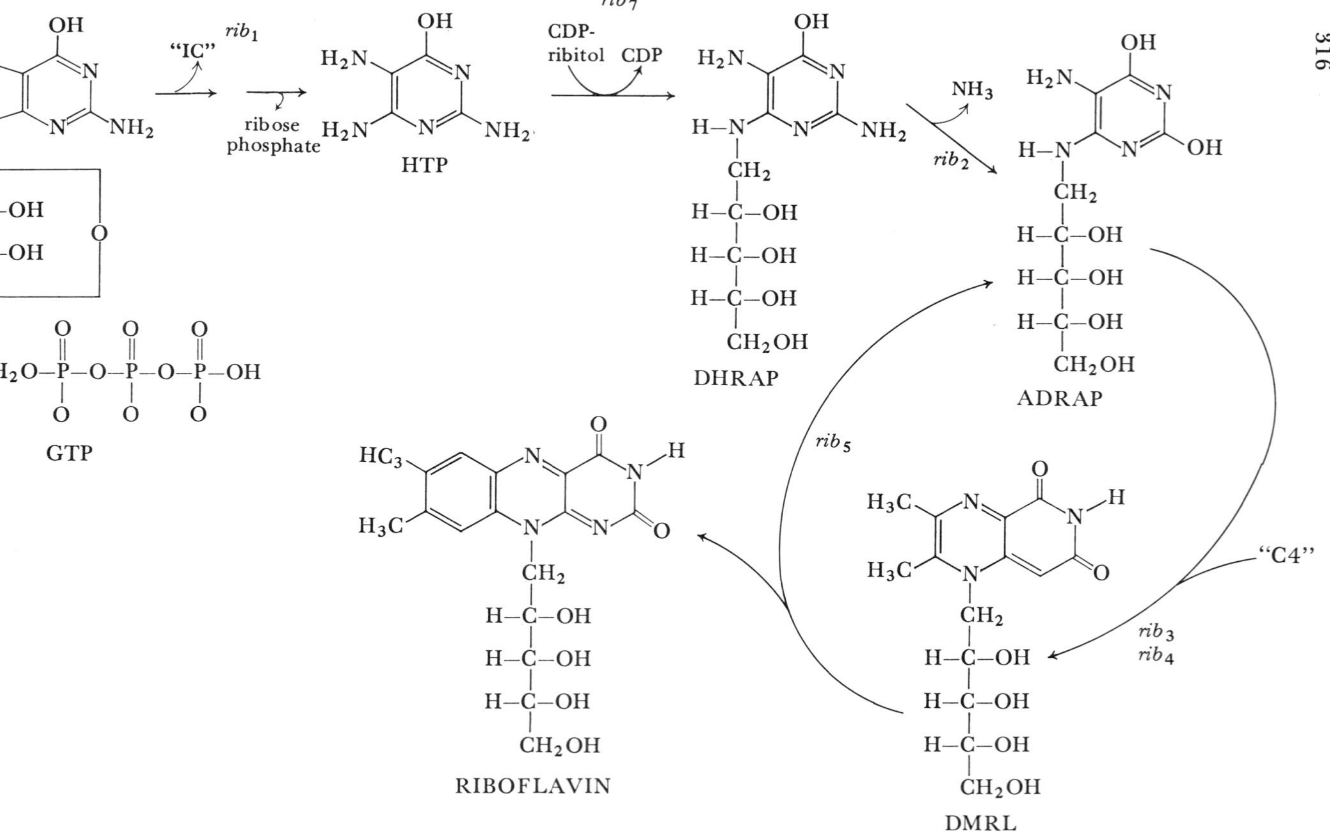

Fig. 3. Probable pathway of riboflavin biosynthesis. The genes (rib_1 to rib_5, rib_7) in *Saccharomyces cerevisiae* which correspond to each reaction are shown. Abbreviations: HTP 6-hydroxy-2,4,5-triaminopyrimidine; DHRAP, 2,5-diamino-6-hydroxy-4-ribitylaminopyrimidine; ADRAP 5-amino-2,6-dihydroxy-4-ribitylaminopyrimidine; DMRL, 6,7-dimethyl-8-ribityllumazine.

D. Biosynthetic Studies

Progress in determining the biosynthetic pathways (mainly in the yeasts and *Ashbya gossypii*) have been summarized by Goodwin (1959) and by Demain (1972). The pathway shown in Fig. 3 is currently accepted as the most likely. Demain (1972) has categorized the 'state of the art' as depending upon the degree of accumulation of the vitamin. He recognizes weak overproducers, moderate over-producers and strong overproducers. Clostridia are representatives of the weak overproducers and are markedly sensitive to the presence of iron in the medium. The moderate overproducers include yeasts, and iron inhibits their overproduction. The strong overproducing group include species of *Ashbya* and *Eremothecium*, and iron has no effect on production. He concludes that iron control of riboflavin over-production must be exercised before the final step of the pathway, since production of DMRL (see Fig. 3, pg. 316) is inhibited by iron in clostridia. If an iron-flavoprotein is the repressor of riboflavin biosynthesis, how can one explain the insensitivity to iron of riboflavin overproduction by ascomycetes? Perhaps these species are constitutive with respect to riboflavin-synthesizing enzymes, or the empirically developed conditions that favour riboflavin over-production inhibit formation of the repressor. One such condition could be the temperature of incubation.

IV. MICROBIAL PROCESSES FOR β-CAROTENE PRODUCTION

A. Historical Evolution

Interest in β-carotene as a product of industrial fermentation lies in its function as a precursor of vitamin A and in its use as a pigment in foods such as margarine. Many micro-organisms produce this com-pound, and Ciegler (1965), in summarizing the occurrence of organisms with ability to produce β-carotene, noted that this capa-bility is spread among bacteria, yeasts, fungi and algae.

Although *Phycomyces blakesleeanus* which has been studied extensively does not produce sufficient quantities to be of economic importance, much of the information obtained with this organism has been applied successfully to the evaluation of *Blakeslea trispora*. These include the finding that β-ionone stimulated β-carotene synthesis (Mackinney *et al.*, 1952) and that temperature of incubation and illumination may be critical factors (Lilly *et al.*, 1960). Related studies with *Choanephora curcurbitarum* showed that the β-carotene yield was increased 15 to 20 times using a combination of + and − cultures, as compared with using either alone.

Commercialization of a fermentation process for β-carotene has not yet occurred, although at one time there were reports that several companies were considering this possibility. The cost of material produced by the fermentation has been estimated at U.S. $31 per kg (Ciegler *et al.*, 1963) which is considerably more than current prices for β-carotene made by chemical synthesis (when corrections for inflation are considered between 1963 and 1975).

B. The *Blakeslea trispora* Process

The initial studies by Hesseltine and Anderson (1957) using + and − strains of *Choanephora curcurbitarum*, *Blakeslea trispora*, *Choanephora conjuncta* and *Blakeslea circinans* showed that combination of equal amounts of cells of *B. trispora* (in contrast to the + and − strains of other moulds) resulted in maximum β-carotene yield. Later studies showed unequal amounts of cells gave higher yields (Zajic, 1964). In practice, both the + and − strains are grown separately in shake flasks in a liquid medium (a typical example contains corn-steep liquor, corn starch, KH_2PO_4, $MnSO_4$, and thiamin) for about 48 hours at 26°C and a 10% (v/v) inoculum of each is used for the fermentation phase (Ninet *et al.*, 1969).

The major improvements in productivity (after the use of mixed cultures) have resulted from supplements to the fermentation medium. β-Ionone, which appears to be toxic when added alone to the basal medium, markedly stimulates β-carotene synthesis when the fermentation medium contains vegetable oils. The compositions of culture media useful in β-carotene fermentations are presented in Table 4, and the effectiveness of various alternatives for the β-ionone component is described in Table 5a. Effects of other supplements are shown in Table 5b. Among the supplements tested, and found somewhat less satisfactory than the combination mentioned in Table 5b, were: a series of aliphatic amides; organic acids; imides; derivatives of succinimide; various lactams and ureides; *N*-heterocyclic compounds; imidazoles; C-2, C-3 and C-4 derivatives of pyridine; derivatives of isonicotinoylhydrazine and other heterocyclic hydrazides (Ninet *et al.*, 1969).

Table 4

Composition of culture media for production of β-carotene by *Blakeslea trispora*
From Ninet *et al.* (1969)

Component	Composition (g/l) in		
	medium A	medium B	medium C
Distiller's solubles	75	75	60
Corn starch	50	0	0
'Fox head' starch	0	70	60
Soybean oil	30	40	35
Cottonseed oil	30	40	35
Yeast extract	1	1	1
KH_2PO_4	0.5	0.5	0.5
$MnSO_4$	0.2	0.2	0.2
Santoquin	0.1	0.5	0.5
Thiamin	0.01	0.01	0.01
Kerosene	20	20	20
β-Ionone	1	1	1

It should be noted that stimulation by ionones is also found with α-ionone (Ciegler, 1965) and mixtures of α- and β-ionone, α-, β-, γ-, and δ-methylionone, and their mixtures (Zajic, 1960). Citrus pulp and grapefruit oil as well as several terpenoid substances may also replace β-ionone (Ciegler *et al.*, 1964; Zajic, 1960).

The lipid components of the medium (see Table 4) are important, and several have found that white grease may be better than cottonseed oil and soybean oil mixtures (Anderson *et al.*, 1958; Ciegler *et al.*, 1959). Addition of detergents (Anderson *et al.*, 1958; Ciegler *et al.*, 1959) has a beneficial effect, with Triton X-100 (octylphenoxypolyethoxyethanol with 9–10 ethoxy residues in the side chain) as an example. One of the major breakthroughs in the development occurred with the finding that addition of kerosene (treated with sulphuric acid) to the medium resulted in marked improvement in β-carotene production (Ciegler *et al.*, 1962). Suitable viscosity may be another important feature of the medium, and a viscosity of 600 centipoise favoured growth of mycelium in one

Table 5

Effects of supplements on β-carotene production by *Blakeslea trispora*

(a) Effect of adding cyclohexane derivatives to medium A[a]. From Ninet *et al.* (1969)

Compound	Concentration (g/litre)	β-Carotene yield (mg/litre)
None	—	1,100
Cyclohexanone	1	870
2-Methylcyclohexanone	1	753
2,2,6-Trimethylcyclohexanone	1	1,287
2,2,6-Trimethylcyclohexanone semicarbazone	1.2	1,375
2,6,6-Trimethyl-1-acetylcyclohexanone	1	1,540
β-Ionone	1	1,804

[a] The composition of medium A is given in Table 4.

(b) Combinations of supplements. From Ninet *et al.* (1969)

Basal medium	Supplement	Nothing	β-Ionone	2,6,6-Trimethyl-1-acetylcyclohexanone
A	None	650	1,100	—
	N,*N*-Dimethylformamide	850	1,350	—
	Succinimide	—	1,550	—
B	None	850	1,550	1,350
	Succinimide	1,350	2,200	1,850
	Isonicotinoylhydrazine	2,200	3,040	2,950
	Iproniazide	2,200	3,200	3,200

study. This was attained in several media, which included 5% distiller's solubles with 5% starch, 10% distiller's solubles with 1% starch, or 6% distiller's solubles together with 2% carboxymethyl-cellulose (Corman, 1959).

Although few details are available on the fermentation conditions, it does seem that aeration at 0.75 volumes of air per volume of liquid (at 8.5–10.2 kg cm^{-2}) with an impeller speed at 300 rev./min and a temperature of 26°C gave reproducible yields (Ciegler *et al.*, 1963).

Although one of the objectives of the studies with kerosene indirectly was to induce excretion of β-carotene into the medium, the reverse situation resulted when the de-odorized product was used (Ciegler *et al.*, 1962). Thus, to recover the feed-grade product, it is only necessary to filter the mycelium and evaporate any residual moisture. One of the major problems with a product of this type is stabilization of the β-carotene. Ciegler *et al.* (1961) found that stabilization could be achieved by addition of ethoxyquin to the fermentation medium or to the dried mycelium. They also noted that addition of chelating agents and encapsulation in gelatin did not improve stability. To be used as food colouring, the β-carotene must also be purified. This can be done by saponification of the dried mycelium and extracting it with a fat solvent such as acetone. The β-carotene may then be preserved in an oil solution.

C. Biosynthetic Studies

The biosynthesis of carotenoids has been studied in some detail, and the scheme shown in Fig. 4, although based on studies on plants and fungi, has been accepted for the organisms producing β-carotene mentioned above (Liaaen-Jensen and Andrewes, 1972; Porter and Anderson, 1962; Qureshi *et al.*, 1974a, b, c).

One of the interesting aspects of the biosynthesis studies is the discovery that mated cultures of + and − strains, but not mated ones, produce a group of compounds which intensely stimulate carotenogenesis by − strains only; + strains were not stimulated in either plate cultures or shake-flask experiments. Prieto *et al.* (1964) and Caglioti *et al.* (1964) found three closely related compounds (labelled β_1, β_2 and β_3) and Sebek and Jäger (1964) found four compounds (named BC factors). These factors are closely related polyunsaturated C_{18} carboxylic acids, and are now known as trisporic acids. The structure of trisporic acid is:

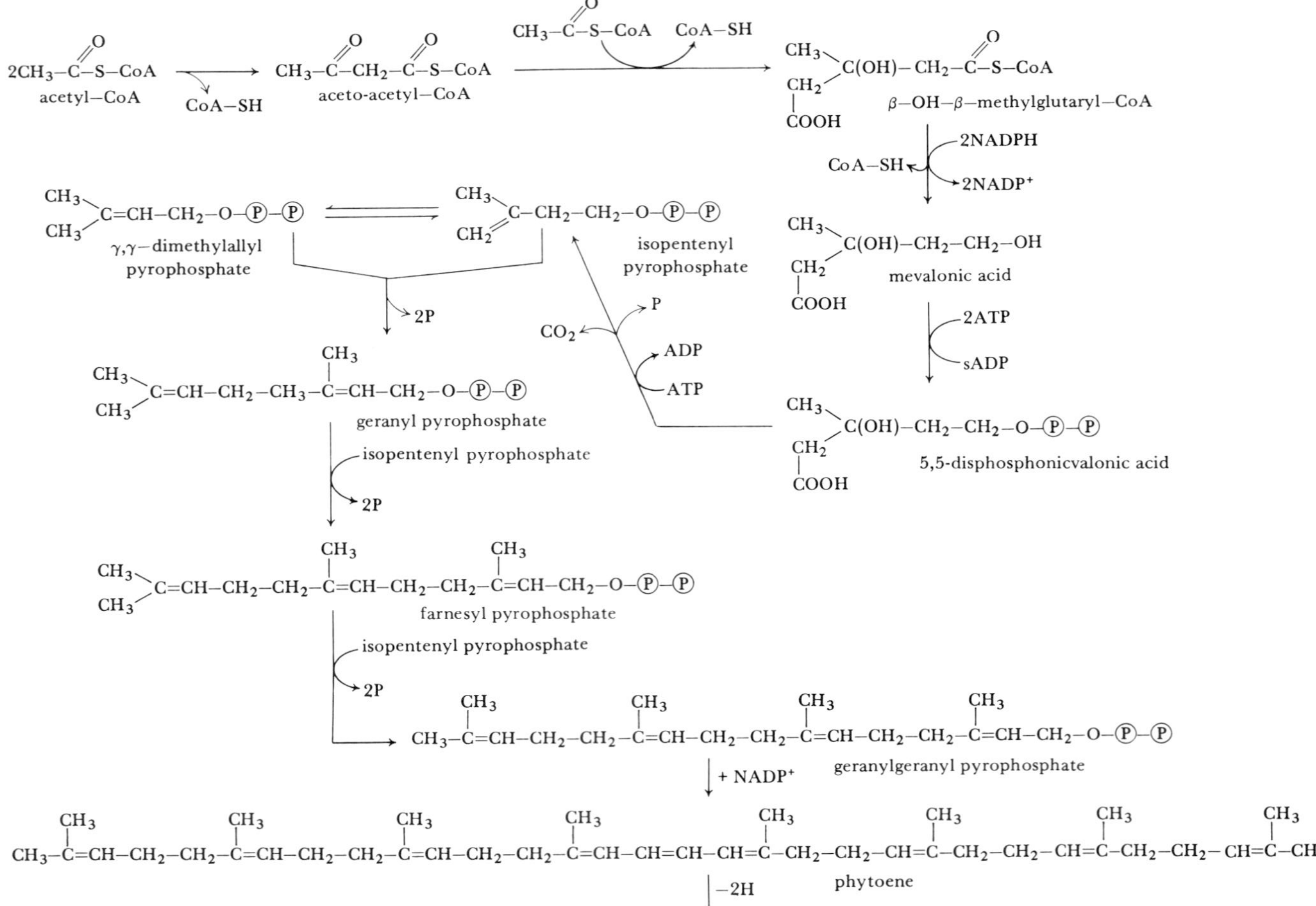

2CH₃−C−S−CoA
acetyl−CoA
CoA−SH
CH₃−C−CH₂−C−S−CoA
aceto-acetyl−CoA
CH₃−C−S−CoA   CoA−SH
CH₃−C(OH)−CH₂−C−S−CoA
CH₂
COOH
β−OH−β−methylglutaryl−CoA
2NADPH
CoA−SH
2NADP⁺
CH₃−C(OH)−CH₂−CH₂−OH
CH₂
COOH
mevalonic acid
2ATP
sADP
CH₃−C(OH)−CH₂−CH₂−O−P−P
CH₂
COOH
5,5-disphosphonicvalonic acid
CH₃−C=CH−CH₂−O−P−P
CH₃
γ,γ−dimethylallyl pyrophosphate
CH₃−C−CH₂−CH₂−O−P−P
CH₂
isopentenyl pyrophosphate
CO₂
P
ADP
ATP
2P
CH₃−C=CH−CH₂−CH₃−C=CH−CH₂−O−P−P
CH₃
CH₃
geranyl pyrophosphate
isopentenyl pyrophosphate
2P
CH₃−C=CH−CH₂−CH₂−C=CH−CH₂−CH₂−C=CH−CH₂−O−P−P
CH₃   CH₃
CH₃
farnesyl pyrophosphate
isopentenyl pyrophosphate
2P
CH₃−C=CH−CH₂−CH₂−C=CH−CH₂−CH₂−C=CH−CH₂−CH₂−C=CH−CH₂−O−P−P
CH₃   CH₃   CH₃   CH₃
geranylgeranyl pyrophosphate
+ NADP⁺
CH₃−C=CH−CH₂−CH₂−C=CH−CH₂−CH₂−C=CH−CH₂−CH₂−C=CH−CH=CH−CH=C−CH₂−CH₂−CH=C−CH₂−CH₂−CH=C−CH₂−CH₂−CH=C−CH₃
CH₃   CH₃   CH₃   CH₃   CH₃   CH₃   CH₃
−2H
phytoene

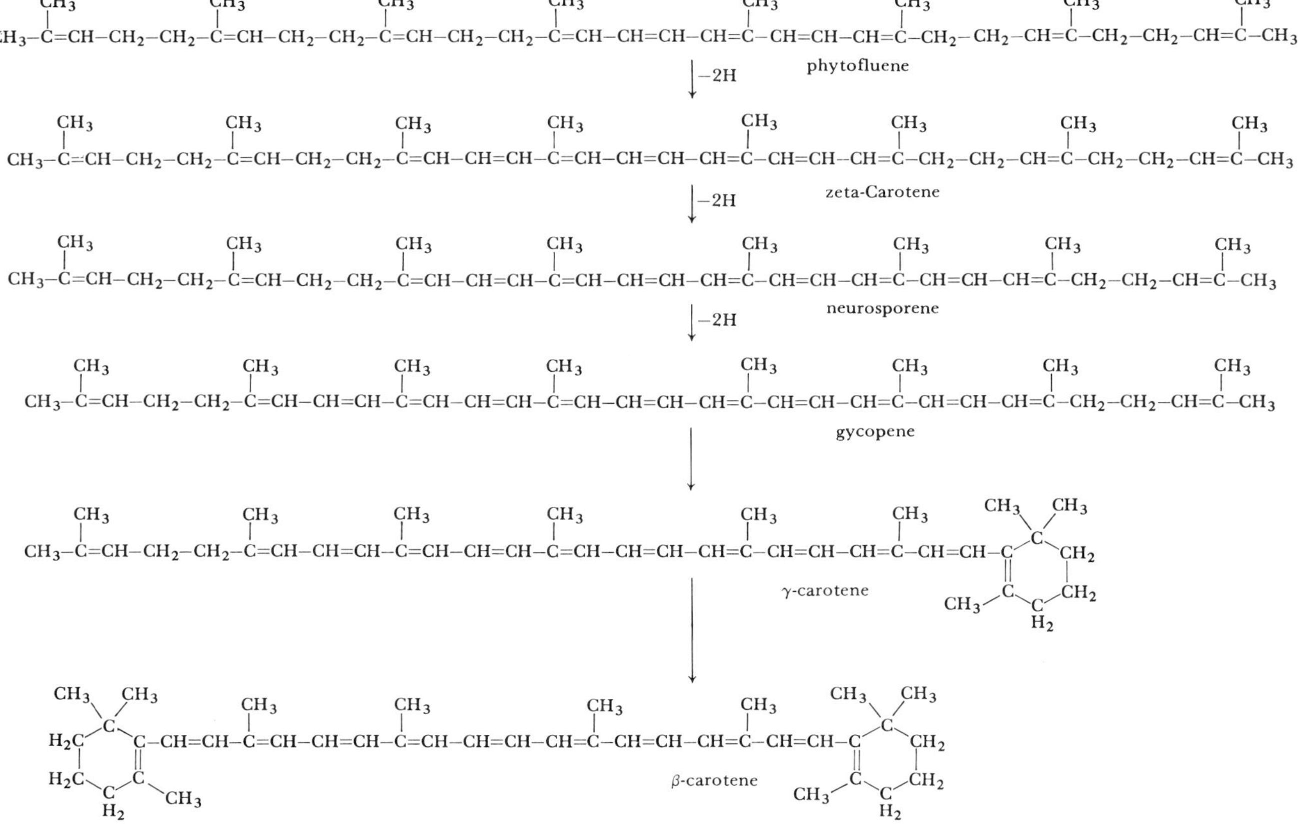

Fig. 4. Biosynthesis of β-carotene via the proposed scheme of Porter and Anderson (1962).

Prieto *et al.* (1964) found that some compounds present in deodorized kerosene were capable of enhancing β-factor synthesis, but that β-ionone inhibited β-factor production. Sebek and Jäger (1964) noted that the trisporic acids are formed from glucose, and that the trisporic acids themselves are not precursors of β-carotene.

There is some possibility of developing a one-organism fermentation in which hormonal factors would take the place of the opposite mating type. Quantitative data on the extent of stimulation by these factors have not been published, although it is believed that addition of the factors alone does not give yields equivalent to those obtained by the two-culture system (Ciegler, 1965).

REFERENCES

Anderson, R. F., Arnold, M., Nelson, G. E. N. and Ciegler, A. (1958). *Journal of Agricultural and Food Chemistry* **6**, 543.

Bungay, H. R., Marsh, M. M. and Peterson, R. C. (1960). *Journal of Biochemical and Microbial Technology and Engineering* **2**, 419.

Caglioti, L., Cainelli, G., Camerino, B., Mondelli, R., Prieto, A., Quilico, A., Salvatore, T. and Salva, A. (1964). *Chimica Industria, Milan* **46**, 1.

Ciegler, A., Arnold, M., and Anderson, R. F. (1959). *Applied Microbiology* **7**, 98.

Ciegler, A. (1965). *Advances in Applied Microbiology* **7**, 1.

Ciegler, A., Lagoda, A. A., Sohns, V. E., Hall, H. H. and Jackson, R. W. (1963). *Biotechnology and Bioengineering* **5**, 109.

Ciegler, A., Nelson, G. E. N. and Hall, H. H. (1961). *Journal of Agricultural and Food Chemistry* **9**, 447.

Cielger, A., Nelson, G. E. N. and Hall, H. H. (1963). *Applied Microbiology* **11**, 128.

Ciegler, A., Pazola, Z. and Hall, H. H. (1964). *Applied Microbiology* **12**, 150.

Ciegler, A., Nelson, G. E. N. and Hall, H. H. (1962). *Applied Microbiology* **10**, 132.

Corman, J. (1959). United States Patent 2,910,410.

Dale, J. K. (1947). United States Patent 2,421,142.

Dale, J. K. (1952). United States Patent 2,603,633.

Dale, J. K. (1957). United States Patent 2,797,215.

Daniels, H. J. (1970). *Canadian Journal of Microbiology* **15**, 809.

Demain, A. L. (1972). *Annual Review of Microbiology* **26**, 369.

Demain, A. L., Daniels, H. J., Schnable, L. and White, R. F. (1968). *Nature, London* **220**, 1324.

Friedmann, H. C. and Cagen, L. M. (1970). *Annual Review of Microbiology* **24**, 159.

Goodwin, T. W. (1959). *Progress in Industrial Microbiology* **1**, 139.

Hall, H. H., Benedict, R. G., Wiesen, C. F., Smith, C. E. and Jackson, R. W. (1953). *Applied Microbiology* **1**, 124.

Hendlin, D. and Ruger, M. L. (1950). *Science, New York* 111, 541.

Hesseltine, C. W. and Anderson, R. F. (1957). *Mycologia* 49, 449.

Hines, G. E., Jr. (1945a). United States Patent 2,367,644.

Hines, G. E., Jr. (1945b). United States Patent 2,387,023.

Kaplan, L. and Birnbaum J. (1973). Abstracts 73rd Annual Meeting American Society for Microbiology, pg. 8.

Liaaen-Jensen, S. and Andrewes, A. G. (1972). *Annual Review of Microbiology* 26, 225.

Levine, H., Oyaas, J. E., Wasserman, L., Hoogerheide, J. C. and Stern, R. M. (1949) *Industrial and Engineering Chemistry* 41, 1665.

Lewis, J. C., Ijichi, K., Snell, N. S. and Garibaldi, J. A. (1949). United States Department of Agriculture Bulletin AIC-254.

Malzahn, R. C., Phillips, R. F. and Hanson, A. M. (1959). United States Patent 2,876,169.

Malzahn, R. C., Phillips, R. F. and Hanson, A. M. (1963). Abstracts 63rd Annual Meeting American Society for Microbiology, pg. 21.

Lilly, V. G., Barnett, H. L. and Krause, R. F. (1960). West Virginia University Agricultural Experiment Station Bulletin 441T.

MacKinney, G., Nakayama, T., Buss, C. D. and Chichester, C. O. (1952). *Journal of the American Chemical Society* 74, 3456.

Mayer, R. L. and Rodbart, R. (1946). *Archives of Biochemistry* 11, 49.

McClary, J. E. (1951). United States Patent 2,537,148.

McCormack, R. B., Langlykke, A. F. and Perlman, D. (1954). United States Patent 2,656,300.

Meade, R. E., Pollard, H. L. and Rodgers, N. E. (1947). United States Patent 2,433,680.

Mervyn, L. and Smith, E. L. (1964). *Progress in Industrial Microbiology* 5, 151.

Michaels, L., Schubert, M. and Smythe, C. V. (1936). *Journal of Biological Chemistry* 116, 587.

Morehouse, A. L. (1958). United States Patent 2,822,361.

Moss, A. R. and Klein, R. (1949). British Patent 615,847.

Ninet, L., Renault, J. and Tissier, R. (1969). *Biotechnology and Bioengineering* 11, 1195.

Pagano, J. F. and Greenspan, G. (1954). United States Patent 2,695,864.

Perlman, D. and Semar, J. B. (1963). *Biotechnology and Bioengineering* 5, 21.

Perlman, D. (1971). *In* "Methods in Enzymology", (S. P. Colowick and N. O. Kaplan eds.), vol. 18C, pg. 75. Academic Press, New York.

Perlman, D. (1974). *Chemical Technology* 4, 210.

Perlman, D. (1977). *Chemical Technology* 7, 434.

Pfeifer, V. F., Tanner, F. W., Jr., Vojnovich, C. and Traufler, D. H. (1950). *Industrial and Engineering Chemistry* 42, 1776.

Phelps, A. S. (1949). United States Patent 2,473,818.

Porter, J. W. and Anderson, D. G. (1962). *Archives of Biochemistry and Biophysics* 97, 520.

Pridham, T. G. and Raper, K. B. (1952). *Mycologia* 44, 452.

Prieto, A., Spalla, C., Bianchi, M. and Buffi, G. (1964). *Chemistry and Industry* 13, 551.

Qureshi, A. A., Andrewes, A. G., Qureshi, N. and Porter, J. W. (1974a). *Archives of Biochemistry and Biophysics* 162, 93.

Qureshi, A. A., Kim, M., Qureshi, N. and Porter, J. W. (1974b). *Archives of Biochemistry and Biophysics* 162, 108.

Qureshi, A. A., Qureshi, N., Kim, M. and Porter, J. W. (1974c). *Archives of Biochemistry and Biophysics* **162**, 117.

Renz, P. (1971), *In* "Methods in Enzymology", (S. P. Colowick and N. O. Kaplan, eds.), vol. 18C, pg. 82. Academic Press, New York.

Riley, P. B., Jackson, P. W., Ross, D. and Savage, P. A. (1961) *Society of Chemical Industry Monograph* No. 12, pg. 127.

Rudert, F. J. (1945). United States Patent 2,374,503.

Sebek, O. K. and Jäger, H. (1964). Abstracts 148th Meeting American Chemical Society, pg. 9Q.

Shive, W. (1953). United States Patent 2,628,186.

Speedie, J. D. and Hull, G. W. (1960). United States Patent 2,951,017.

Sudarsky, J. M. and Fisher, R. A. (1957). United States Patent 2,816,856.

Szczesniak, T., Karabin, L. and Wituch, K. (1973). Polish Patent 66,611.

Tanner, F. W., Jr., Vojnovich, C. and Van Lanen, J. M. (1949). *Journal of Bacteriology* **58**, 737.

Vogelmann, H. and Wagner, F. (1974). *Biotechnology and Bioengineering Symposium* No. 4, pg. 969.

Wagman, G. H., Gannon, R. D. and Weinstein, M. J. (1969). *Applied Microbiology* **17**, 648.

Wood, T. R., and Hendlin, D. (1952). United States Patent 2,595,499.

Zajic, J. E. (1960). United States Patent 2,959,522.

Zajic, J. E. (1964). United States Patent 3,128,236.

9. Polysaccharides

C. J. LAWSON AND I. W. SUTHERLAND

Tate & Lyle Ltd., Group Research Laboratory, Reading, Berks, England, and Department of Microbiology, University of Edinburgh, Edinburgh, Scotland

I. INTRODUCTION

Polysaccharides have been recognized and exploited by Man for centuries. As far as is known, they occur as energy reserves and structural materials in the tissues of all living things. In animals, their structural role is performed in connective tissue, in plants in the cell walls, and in micro-organisms as cell-wall materials and extracellular capsules. A large number of polysaccharides obtained from plant tissues, including certain seeds, seaweeds, fruits and trees, have been developed into commercially important products known collectively as industrial gums and, in Table 1, the more important of these are listed and categorized according to source.

The commercial usefulness of gums is based upon their ability to alter the rheological properties of water. They do this by performing two broadly interconnected functions, namely by gelling aqueous solutions or by modifying their flow characteristics, often producing marked non-Newtonian behaviour such as pseudoplasticity or thixotropy. The manner in which polysaccharides interact with themselves and with their aqueous environment is both complex and diverse. This can be exemplified by the carrageenans, an important group of polysaccharides extracted from the red seaweeds. Certain carrageenans form strong gels, and it has been shown that this is caused by polysaccharide chains coming together in an ordered association which is a double helix or group of aggregated double helices. Other types of gel-forming polysaccharides are pectins and alginates which are extracted from citrus fruits and brown seaweeds, respectively. When these polysaccharides form gels, the resultant ordered association between chains has been called the 'egg box'. Buckled chains pack in such a way as to leave cavities or nests within which are packed the divalent cations (e.g. Ca^{2+}) which cause gelation (Fig. 1, Rees, 1975).

Table 1

Classification of water-soluble polymers

Origin			Examples
(i) *Chemically unmodified natural gums*			
		trees	Gum arabic
			Gum karaya
			Gum tragacanth
		seeds	Locust bean gum
			Guar gum
			Psyllium gum
			Quince gum
		seaweeds	Agar
			Alginate
	Plants		Carrageenan
			Furcellaran
		cereals	Corn starch
			Wheat starch
			Maize starch
			Sorghum starch
		tubers	Potato starch
			Arrowroot starch
			Tapioca starch
		citrus fruits	Pectin
	Micro-organisms	bacteria	Dextran
			Xanthan gum
	Animals	milk	Casein
		bones	Gelatin
(ii) *Chemically modified natural gums*			
	Plants		
		trees	Carboxymethyl cellulose
		cottons	Methyl cellulose
			Hydroxymethyl cellulose
		cereals	Dextrins
		tubers	Carboxymethyl starch
			Hydroxypropyl starch
		seaweeds	Propylene glycol alginate
		citrus fruits	Low methoxy pectin
	Micro-organisms	bacteria	Diethylamino-ethyl (DEAE) dextran
(iii) *Synthetic water-soluble polymers*			
	Chemical synthesis	Vinyl polymers	Polyvinyl pyrrolidone
			Polyvinyl alcohol
			Carboxyvinyl polymer
		Acrylate polymers	Polyacrylamide
			Polyacrylic acid
		Ethylene oxide polymers	Polyethylene oxide (Polyox)
		Inorganic polymers	Laponite (synthetic hectorite)

Due to their extremely diverse physical properties, polysaccharides have found many applications in, amongst others, the food, pharmaceutical, cosmetic, paper, oil and textile industries, and in Table 2 some examples of this diversity of use are shown. A comprehensive review of applications for these water-soluble polymers is not

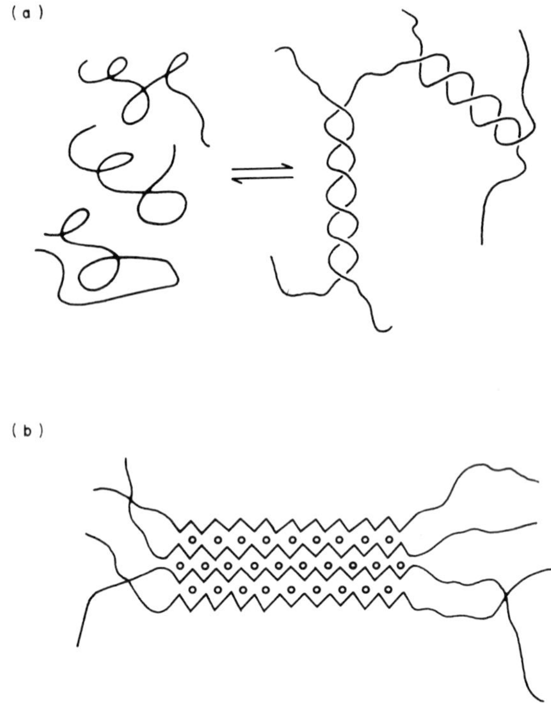

Fig. 1. Structures of polysaccharide gels. (a). Schematic representation of the mechanism proposed for gelation of carrageenans and some related polysaccharides. (b). Proposed structure of alginate and pectinate gels.

possible within the confines of the present discussion, but several treatises on the subject are available to the reader wishing a more detailed account (Whistler, 1973a; Glicksman, 1969; Lawrence, 1973; Meltzer, 1972). Traditionally, polysaccharides have been exploited for their thickening, stiffening and setting characteristics, and any specialized uses were a result of alchemy rather than by the application of detailed structural knowledge.

Table 2

Some important physical properties of industrial gums and their applications

Physical property	Gum	Application
Cold set, clear, gel formation with divalent cations	Alginate	Reformed fruit pieces, dental gels
Gel formation with sucrose	Pectin	Jam manufacture
Heat-reversible gel formation	Agar	Microbiological solid media, synthetic meat gels, diatetic jellies
Heat-reversible gel formation in the presence of potassium ions	Carrageenan	Synthetic meat gels, instant desserts
Non-reactivity with 'reactive' (Procion) dyestuffs	Alginate	As a textile print-paste thickener in conjunction with reactive dyes
Smooth 'short' texture	Fully pregelatinized corn starch	In many proprietary puddings
Cold-set soft gel formation with water or milk	Pregelatinized and/or oxidized potato or tapioca starch	Instant puddings
Stability in the presence of strong acids	Xanthan gum	In rust-curing gels containing phosphoric acid
Pseudoplastic behaviour under conditions of high shear	Xanthan gum	As a lubricant for the bentonite muds used to drill oil wells
Synergistic gel formation with carob gum	Xanthan gum	Synthetic meat gels
Retardation of sugar crystallization at low moisture contents	Gum arabic	In pastilles and jujubes
Complex formation with milk protein	Carrageenan	Milk drinks
Stability at low pH values in food products	Propylene glycol alginate	French dressings, salad dressings and pharmaceutical jellies
Retardation of ice-crystal size	Propylene glycol alginate	Ice cream and ice lollies
Gel formation with heat	Curdlan	Not known

A much greater understanding of the relationships between the demonstrable physical properties of several of the polysaccharides and the primary, secondary and, in a few cases, tertiary chemical structures has been obtained. A benefit of this greater knowledge has been the ability to write more objectively defined and precise specifications for existing products, and it should be possible to develop an increasing number to suit particular applications. However, an important question for the future development of industrial polysaccharides is in the identification of sources yielding gums having physical properties not available from existing products, and the subject of this chapter is concerned with one such source, namely micro-organisms. The unique physical properties shown by many microbial gums have stimulated industry throughout the World to investigate the possibilities of exploiting micro-organisms and at least two products, dextran and xanthan gum, are currently in commercial production, with several more in development.

If the great technical challenges which the development of these products pose can continue to be met with success, they will penetrate new and exciting market sectors, assuming that other influential factors such as continuity of supply and price can be maintained within competitive limits.

II. COMMERCIAL IMPORTANCE OF POLYSACCHARIDES

A. Market for Water-Soluble Gums

The present World consumption of gums is difficult to calculate precisely, but some estimates are given in Table 3 and, in Table 4, the United States demand in 1973 is indicated. In Table 5, the imports of some gums into the United States is given, and it is interesting, for example, to compare consumption of gums with the import levels as this gives some indication of the use of 'home-grown material'. Also, a decline in the import levels of some products, particularly gum arabic, should be noted as this indicates a shift in use patterns towards other water-soluble polymers. Market prices for gums tend to be influenced by quality, source and the market in which they are sold, and in Table 6 prices as listed in The Chemical Marketing Reporter of 1971, 1974 and 1975 are shown.

Table 3

World production of some industrial gums

Gum	Year	Production (tonnes)
Agar[a]	1968	5,500
Alginate[b]	1973	17,000
Arabic[a]	1966	60,000
Carrageenan[b]	1973	10,000
Locust bean[a]	1970	15,000
Methylcellulose[a]	1972	25,000
Pectin[a]	1971	9,000
Carboxymethylcellulose[a]	1969	60,000 (33,000 U.S.A.)
Xanthan gum[c]	1975	5,000

[a] Whistler (1973); [b] E. Booth (personal communication); [c] This value is an estimate.

Table 4

Consumption of industrial gums in the United States in 1973.
Unpublished results of R. L. Whistler

Gum	Quantity (tonnes) used		
	Food uses	Industrial	Total
Corn sugars	2,232,142	?	—
Corn starch	223,214	1,116,071	1,339,285
Carboxymethylcellulose	6,696	43,303	50,000
Methylcellulose	900	23,660	24,553
Guar	6,696	15,625	22,321
Arabic	10,267	3,125	13,392
Pectin	5,357	0	5,357
Locust bean	4,017	1,785	5,803
Alginate	4,017	4,017	8,034
Ghatti	4,464	446	4,910
Carrageenan	4,017	89	4,106
Xanthan	1,000	3,678	4,678
Karaya	446	3,125	3,571
Tragacanth	580	89	669
Agar	133	178	311
Furcellaran	89	0	89

Table 5

Imports of water-soluble gums into the United States.
Data from the United States International Trade Commission

| | Import (tonnes) in | | | | |
Water-soluble gum	1970	1971	1972	1973	1974
Arabic	12,000	13,300	14,240	7,500	—
Karaya	3,586	3,647	3,452	3,274	2,959
Tragacanth	801	728	658	542	730
Guar	12,346	20,817	13,150	12,455	26,186
Locust bean	5,111	3,784	4,215	3,465	2,441
Carrageenan	180	180	180	340	340
Agar	379	352	526	468	—
Pectin	348	424	638	754	—
Starch (corn and potato)	101,877	90,610	77,934	70,892	—

Table 6

Retail prices of industrial gums.
Data obtained from The Chemical Marketing Reports

| | Price (pound sterling $\times 10^{-3}$) in | | |
Gum	1971	1974	1975 (Dec.)
Agar	2.9	9.1	7.7–8.3
Alginate	—	0.9–1.9	1.9–3.7
Arabic	0.56	2.0	1.0–1.2
Gelatin	0.64	1.9–3.0	1.6–2.7
Guar	0.50	0.50	0.66–0.85
Karaya	0.90	0.95	0.84–1.1
Locust bean	0.63	1.10	1.2–1.3
Xanthan	—	2.0–3.3 (U.K.)	2.8–4.0 (U.S.A.)[a]
			3.6–5.1 (U.K.)
Pectin	2.6	2.3	—
Carboxymethylcellulose	0.50	0.67	0.73–1.0
Tragacanth	2.5–7.8	11.4–15.6	15.4–16
Starch (modified)	—	0.1–2.0	—
Starch	—	—	0.13–0.17[b]

[a] Direct quotation; [b] Data obtained from The Public Ledger.

The United States of America has an annual growth in production of all water-soluble gums of about 1.3%. However, in the case of cellulose derivatives, microbial gums and synthetic polymers, such as polyvinyl alcohol and polyacrylic acid, the growth rate is much higher at around 8%. As already mentioned, the increase in consumption of these latter product groups has been at the expense of other gums such as gum arabic, locust bean and Karaya. Several reasons for this decline in the use of these more 'traditional' gums will be given later in this chapter.

B. Problems Associated with the Production of Plant Polysaccharides

Many higher plant and algal polysaccharide suppliers face several difficulties of a highly intractable nature when attempting to maintain guaranteed supplies of products with stable specifications. The first problems arise within the plant tissues themselves. Polysaccharides are produced either as storage products or as structural materials and often vary in chemical composition as a response to metabolic requirements within the plant. These requirements may be caused by events such as environmental changes from season to season or through the aging cycle of the plant. Plants showing typical behaviour of the kind described are the brown seaweeds; these contain the alginates which have been found to vary in uronic acid content. Examination of different parts of the plants has shown variations related to some extent to the age of the tissues as indicated by values for mature and new fronds (Table 7) and for different parts of the stipe of *Laminaria hyperborea* (Table 8). The stipe of *Laminaria digitata*, a seaweed from which commercially useful alginates are collected, has been shown to contain alginate with a higher guluronic acid content than that from the frond, and the manufacturers of alginate attempt to exploit this by separating stipe from frond in order to create products having different specifications.

In the case of starch, development work on a number of the common sources such as corn and potatoes has effected a substantial degree of product improvement. The newer amylopectin starches from agronomically modified corns, such as waxy maize and waxy

Table 7

Composition of alginates from different species of algae

Species	Ratio of mannuronic acid to guluronic acid residues	
	Fischer and Dörfel (1955)	Haug (1964)
Ectocarpales		
Ectocarpus confervoides	0.4 (0.3)	—
Ectocarpus sp.	—	0.45
Sphacelariales		
Sphacelaria bipinnata	0.6 (0.4)	—
Dictyotales		
Dictyota dichotoma	0.6 (0.4)	1.05
Dictyopteris polypodioides	0.6 (0.4)	—
Chordariales		
Mesogloia vermiculata	—	0.25
Chordaria flagelliformis	—	0.90
Spermatochnus paradoxus	—	1.30
Dictyosiphonales		
Scytosiphon lomentaria	—	1.15
Dictyosiphon foeniculaceus	—	0.85
Desmarestiales		
Desmarestia aculeata	—	0.85
Laminariales		
Chorda filum	1.1 (0.8)	—
Laminaria digitata	3.1 (2.1)	1.45–1.6
Laminaria digitata (f)		1.2–1.85
Laminaria digitata (nf)		
Laminaria hyperborea	1.6 (1.1)	
Laminaria hyperborea (s)	—	0.4–1.0
Laminaria hyperborea (f)	—	1.05–1.65
Laminaria hyperborea (nf)	—	1.90
Laminaria saccharina (f)	—	1.25–1.35
Alaria esculenta		1.20–1.70
Fucales		
Ascophyllum nodosum	2.6 (1.7)	1.40–2.25
Fucus vesiculosus	1.3 (0.9)	0.75–1.20
Fucus serratus	2.7 (1.8)	1.15
Pelvetia canaliculata	1.5 (1.0)	1.30–1.50
Himanthalia elongata	2.7 (1.8)	1.00–1.80
Halidrys siliquosa	1.1 (0.8)	0.75
Cystoseira barbata	0.7 (0.5)	—
Cystoseira abrotanifolia	1.9 (1.2)	—
Sargassum linifolium	0.8 (0.6)	

Values in brackets are corrected by Haug's (1964) factor for the higher rate of destruction of guluronic acid. Whole plants were used for analysis except in the cases indicated as follows: f, fronds; nf, new fronds; s, stipes.

sorghum, have yielded many products with well defined and reproducible specifications which have added a new dimension to starch
technology. These compounds have better optical and rheological
properties, and form gels in ways which allow products to be
developed which were impossible with the mixed amylose-
amylopectin starches previously available. For many other industrial
gums, however, the manufacturer has had little control over his raw
material. Very little success has been achieved in attempting control

Table 8

Composition of alginate from stipes of *Laminarea hyperborea*
From Hang (1964)

	Ratio of mannuronic to guluronic acid residues	
Part of stipe	Lower part	Upper part
Medulla	0.70	0.92
Inner cortex	0.55	0.54
Outer cortex	0.44	0.43
Peripheral tissue	0.54	0.69
Weight average	0.46	0.57

over the metabolism of seaweeds, for example, and the type of
development work used in the formation of starch products has not
been possible.

Another type of uncontrolled situation is in the production of
gum arabic which is obtained as the exudate from the gum nodule
which forms when the bark of *Acacia senegal* is tapped. The gum is
harvested by nomadic tribesmen from trees growing in the wild. The
nodules are bleached in the hot sun and then brought to central
collection points. The nodules are graded and sold to the gum
supplier by government agencies. Thus, the supplier has no control
over the product specification before or after harvesting.

Another difficulty often encountered in the manufacture of plant
and algal gums is modification and degradation of products as they
are processed. Many harsh processing conditions are encountered,
alkali extraction, acid precipitation, hot-water leaching, and treat-

ment with bleaching agents. In the manufacture of sodium alginate, the dry seaweed is first milled and shredded, then digested with sodium carbonate to extract the crude sodium alginate. Small amounts of cellulosic material are removed at this stage, and the polysaccharide precipitated as the calcium salt, or if appropriate as alginic acid using hydrochloric acid. The calcium alginate must be converted to the acid form of the polysaccharide by leaching with hydrochloric acid and then converted to sodium alginate by titration. Two alternative titration techniques may be employed. The first is a heterogeneous reaction using aqueous sodium hydroxide followed by aqueous sodium carbonate in the presence of iso-propanol, which prevents the polysaccharide from redissolving. The aqueous isopropanol is then filtered off and the product dried. The alternative technique involves addition of an approximately stoicheio-metric amount of dry sodium carbonate to the wet cake of alginic acid and mixing the viscous dough in a high powered mixer. The pH value is then adjusted with small quantities of sodium carbonate and the semisolid mass is drum or roller dried. Before final drying, hypochlorite may be used to bleach the polysaccharide (E. Booth, personal communication). These processing conditions can cause alkaline degradation by β-elimination reactions, acid-catalysed hydrolysis and possible biological and physical degradation. Other undesirable qualities such as colour, odour and chemical impurities may also be introduced. To obtain a consistent product specification, having the desired viscosity or gel-forming characteristic, the manufacturer often relies on blending different samples. However, when the user requires the polysaccharide to perform in a specific manner utilizing more than one physical property, e.g. to have a specific viscosity and at a subsequent stage to form a defined gel, blending will render the definition of the two rheological characteristics mutually incompatible. Consequently, users may be obliged to reformulate products.

A final problem may well be difficulty of supply. For any product derived from plant and algal sources, supply in both the long and short term can be difficult. Poor harvests of the economically important crops, such as corn, citrus fruit and legumes, due to climatic conditions and disease are commonplace and often cause temporary shortages. The plant species not under controlled cultivation can present even more serious problems. Recent severe droughts in the Sudan have caused a shortage of gum arabic due to defoliation and death of many acacia trees. The supply of seaweed polysaccharides has been affected by severe climatic conditions and changes in ocean currents. Over a period of years, growth areas change and a species of weed may have its territory invaded by more vigorous types. This latter situation is often caused by pollution, when green seaweeds, preferring to grow in a more richly organic environment, invade sites occupied by the more delicate red and brown varieties. Over-harvesting of certain species of seaweed can cause supply problems, particularly species of *Gelidium*, a common source of agar. In order to improve supplies, it has been suggested that certain species of algae be cultivated in areas where they are not normally found. One species of brown algae, *Macrocystis pyrifyra*, has been the subject of much recent debate in this respect. This weed grows naturally in several parts of the world, and it is harvested off the coast of California as a source of alginate. This weed does not grow off European coasts, and the suggestion was made that, in order to increase the supply of alginate-bearing weeds, *Macrocystis* sp. should be cultivated off the coast of Brittany in France. As the weed has stipes up to 66 m long which can be more than 2.5 cm in diameter and grows in water between 750 and 2,400 cm in depth, various objections were made. It was pointed out that problems with coastal shipping would occur due to the fouling of propellers and rudders, and the presence of a seaweed with such large dimensions would almost certainly upset the ecological balance of the area.

C. Production of Polysaccharides of Commercial Significance *via* Chemical Modification

One approach to obtaining polysaccharides with modified specifications which has been practised for many years is chemical derivative formation. Chemical modification of polysaccharides has often proved successful, and important commercial products have been developed, such as the starch and cellulose ethers and esters. The approach to chemical-derivative formation has stemmed from two interconnected concepts. The first concept is through a desire to improve the performance of a product for a particular application. An example of derivative formation which fits this category is the reaction of propylene oxide with partially neutralized alginic acid resulting in the formation of the propylene glycol ester derivative of this polysaccharide. Substitution of a number of the carboxyl groups in alginic acid yields a product which is more compatible with acids and multivalent metal cations than the unsubstituted compound. An application of propylene glycol alginate is in soft drinks where the polymer performs the function of modifying texture, foam stabilization and penetration control. This application is unsuitable for unmodified alginate because the acidic nature of soft drinks would render the polysaccharide insoluble.

The second broad concept behind the chemical modification of polysaccharides is that of forming entirely new polysaccharide products with properties completely different from the parent polymer. An example of this technique is in the formation of cellulose ethers. Native celluloses are completely insoluble in water and this restricts their use. The methyl and hydroxypropyl derivatives are water soluble and have been developed for their non-ionic, thickening, film-forming, gelling and other properties. They have become major commercial products in their own right and bear little resemblance to the compounds from which they were formed, in contrast to the propylene glycol alginates which still retain facets of native alginate.

There are, however, limitations to the extent that chemical modification can be exploited to produce the range of products of actual and potential use to industry. The nature of the chemically modified polysaccharide is partly dependent upon that of the native polysaccharide. If the latter has undesirable characteristics, such as a low molecular weight or acid-labile glycosidic bonds, so may the derivative and this may adversely affect its performance.

A further point for consideration is that chemical processes are relatively crude when compared with the enzyme-catalysed reactions employed in biosynthesis of polysaccharides. Many modifications are impossible to achieve with the technology available—the specific substitution of hydroxyl groups in a polysaccharide is not yet possible by chemical means, although one cell-free, enzymic process is commercially viable in the production of dextran (Whistler, 1973b).

As the production of polysaccharides from micro-organisms is achieved through the mediation of enzyme systems, this approach has been the subject of much attention. It offers a more or less controllable synthetic approach which, although not as well defined as many purely chemical reactions, is much more satisfactory than many of the traditional processes.

As can be seen from the foregoing discussion, the production of native and chemically modified polysaccharide gums is, in many instances, a highly successful and profitable business. However, areas of difficulty remain, many of an intractable nature.

III. BIOPOLYMERS, POLYSACCHARIDES FROM MICRO-ORGANISMS

A. Definition

The discussion has so far been limited to a consideration of the present status of plant and algal gums in order to set the scene for this chapter on the production of microbial gums. As has been pointed out, the development of chemical processes for the primary production of defined polysaccharide products has been the subject of much research, but limited practical success has been achieved apart from starch, cellulose and a few other gums. The failure of purely chemical reactions to extend greatly the available range of industrial gums has led to the current interest in the exploitation of micro-organisms. The micro-organisms involved include bacteria, fungi and yeasts, many of which elaborate polysaccharides as exocellular capsules and slimes. Many micro-organisms secrete into the aqueous environment upon which they are grown, copious quantities of slimy or gelatinous material which, upon analysis, are usually found to be carbohydrate polymers.

It will be useful to consider polysaccharides from micro-organisms by first defining them in terms which enable a logical comparison to be made with existing polysaccharide types of industrial importance. The polysaccharides in question are those carbohydrate polymers which appear outside and normally unbonded to the cell walls of many bacteria, fungi and yeasts, and exhibit rheological and gel-forming properties which are, by definition, of interest by comparison with industrial polysaccharides already in use (Table 2, pg. 331).

Bacterial exopolysaccharides have been shown to contain a variety of monosaccharide residues including neutral hexoses and methyl-

Table 9

Processes used in production of microbial polysaccharides of commercial importance

Name	Organism	Type of process	Substrate	Sugar residues in polymer
Dextran	*Leuconostoc mesenteroides*	Cell-free enzyme	Sucrose	Glucose
Xanthan gum	*Xanthomonas campestris*	Bacterial (batch)	Glucose, glucose syrup	Glucose (acetate), glucuronic acid, mannose (pyruvate)
Pullulan	*Aureobasidium pullulans*	Fungal (batch)	Glucose syrup	Glucose
Erwinia exopolysaccharide	*Erwinia tahitica*	Bacterial (batch)	Glucose? glucose syrup?	Glucose, galactose fucose, uronic acid, (acetyl)
Scleroglucan	*Sclerotium glucanicum*	Fungal	Glucose	Glucose
Microbial alginate	*Azotobacter vinelandii*	Bacterial (batch and continuous)	Sucrose	Mannuronic acid, guluronic acid, (acetate)
Baker's yeast glycan	*Saccharomyces cerevisiae*	Yeast (batch)	Glucose	Glucose, mannose
Curdlan	*Agrobacterium* sp., *Alcaligenes faecalis*	Bacterial (batch)	Glucose	Glucose

pentoses, oxo sugars, uronic acids and amino sugars. Other components include acetyl, pyruvate, succinate and phosphate groups. The most commercially developed polymers are listed in Table 9, which shows the micro-organisms responsible, component sugar residues, type of fermentation process employed, and the carbohydrate substrate used in the fermentation.

B. Technical Advantages and Disadvantages of Industrial Polysaccharide Production by Fermentation

There are several inherent advantages in the fermentation approach to polysaccharide production compared with more traditional approaches. These are: (i) Through the use of well characterized, cheap and plentiful raw materials, particularly if fully defined media can be developed. (ii) A potential ability to exercise control over the synthetic process via ongoing fermentation parameters such as pH value, temperature, medium composition, inoculum composition, inoculum size, broth mixing and aeration, and fermentation time. Two separate objectives can be identified as having importance in control of fermentations. The first aspect is in the maintenance of product specification through process control, i.e. the ability to reproduce fermentations for synthesis of products within defined limits. The second aspect is manipulatory control of product type, e.g. the ability to produce a range of microbial gums having different molecular structures and weights through specific changes in fermentation conditions. Mutation can be an important factor in achieving this later objective. (iii) The possibility of using mild recovery techniques (for example, solvent precipitation) thus preventing unwanted product degradation with concomitant loss in quality. (iv) A greater flexibility in the choice of production sites. (v) Use of continuous polymer production giving high productivities.

On the other hand, fermentation presents several technical problems or disadvantages over traditional methods. These are mainly due to the highly viscous fermentation broths leading to: (i) Low concentrations of product in most instances below 5% w/v, leading to: (a) the need for very large fermenters (50–200 m^3); (b) large water usage; (c) extensive use of recovery solvents. (ii) High power requirements (because of the high viscosities in cultures). (iii)

Difficulties in removing cells from the cultures. (iv) Bad mixing and aeration leading to oxygen limitation and lowered productivities.

C. Commercial Potential of Polysaccharides from Micro-Organisms

In discussing the commercial potential for microbial polysaccharides, it will be useful to examine their development to date and, in Table 10, an indication of the commercial status of the most advanced microbial gums is given. In contemplating the production of microbial gums, the potential manufacturer must answer a number of important questions. (i) What annual production capacity will the plant have? To answer this question the following information must be available: (a) the present and future market size for both products of specific interest, and for likely competing products: (b) the number of competing products and their manufacturers: (c) the sensitivity of direct manufacturing cost to plant size. (ii) If the product is a new one, whether the properties are sufficiently unique to merit a development programme. (iii) If the product is: (a) either to be the same as a microbial gum already in production, e.g. dextran; or (b) the equivalent to a plant gum, e.g. microbial alginate, whether the new version can fully match or even supersede the properties of the available material. (iv) Whether the new polysaccharide can be made at a cost and sold at a price which will provide a reasonable return on investment, taking into account the possibility of price-cutting tactics by incumbent manufacturers. (v) The likely product life, and whether new, superior products are contemplated. Some of the above considerations dealing with market size and structure will be extremely difficult, if not impossible, to clarify, as the World gum market is extremely complex and much market data are not available, but it is worth the potential microbial gum producer making a serious attempt to obtain answers.

Many publications have appeared, mostly since 1945, on the subject of microbial gum production, but the vast majority of these have not been exploited commercially. Much of the early work leading to results with commercial significance was performed at The Northern Regional Research Laboratories (N.R.R.L.) of the United States Department of Agriculture at Peoria, Illinois, U.S.A. in a project directed towards the utilization of surplus corn starch. This

Table 10

Commercial information on microbial polysaccharides of commercial importance

Name	State of development	Trade name	Company involved
Dextran	Present—in production; future —static	Various	Dextran Products, Polydex
Xanthan gum	Present—in production; future —expanding	Keltrol, Rhodopol 23	Kelco, Rhône Poulenc/General Mills
Pullulan	Present—in development; future —commercialization announced	Pullulan	Hayashibara Corp.
Erwinia exopolysaccharide	Present—in production (U.S.A.); future —not known	Zanflo	Kelco
Scleroglucan	Present—in development; future —not known	Polytran F.S.	Pillsbury
Microbial alginate	Present—in development; future —no decision yet to commercialize		
Baker's yeast Glycan	Present—in development; future —not known	BYG[R]	Anheuser-Busch Inc.
Curdlan	Present—in development; future —not known		Takeda Chemical Ind.

starch surplus stemmed from an overproduction of corn in the American mid-west and, as a source of glucose, it was considered to be a useful potential fermentation substrate.

D. Dextran

Dextrans are polyglucans produced by a wide range of bacterial species. The nomenclature of many of the strains cited is possibly dubious, but the list of dextran-synthesizing bacteria includes Gram-positive and Gram-negative species: *Aerobacter* spp., *Aceto-*

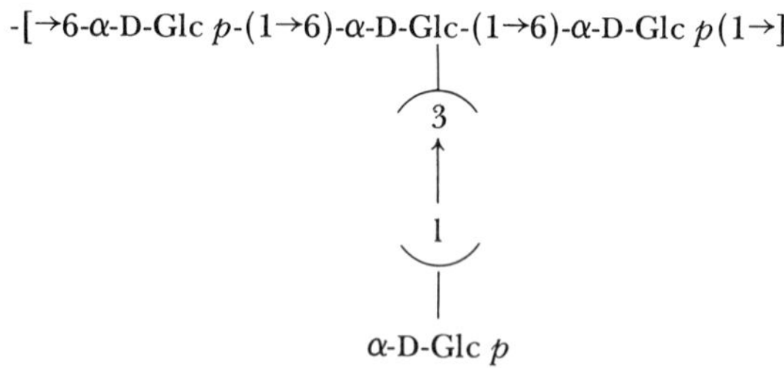

$$-[\rightarrow 6\text{-}\alpha\text{-D-Glc }p\text{-}(1\rightarrow 6)\text{-}\alpha\text{-D-Glc-}(1\rightarrow 6)\text{-}\alpha\text{-D-Glc }p(1\rightarrow]$$

3
↑
1

α-D-Glc p

Fig. 2. Structure of dextran from *Leuconostoc mesenteroides* N.R.R.L.-B-512.

bacter spp., *Betabacterium dextranicum, Streptococcus bovis, Strep. viridans, Streptobacterium dextranicum* and *Leuconostoc mesenteroides*. Most work has used strains of the last named species. The products are normally high molecular-weight polysaccharides containing up to 90–95% of α 1 → 6 linked glucose residues, the remaining linkages being either α 1 → 4 or α 1 → 3. The proportion of different linkages varies considerably (Fig. 2), and the content of α 1 → 6 links may be as low as 50% (Jeanes *et al.*, 1954). The differentiation of dextrans into three different groups, A, B and C, has been based on the presence of certain proportions of the three types of linkages, but separation has also been made into water-soluble polysaccharides (Table 11). The molecular-weight range of the dextran formed by one bacterial strain, under given conditions, reveals considerable differences in the size of the products. In extreme cases, the range might be from 50,000 to 3×10^8 daltons. Claims have been made for conditions under which specific strains of *L. mesenteriodes* or *Streptococcus* sp. are capable of producing

Table 11

Properties of dextrans.
From Jeanes *et al.* (1954)

Group	Linkage class	Intrinsic viscosity (at 25°C)	Nature of polymer	Appearance of 1-2% (w/v) aqueous solution
1	A, B	1.2-0.6	Very cohesive, tough gum or flocculant	Very turbid
2	A, C	0.5-0.2	Fine or flocculant precipitate	·Opalescent solution
3a	A, B	0.9-0.5	Fine or flocculant precipitate or dense gum	Very turbid
3b	C	1.4-0.5	Flocculant precipitate or dense gum	Very turbid
4a	A, B	1.3-1.0	Soft gum	Slightly opalescent
4b	A, B	2.0-1.6	Cohesive, stringy gum	Slightly opalescent
4c	A, C	1.4-0.4	Stringy or fluid gums	Clear or slightly turbid
5a	A, B	1.2-0.6	Short or stringy gums	Turbid or slightly opalescent
5b	A, B	1.0-0.9	Flocculant precipitate or short gum	Slightly to very turbid

material with a given molecular weight range—as for example dextrans of molecular weight 54,000–60,000 daltons (Hehre, 1956), but it is more usual to obtain high molecular-weight material and to degrade it by mild acid hydrolysis. Some strains produced polymer which retained the molecular-weight characteristics if left in the culture fluid for prolonged periods, while others were partially degraded to material of smaller size (Jeanes *et al.*, 1957). Smith (1970) suggested that strain B-1229 of *L. mesenteroides* produced a complex of dextransucrase and insoluble dextran which could be used as a precursor for soluble dextran. The soluble polymer was only released from the complex on completion of synthesis.

Polysaccharide production is obtained in fermenters inoculated with 10% of a seed culture. The enzyme dextransucrase catalysing the reaction:

$$n \text{ Sucrose} \rightarrow (\text{Glucose})_n + n \text{ Fructose}$$

is rapidly produced and the culture pH value falls. Adjustment is made with alkali, and further sucrose is added at intervals. As production of dextran is rapid and is basically a process involving the use of bacterial extracellular enzymes present in the culture fluid, contamination presents less of a problem than is found in more prolonged fermentations. Cultures of *Leuconostoc mesenteroides* do not require vigorous aeration.

The product is normally a high molecular-weight polysaccharide which is not suitable for use in the role of blood extenders for which most dextran is produced. Addition of low molecular-weight dextran to the culture fluid prior to inoculation has been found to provide numerous receptor molecules on which dextran molecules are built up. The product from such preparations has a lower average molecular weight in the range $10^5 - 10^6$ daltons. Harvesting of either low or high molecular-weight polymer follows precipitation by addition of alcohol or acetone.

High molecular-weight dextran is normally submitted to degradation with acid or heat, but enzymic methods may also be developed to produce the right range of molecular sizes. Some dextranases such as those described by Zevenhuizen (1968) and by Sawai *et al.* (1973) are exo-enzymes releasing glucose and maltose, respectively. Other enzymes are endodextranases breaking the linear polymer chains at random; under carefully controlled conditions, these enzymes might replace other methods of dextran fragmentation. In terms of significant commercial success dextran has been limited in that production has been confined to several hundred metric tonnes per annum. In technical terms, production of dextran is the nearest synthesis to a chemical reaction for polysaccharide production that exists as, although early production was via cell cultures, a cell-free enzymic synthesis was developed and forms the basis of dextran production today (Whistler, 1973).

Other glucans showing some similarities to dextrans are formed by *Streptococcus mutans*, one of the species of oral streptococci associated with dental caries. In one polymer, the glucose residues are α $1 \rightarrow 3$ linked as a high molecular weight, water-insoluble polysaccharide (Guggenheim, 1970); other polymers containing a mixture of α-glucosyl linkages are also formed. The enzymes involved in synthesis of these polysaccharides can be isolated in an extracellular cell-free system capable of producing the insoluble polymer and a water-soluble dextran (Baird *et al.*, 1973). A third product, a water-soluble β $2 \rightarrow 1$ linked fructan, was also identified. The ratio of product formation was approximately 1:3:5 and 13% of the sucrose was converted to polymeric material. This low value was ascribed to the possible presence of invertase-like enzymes. Although the water-insoluble polymer presumably plays an important role in cariogenesis, it might also be suitable for commercial applications as a binding agent. The adherence of *Streptococcus mutans* cells through the mediation of the polymer has, however, been thought to be due to the combination of cell-bound enzymes and a 'binding-site' on the bacterial surface (Mukasa and Slade, 1973).

E. 'Xanthan Gum'

A strain of *Xanthomonas campestris* examined by Sloneker and Jeanes (1962) in the Northern Regional Research Laboratories at Peoria, Illinois, U.S.A. produced an exopolysaccharide composed of D-glucose, D-mannose and D-glucuronic acid residues and was both acetylated and pyruvylated. The monosaccharides were present in the approximate molar ratio 3:3:2. On the basis of further studies involving periodate oxidation and partial acid hydrolysis (Sloneker *et al.*, 1964), a repeating-unit structure with a degree of polymerization of 16 was proposed, the acetylated and pyruvylated residues being mannose and glucose, respectively. However, this structure has recently been extensively revised and will be discussed more fully later in this chapter (pg. 357).

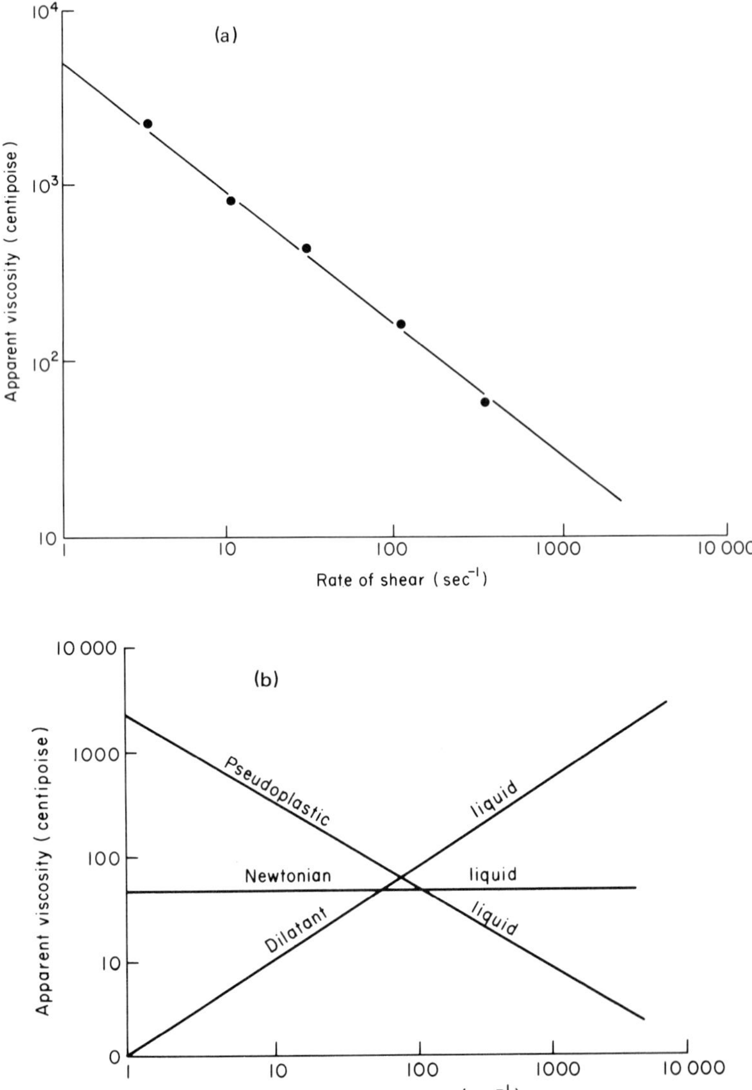

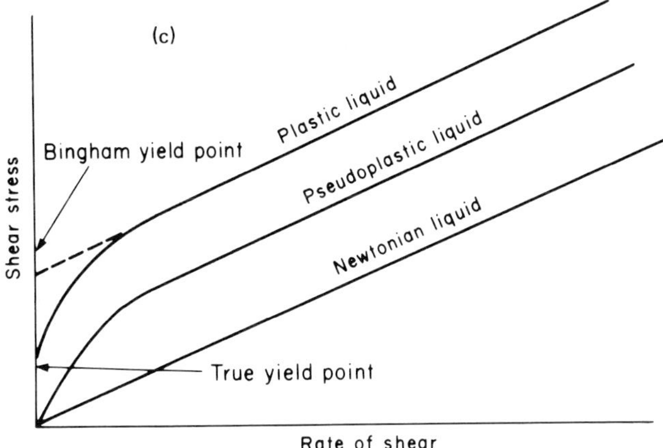

Fig. 3. Flow behaviour (a) of xanthan gum solutions using power law rheology, which is described as follows:

$$\text{For a Newtonian liquid:} \quad T = \eta a(\mathrm{d}v/\mathrm{d}x) \tag{1}$$

where T equals the shear stress in dynes cm^{-2}, $\mathrm{d}v/\mathrm{d}x$ equals the shear rate in sec^{-1}, and ηa is the apparent viscosity in $\mathrm{g\ cm}^{-1}\ \mathrm{sec}^{-1}$. For a large number of viscous substances, the following empirical relationship holds:

$$T = k(\mathrm{d}v/\mathrm{d}x)^n \tag{2}$$

This is the *power law equation*, in which k is the consistency index, and n is defined as the flow behaviour index. For Newtonian liquids, n equals unity; for plastic or pseudoplastic liquids, the value is less than unity, and for dilatant liquids greater than unity. Combining equations (1) and (2):

$$\eta a = k(\mathrm{d}v/\mathrm{d}x)^{n-1}$$

which, on taking logarithms, gives:

$$\log \eta a = \log k + (n-1) \log (\mathrm{d}v/\mathrm{d}x) \tag{3}$$

It follows that a plot of apparent viscosity (ηa) against shear rate ($\mathrm{d}v/\mathrm{d}x$) on logarithmic paper (b) will give a straight line, or a number of straight lines, depending upon whether the value for n remains constant over the full shear range examined. In the data shown for xanthan gum, the value for k is 5,800, and for n 0.19.
The value for k is obtained from the graph at the intercept, i.e. at a shear rate of one sec^{-1}. The value for n is obtained from the slope of the line, i.e. the slope is $n-1$, and $n = 1 + \text{slope}$.

Two types of shear-thinning behaviour have been categorized. In the first instance, when the shear stress is applied to solutions of some polysaccharides, e.g. xanthan gum, a certain value must be reached before shear commences (plastic flow). The point where this occurs is called true yield point (c).
If the straight-line portion of the curve is extrapolated back to the shear stress axis, the intercept that results is known as the Bingham Yield point.
In the second instance no critical yield stress exists (pseudoplastic flow).
The reader is referred to (Sherman, 1975) for further information on rheology.

At Peoria, laboratory culture of the bacteria and isolation of the xanthan gum were scaled up using fermenters of varying size (90–900 litres) containing a growth medium composed of corn sugar, distiller's solubles and salts (Rogovin *et al.*, 1961). Recovery of the product was accomplished by centrifugation to remove the bacteria, followed by methanol precipitation in the presence of potassium chloride, and vacuum drying. In this early work, highest conversion of substrate to polymer was obtained when the initial concentration of glucose in the medium was limited to 1%; an increased glucose concentration led to maximal polysaccharide formation without complete utilization of substrate although various attempts were made to alter the culture conditions. Although polysaccharide production continued over a period lasting up to 8 days, most of the polymer was formed in the first 72 h, by the end of which time the viscosity of the culture fluid had reached almost 4,000 cP. Longer fermentation led to higher viscosity (c. 7,000 cP) and problems in removing the bacteria.

1. Commercial Development

The work at Peoria demonstrated that: (i) xanthan gum was produced in high yield; (ii) solutions of the gum showed high viscosity at low concentration (Fig. 3); (iii) aqueous solutions of the gum were highly pseudoplastic (Fig. 3); (iv) in terms of practical use the viscosity of xanthan gum solutions was independent of temperature (the detailed behaviour of aqueous solutions of xanthan gum is complex; see Rees, 1975); (v) the gum was stable over a wide range of pH values; (vi) aqueous solutions of the gum formed stable gels in the presence of locust bean gum.

On the basis of the behaviour described, and the high yields of gum, commercial development of xanthan gum was considered appropriate.

Laboratory and pilot-plant development work continued at Peoria into the early 1960s, when the U.S. Department of Agriculture approached various industrial companies which they considered might have an interest in commercializing the process. A number of groups originally expressed interest. Of these Kelco, General Mills and Exxon Production Research have maintained a significant interest. Kelco took the process through a pilot-plant evaluation, took out several United States production patents, and commenced commercial production in 1967. At present, the company is the major manufacturer of the product, and it is estimated that production in 1975 was the order of 5,000 tonnes at prices as set out in Table 12. Local expansion of the plant has been announced, and a further

Table 12

Retail prices quoted for xanthan gum in November 1975

Product trade name	Source of price	Product manufactured in	Quoted Price (tonnes £)
Keltrol (food)	Associated British Maltsters (U.K.)	America (Kelco)	5,010[a]
Kelzan (industrial)	Associated British Maltsters (U.K.)	America (Kelco)	3,865[a]
Keltrol	Kelco (Chicago)	America (Kelco)	4,000
Kelzan	Kelco (Chicago)	America (Kelco)	3,000
XC Polymer (oil drilling)	Milchem Inc. (Houston)	America (Kelco-Xanco)	6,000[b]
Rhodigel 23 (food)	Rhodia U.K.	France (Rhône Poulenc)	5,000[c]
Rhodipol 23	Rhodia U.K.	France (Rhône Poulenc)	3,650[c]
XC Polymer (oil drilling)	Imco Services (Houston)	America (Kelco-Xanco)	6,000[b]

[a] Price will include: (i) U.K. import duty, (ii) transport costs from U.S.A. to U.K. and (iii) profit to the producing firm.
[b] Xanthan gum is sold to the oil industry as part of a 'service' package, and this price reflects the considerable personnel and expertise involved.
[c] Price will include: (i) transport costs from France to U.K., and (ii) U.K. import costs.

plant to produce xanthan and other biopolymers is under construction at a cost of U.S. $35m. This plant is planned to go on-stream in late 1976. General Mills formed a joint venture group with the French chemical company, Rhône Poulenc, to produce xanthan gum in a plant at Melle in the département of Deux-Sevres, reported to go on-stream during early 1975, product prices quoted being those in Table 12. In the United States of America, General Mills have announced the construction of a plant in Iowa to come on-stream in 1978. Little information is available as to plant capacity, but plants having a capability of producing much less than 2,000 tonnes per annum are unlikely to be viable. The World production of xanthan gum in 1975 was estimated to be approximately 5,500 tonnes and it is possible that, by the end of the decade, consumption could total 12,000 tonnes.

Exxon production research pursued its own development of the gum, more from a user aspect as an oil-well drilling mud viscosifier in primary oil-well drilling applications, and in enhanced recovery from partially depleted reservoirs as a mobility control agent in conjunction with other chemicals such as surfactants. The latter application will be crucial in the future development of xanthan gum and possibly other biopolymers, and will be discussed more fully later in this chapter. Exxon have obtained a number of both user and producer patents, and now issue licences for the Kelco product to be used for oil drilling and flooding, and collect royalties reported to be U.S. 0.2$/kilo. Although Exxon have conducted an extensive pilot-plant programme into the production of xanthan, commercial production of the gum has not yet been announced.

2. Applications for Xanthan Gum

The unique physical properties shown by aqueous solutions of xanthan gum have found application in many non-food industries, such as textile printing and dyeing, oil-well drilling and flooding, ceramic glaze manufacture, cleaners and polishes, latex emulsion paints and rust-curing agents. The most important single use for the gum is in the production of crude oil. It was found that the pseudoplastic flow behaviour combined with its peculiar stability towards acid, alkali, heat and many cations gave xanthan gum a technical advantage over other polymer lubricants in the bentonite muds used to drill oil wells. Kelco in conjunction with Exxon set up a subsidiary company, Xanco, in Houston, Texas, to exploit oil-recovery applications.

A more recent opportunity for the gum stems from research being undertaken by many oil companies throughout the World, namely enhanced oil recovery. For the purpose of discussion, enhanced recovery is the general technique of winning further oil from a well after the natural pressure has ceased to force the oil out. In practice, substantial amounts of either thermal or chemical agents are introduced, which before their injection were not part of the reservoir. Thermal approaches could include steam flooding and fire flooding; chemical methods would include injection of carbon dioxide or a combination of surfactants, hydrocarbons (often alcohol) and a mobility control agent such as a polymer. It is this latter 'surfactant flooding' technique which has been most successful to date, and xanthan gum is one of the polymers of current interest (the other most important polymer being polyacrylamide). Theoretically, a very large volume of crude oil remains to be recovered by enhanced techniques. Of approximately 434 billion barrels of crude oil discovered to date in U.S.A., only about 32% (140 billion barrels) has been produced, or is regarded as a producible reserve, based upon economics and technology available at the end of 1974 (American Gas Association, 1975). It is estimated that 40–60 billion barrels may be producible by enhanced methods, with 70–90 billion barrels to be produced by improved enhanced techniques. This will still leave 165–170 billion barrels unrecoverable with known or used technology. Many U.S. companies are now actively interested in this area with an eye towards supplying the large volumes of materials which will be required, as it has been estimated that to obtain 100,000 barrels of oil per day would require 623×10^6 lbs of petroleum sulphonates per year, 29.7×10^6 litres of alcohols per year and 37.4×10^6 lbs of polymers such as xanthan gum or polyacrylamide (Sharp, 1975; Umland, 1974).

The breakthrough for xanthan gum in human-food markets came when acceptance was published in the United States (Federal Register, 1969). Since that time, the gum has been developed for use in fruit-flavoured beverages, canned and frozen foods, relishes, french dressings, instant foods and toppings and whips. These products are all improved in terms of mouthfeel and texture by inclusion of small quantities of the gum into the product often in conjunction with locust bean gum with which it forms a stable gel. The gum has recently been the subject of draft approval in Category II of the E.E.C. food regulations (R/1233/73(Agri 404)). Another sector with a strong interest in xanthan gum is the animal-feed industry because use of xanthan in conjunction with locust bean gum is extremely competitive with agar as a jellying agent in canned pet foods, or as a suspending agent for low-solids liquid animal feeds.

3. Fermentation Process

Xanthan gum is produced commercially in a conventional batch process using, as carbohydrate substrate, commercial grade D-glucose or starch thinned using a combination of acid and enzymic processes. Nitrogen sources such as corn-steep liquor (from the corn wet-milling industry), casein hydrolysate (from dairy processors) and distiller's dried solubles (from the production of alcoholic beverages) have been used at one time or another but, due to their rather undefined and variable composition, have been largely superseded by yeast and soy extracts. The inorganic requirements of the fermentation are dipotassium hydrogen phosphate, for essential mineral requirements and buffering, and magnesium chloride.

Growth conditions are carefully controlled, the important variables being temperature, pH value and fermentation time. Aeration and mixing of the fermentation broth are critical variables as, due to the extremely high viscosities encountered, oxygen-transfer rate is affected and will become limiting unless the fermenter baffle and impeller geometry are carefully designed for optimum gas transfer.

When fermentation is terminated, the broth has a pH value of about 6 and a viscosity of greater than 30,000 cP (measured on the Contraves viscometer at $25°C$ and a shear rate of one sec^{-1}). There are no reports that, in industrial fermentations, the cells are removed from the fermentation broth, although recent patents have claimed the use of enzymes for cell digestion (Kelco, 1975a), products normally being precipitated using methanol or isopropanol in the presence of potassium chloride. The wet cake can then be recovered by filtration or centrifugation, and the product shredded before drying on a moving band drier. Alcohol is normally recovered from the spent liquor by distillation and from the cake by passing the hot gasses from the drier into charcoal recovery columns. The yield of product in such a fermentation can be expected to be between 75 and 80%. However, for reasons already mentioned, the concentration of gum is limited to less than 5% in the fermenter, mainly due to the high broth viscosity encountered.

Continuous production of xanthan gum has been investigated in the laboratory (Rogovin, 1969), but there is no evidence of industrial fermentations which employ this mode of culture. If continuous culture of xanthan could be achieved on an industrial scale, production costs would be greatly lowered as batch processes run to approximately 80 h, whereas a continuous process with a dilution rate of 0.05 h^{-1} would give a 20 h fermentation. A continuous process, therefore, could improve the efficiency of plant utilization with a consequent lowering of capital costs.

4. Chemical Structure

As already stated, the chemical structure of xanthan gum as originally conceived was composed of a 14 or 16 unit repeat structure made up from residues of glucose, mannose and glucuronic acid and the substituents pyruvate and acetate (Sloneker *et al.*, 1964;

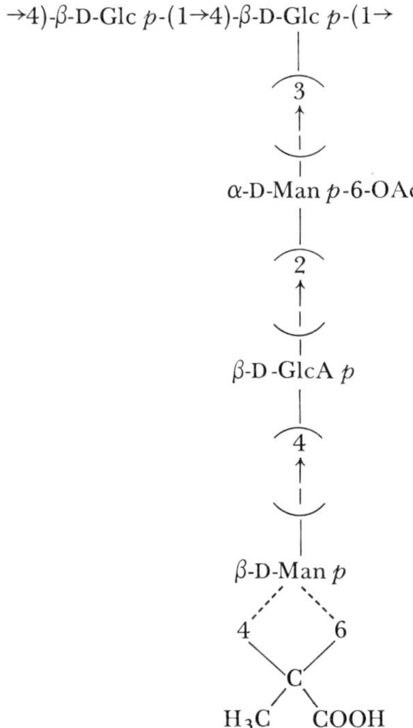

Fig. 4. Proposed covalent structure for xanthan gum.

Siddiqui, 1967). This early structure has now been extensively revised, and the gum has been shown to consist of a cellulose backbone with a side chain, containing residues of mannose, glucuronic acid and acetate, on every second main-chain glucose unit; the pyruvate residues are attached non-stoicheiometrically as a ketal to the terminal side-chain mannose unit (Fig. 4; Jansson *et al.*, 1975; Melton *et al.*, 1976; Lawson and Symes, 1977).

Measurements of molecular weight using light-scattering techniques gave values of 13×10^6 and 50×10^6 daltons. A dispersion of the polysaccharide made in 4 M urea and heated gave a value of 2×10^6 daltons. These differences were explained as being a consequence of differences in fermentation conditions, and micro-gel formation (Dintzis *et al.*, 1970).

F. Pullulan

1. Commercial Development

Pullulan, the exocellular polysaccharide elaborated by *Aureobasidium pullulans*, is an example of a polysaccharide at present at the development stage. The material was originally of interest solely as a substrate for pullulanase, an enzyme which breaks $\alpha 1 \rightarrow 6$ glucosyl linkages. However, recent interest has been shown in the gum through its ability to form strong resilient films and fibres, and to be moulded into 'shaped bodies'. The Hayashabara Company of Japan have been pursuing the commercial development of pullulan, and were in pilot-plant production in 1974 with a plant producing 12 tonnes per annum. The construction of an intermediate-scale commercial plant with a capacity of several hundred tonnes will commence in June 1975 (Yuen, 1974).

2. Applications for Pullulan

Patents have been issued claiming unique use for pullulan in several areas of food and industrial applications. However, as products are not yet on sale, little information is available as to the practical value of the claims made.

3. *Chemical Structure*

The gum is composed of maltotriose units or occasionally malto-tetraose units linked together with $\alpha 1 \rightarrow 6$ bonds (Fig. 5; Catley and Whelan, 1971).

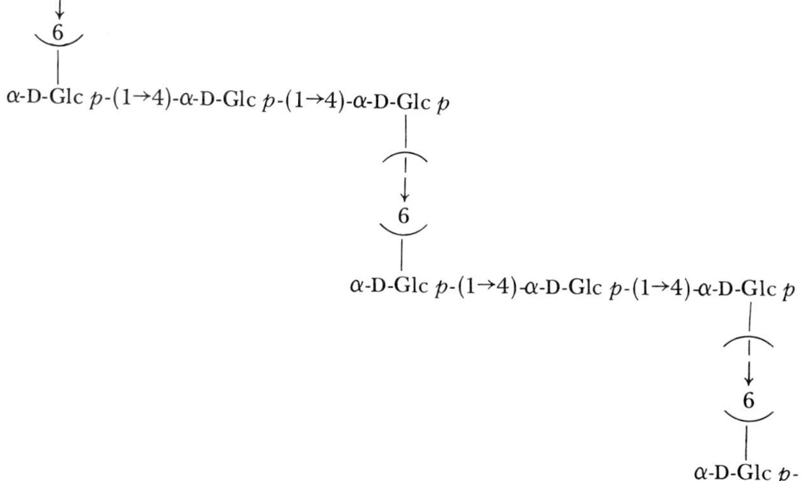

Fig. 5. Covalent structure of pullulan.

G. Microbial Alginate

1. *Technical Development*

Alginic acid is a commercially important gum with gelling and visco-elastic properties of use in the food, textile, pharmaceutical and paper industries (Percival and McDowell, 1967). Traditionally, alginates have been regarded as products found in the cells of *Laminaria* spp. and other related members of the Phaeophyceae, where they are important structural polymers. In these seaweeds, the alginate content varies considerably, and the extracted polymer shows variations in the ratio of mannuronic acid to guluronic acid residues. Two species of Gram-negative bacteria, *Pseudomonas aeruginosa* and *Azotobacter vinelandii*, were found to produce exopolysaccharides closely resembling the algal products in being heteropolysaccharides containing residues of D-mannuronic acid and L-guluronic acid (Gorin and Spencer, 1966; Linker and Jones, 1964, 1966). The polysaccharides from *A. vinelandii* contained mannuronic acid as the major component, and acetate was present as well as guluronic acid. Comparison with algal alginate showed similar periodate uptake, but a higher ratio of mannuronate to guluronate residues and the added presence of acetate. It should be remembered, however, that algal material is traditionally extracted with weak alkali, a procedure which would remove any acetyl groups which might have been originally present. A later study of alginate from *Azotobacter vinelandii* (Larsen and Haug, 1971) indicated a similar ratio of mannuronate to guluronate residues in several preparations and an acetyl content to give one substituent group on every five monosaccharide residues. The product from *Ps. aeruginosa* was found in strains isolated from cystic fibrosis patients and shown independently to resemble alginate (Linker and Jones, 1966; Carlson and Matthews, 1966).

Convincing evidence for a polymannuronate was provided by Linker and Jones (1966) for a single preparation from *Pseudomonas aeruginosa*. Most preparations resembled seaweed alginate in their range of uronic acid ratios and their infrared spectra. This study was extended (Evans and Linker, 1973) to polysaccharides prepared from a wide range of *Ps. aeruginosa* isolates. All of the polysaccharides were similar, being alginate-like polymers containing residues of D-mannuronic acid and L-guluronic acid as well as acetyl groups, the D-mannuronic acid residues being β-linked. The ratios of the component uronic acids varied, as did the acetyl contents. The polysaccharides were also compared as substrates for two enzyme preparations from a bacterium and abalone, respectively. Some differences were noted in the oligosaccharides produced from different substrates and, in general, the pattern of fragments from bacterial enzyme digests was much simpler than that from abalone enzyme treatment.

Although polysaccharides from *A. vinelandii* and *Ps. aeruginosa* have been identified as alginates, there have been few reported studies on the conditions necessary for the production of these polymers. *Azotobacter* isolates are usually mucoid and appear to produce alginate when grown in nitrogen-free medium. There is, presumably, an effective nitrogen-limiting, carbohydrate-rich situation favouring polysaccharide synthesis in a manner similar to that found in the Enterobacteriaceae (Wilkinson, 1958). The composition of the polymers is unaffected by the carbohydrate source used, which may be mannitol, sucrose or other sugars.

Evans and Linker (1973), in their study of alginate-synthesizing strains of *Ps. aeruginosa*, noted that the conditions required for maximal polysaccharide production also resulted in minimal reversion to the non-mucoid state. Very little polysaccharide was produced at 37°C, and more was formed at 12°C than at 25°C, although bacterial growth was very much slower at the lower temperature. Optimal production was obtained on supplemented MacConkey bile-salt agar. We have also noted (I. W. Sutherland, unpublished results) that alginate from *Pseudomonas aeruginosa* is formed in greater amounts on MacConkey agar rather than on a nitrogen-deficient medium with excess carbohydrate which was earlier shown to be suitable for producing large amounts of the polysaccharide colanic acid from the strains of Enterobacteriaceae (Sutherland and Wilkinson, 1965). Factors favouring alginate production on MacConkey agar included addition of 5% glycerol or 0.25 M sodium chloride. Higher concentrations of either sodium chloride or glycerol inhibited both bacterial growth and polysaccharide production at 25–37°C, but not at 12°C. Good production of polysaccharide can also be obtained in shake flasks using liquid medium, but we have not yet defined conditions for optimal production (I. W. Sutherland and N. Piggott, unpublished results).

Pseudomonas aeruginosa was rejected for commercial study because of its association with pathogenic conditions in man, and *A. vinelandii* was selected for examination, the company concerned being Tate and Lyle Ltd. in the U.K. Research commenced in 1969, and by 1975 a process had been developed which was being operated on pilot plant (1 m^3)-scale fermenters.

The term 'alginates' is given to the group of acidic polysaccharides normally extracted from the brown seaweeds, alginates from different sources having been shown to vary within a structural framework based upon a linear block copolymeric arrangement of α-L-guluronic acid and β-D-mannuronic acid residues (Figs. 6, 7). Initial studies based upon small-scale fermentation experiments showed that a number of polysaccharide types could be obtained from cultures of *Azotobacter* sp., all falling within the structural framework already

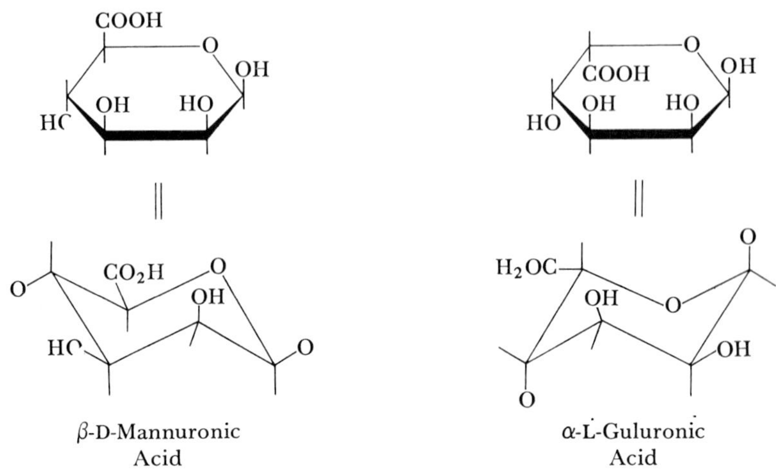

Fig. 6. Structures of component uronic acids in algal and bacterial alginates.

mentioned except that the bacterial product was partially acetylated, and viscosity measurements indicated that molecular weight distributions appeared to be somewhat lower than for most algal alginates. It was considered that, if a number of objectives could be fulfilled, a feasible process for production of microbial alginate might be developed. The objectives were: (a) to increase the yield of product from the 5% found in early experiments to at least 50%; (b) to raise and control the molecular weight over as wide a range as possible. An important determinant of the applications and selling prices of alginates is the viscosity generated by the material. Applications exist where a low viscosity material is preferred but, in general, the higher viscosity products command the highest prices; (c) to achieve control over the mannuronic: guluronic acid ratio, and block structure. The gelling properties of the polysaccharide are determined primarily by these criteria and, as the bacterial product would be required to compete with as wide a range of algal products as possible, this

-G-G-G-G-G-G-G-G-	Poly G
-M-M-M-M-M-M-M-M-	Poly M
-G-M-G-M-G-M-G-M-	Poly MG

Fig. 7. Idealized structures of heteropolymeric and homopolymeric blocks found in algal and bacterial alginates.

objective was considered as important as achieving objective (b); (d) to develop a continuous process.

Of the many fermentation parameters examined in batch culture, e.g. medium composition, culture pH value, temperature, oxygenation of broth, inoculum age and inoculum composition, using the parent strain of *A. vinelandii*, that which produced the single most important increase in both yield and molecular weight of microbial

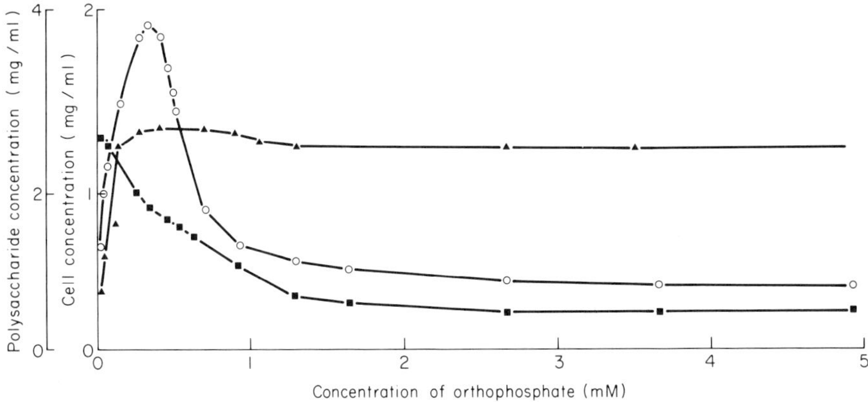

Fig. 8. Effect of phosphate concentration on synthesis of alginic acid by *Azotobacter vinelandii* in batch culture. ▲ indicates change in cell concentration, ○ change in alginate concentration, and ■ the ratio of alginate produced to cell concentration.

alginate was lowering of the concentration of phosphate in the medium to one-twentieth of that in the original growth medium (Imrie, 1973). Because of the decreased buffering capacity of the low-phosphate medium, pH control became necessary to maintain the level of polysaccharide production as, because of alginate and carbon dioxide production, the culture pH value fell until cell growth ceased. These improvements raised the yield of product so that approximately 25% of the sucrose supplied was converted into alginate (Fig. 8). A second result of the changes in medium composition was that solutions of bacterial alginate were found to exhibit higher viscosities. This was due to a narrowing of the molecular-weight distribution together with an increased weight average molecular weight. The visco-elastic properties available from bacterial alginates now fell into the range covered by many low- and medium-viscosity algal products (Fig. 9), and by subsequent process improvements it became possible to match even high-viscosity alginate.

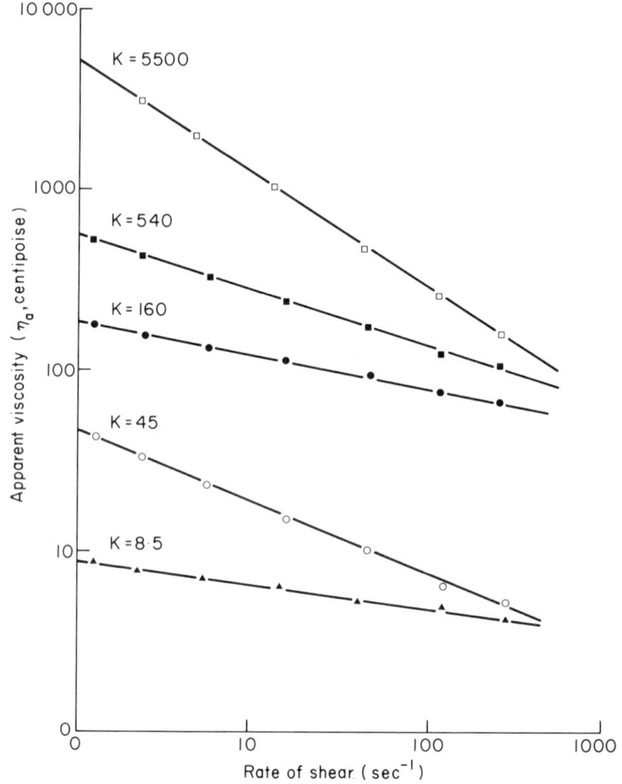

Fig. 9. Improvement in the rheological behaviour of bacterial alginate resulting from process modification. The bottom line describes behaviour obtained using the conditions described by Gorin and Spencer (1966), the next two lines (○, ●) as a result of lowering the phosphate concentration in the medium, and the top two lines (■, □) from further process modification.

Further improvements in the efficiency of conversion of sucrose into alginate by the bacteria were achieved using continuous culture. The maximum efficiency was increased to 50% compared to the 25% in batch cultures already described, although the product concentrations remained low. Fermentation productivity was also increased using continuous culture, giving a five-fold increase in similar alginate concentration to the batch culture (from 0.044 to 0.20 kg m^3 h).

Table 13

Effect of calcium ion concentration on the ratio of mannuronic to guluronic
acid residues in alginate from *Azotobacter vinelandii*

Concentration of $CaCl_2$ (mM)	0	0.30	0.50	0.75
Percent mannuronic acid residues	78	73	55–46	31

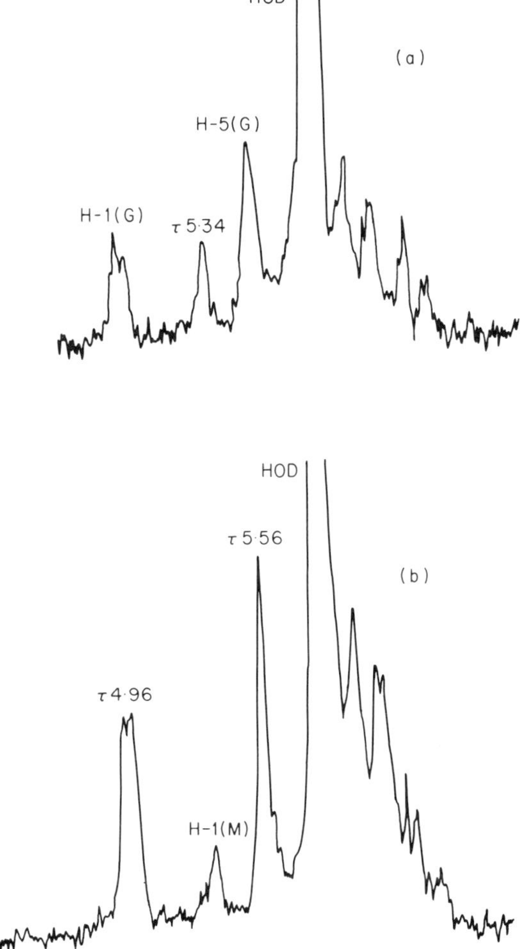

Fig. 10. Paramagnetic resonance spectra (100 MHz) of homopolymeric blocks in (a)
bacterial alginate, and (b) in algal alginate, both recorded at 80°.

It has been shown in the Tate and Lyle Laboratories and elsewhere (Haug and Larsen, 1971) that the concentration of calcium ions in the medium affects the epimerization of mannuronic acid to guluronic acid, and a range of materials with widely different mannu-

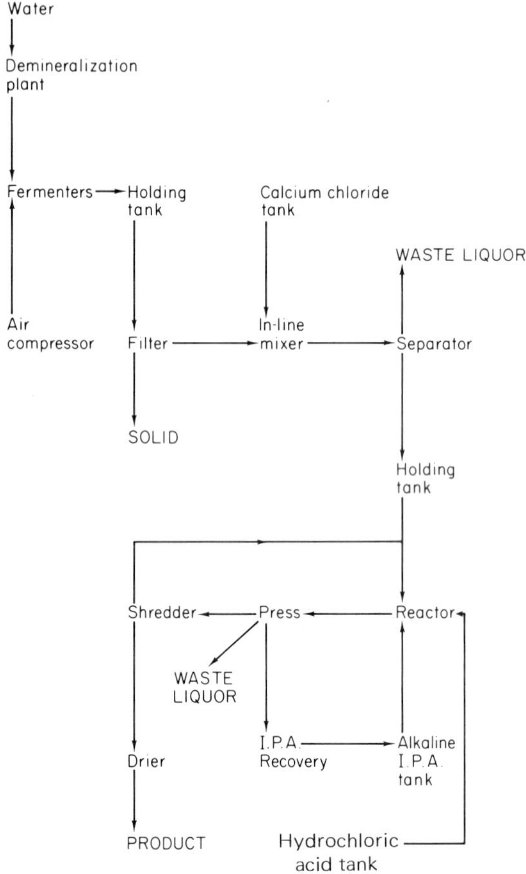

Fig. 11. A possible recovery process for microbial alginate.

ronic: guluronic acid ratios has been produced (Table 13). Knowledge of the effect of calcium ion concentration on alginate block structure is, at present, limited but, using the nuclear magnetic resonance (nmr) technique of Penman and Sanderson (1972), it should be possible to follow variations in this parameter. A nmr spectrum of blocks derived of algal alginate is compared with those from bacterial alginate in Fig. 10 showing a marked similarity in the spectra. The recovery of microbial alginate could be based upon conventional solvent-precipitation processes as for xanthan gum, but an analogous process to that described for the algal product might be considered (Fig. 11).

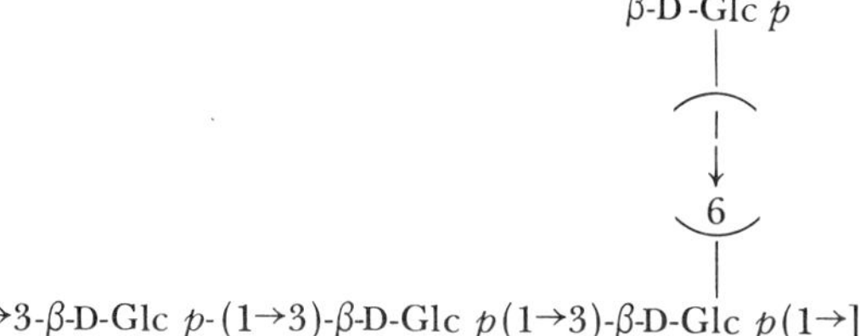

$$\beta\text{-D-Glc } p$$

$$-[\rightarrow 3\text{-}\beta\text{-D-Glc } p\text{-}(1\rightarrow 3)\text{-}\beta\text{-D-Glc } p(1\rightarrow 3)\text{-}\beta\text{-D-Glc } p(1\rightarrow]_n$$

Fig. 12. Covalent repeating unit proposed for the scleroglucan from *Sclerotium glucanicum*.

H. Scleroglucan

Scleroglucan from the fungus *Sclerotium glucanicum* is a β-linked glucan with the covalent structure shown in Fig. 12 (Johnson *et al.*, 1963). This polysaccharide was commercially developed by The Pillsbury Company who market it under the trade name of Polytran. The properties claimed for Polytran are those of pseudoplasticity over a broad range of pH values and temperature, its performance being largely unaffected by the presence of various salts. In application, it is claimed that Polytran stabilizes the rheology of bentonite clays during storage, exposure to high temperature and, when mixed with acids and alkalis, it also improves ceramic glazes, latex paints, drilling muds, printing inks and seed coatings. However, little information is available concerning uses for Polytran. In 1975, prices were in the region of £4,750 per tonne indicating that, for a substantial market to have developed, unique end-use applications would have been developed precluding other cheaper gums.

I. Curdlan

Curdlan is the polysaccharide elaborated by a chemical mutant of *Alcaligenes faecalis* var. *myxogenes* strain 10C3. Curdlan, in comparison with the polysaccharides callose and pachyman which are known to contain high proportions of β-1,3-glucosidic links, is almost entirely a β-1,3-glucan (Harada, 1974). The commercially important property claimed for curdlan is non-reversible gel formation on heating aqueous solutions. Curdlan gels have been studied (Kimura, 1972) and found to be intermediate in strength between typical examples of agar and gelatin gels. These properties, it is claimed, could be useful in, for example, foods which during production, or cooking, are heated with water (noodles, spaghetti, jellies or mayonnaise). The Takeda Chemical Company of Japan has been developing curdlan but, as in the case of Polytran, little information is available concerning the commercial status of the material.

J. *Erwinia* Polysaccharide

This microbial polysaccharide, 'Zanflo', has been developed specifically for use in carpet-printing applications. The polymer is thought to be produced from a strain of *Erwinia tahitica* and is a product of the Kelco Company (Kelco, 1975b). No detailed study of the covalent structure of the gum has been published. It is, however, reported that the gum is composed of residues of glucose, galactose, uronic acid and fucose in an approximate ratio of 3 : 2 : 1.5 : 1, and has an acetyl content of 4.5% (Kang and Kovacs, 1974).

The physical properties claimed appear to be similar to those of xanthan gum, e.g. high pseudoplasticity, good acid and alkali stability, compatibility with most cationic agents and freeze-thaw stability, the main difference being that solution viscosities undergo thermally reversible changes (Kelco, 1971). No market or price information is available.

IV. DEVELOPMENT OF NEW MICROBIAL POLYSACCHARIDES

The laboratory interested in producing exopolysaccharides might initiate a development programme in which the work follows the pattern: (a) screening material from various sources for suitable microbial species; (b) identification of mucoid (slime- or capsule-producing) micro-organisms; (c) selection of a number of strains based upon a qualitative assessment of the chemical and physico-chemical properties of the polymers; and (d) further development of a rigorously selected group of isolates to pilot-plant scale in order to obtain a comprehensive background knowledge of the growth characteristics of the micro-organisms, the conditions essential for polysaccharide production, and a detailed knowledge of the physical and chemical characteristics of the polysaccharide.

A. Sources and Isolation

A decision to attempt isolation of polysaccharide-producing micro-organisms can be based on one of two philosophies. One may either

Table 14

Some polysaccharide-producing micro-organisms

Yeasts	*Rhodotorula* spp.
	Pichia spp.
	Pachysolen tannophilus
	Lipomyces spp.
	Hansenula capsulata
	Hansenula holstii
	Cryptococcus spp.
	Torulopsis molischiana
	Torulopsis pinus
	Aureobasidium pullulans
Other fungi	*Penicillium* spp.
	Tremella mesentaria
Gram-positive bacteria	*Bacillus* spp.
	Leuconostoc spp.
	Streptococcus mutans
	Streptococcus spp.
Gram-negative bacteria	*Azotobacter* spp.
	Rhizobium spp.
	Escherichia coli
	Klebsiella aerogenes
	Acetobacter xylinum
	Arthrobacter viscosus
	Pseudomonas aeruginosa
	Xanthomonas campestris
	Xanthomonas phaseoli
	Achromobacter spp.
	Alcaligenes faecalis var. *myxogenes*
	Agrobacterium spp.
	Erwinia spp.
	Sphaerotilus natans

isolate and examine micro-organisms known to produce large amounts of exopolysaccharide, or one can examine isolates from certain carbohydrate-rich or other environments and determine whether any have the desired culture characteristics.

Many species of bacteria and yeasts are capable of producing extracellular polysaccharides either in the form of capsules or of slime. The chemical composition of these polysaccharides has been reported for many preparations, but detailed structural studies have been made on only a few microbial exopolysaccharides. They may be heteropolysaccharides containing up to four monosaccharides in repeating unit structures ranging in size from disaccharides to hexasaccharides, and in at least one instance heteropolysaccharides with no obvious ordered repeating structures, namely, microbial alginates, or they may be homopolysaccharides formed from a single type of monosaccharide residue. Any of these types of polymer may contain various types of acyl substituents. Some of the microbial species which normally produce exopolysaccharides are listed (Table 14). They include yeasts, Gram-positive and Gram-negative bacteria. Very little is known about the properties and potential of polysaccharides from the surfaces of green and blue-green algae, but it is probable that the slower growth rates of these micro-organisms makes commercial production of any polymers produced by them less attractive.

B. Carbohydrate-Rich Environments

One approach is based on the fact that many habitats contain large amounts of carbohydrate waste materials from industrial processes, while only relatively small amounts of nitrogeneous substrates are available. These conditions favour polysaccharide production in many micro-organisms. Investigation of such environments may thus provide useful exopolysaccharide-producing micro-organisms which, after isolation in pure culture, can be screened, identified and evaluated. The added advantage of obtaining microbial species from such environments is that they should be capable of converting the waste (i.e. cheap) carbohydrate to an extracellular polymer which is

hopefully of commercial value. Possible locales and their carbohydrate sources are:

Sugar (beet or cane) refineries—sucrose
Paper and pulp mills—cellulose
Potato crisp or chip-processing plants—starch
Brewery or distillery effluents—various
Corn or vegetable-processing plants—starch, cellulose
Activated sludge—various

While improved treatment methods should ensure that water
discharged from such processes ought to have low carbohydrate
content and a low Biological Oxygen Demand, the liquors prior to
treatment or in settling tanks or lagoons could prove a source of
exopolysaccharide-producing micro-organisms. Samples taken from
machinery, or the buildings housing it, should also be examined
when processing plants are considered as a source of micro-
organisms. More detailed information on selective culture methods
and culture conditions, can be found in Norris and Ribbons (1969,
1970).

C. Identification

Identification may well present problems unless the micro-organisms
belong to a well characterized group such as the nitrogen-fixing
bacterial genus *Azotobacter* or the plant-pathogenic genus *Xanthomonas.* Many of the bacteria isolated possess general attributes of the
family Pseudomonadaceae and the approach to identification of such
cultures employed by Hendrie and Shewan (1966) is to be recommended. This necessitates microscopic examination to determine the
Gram reaction and cell morphology, and to ascertain whether the
bacteria are motile. Electron-microscopic examination is made if
flagella are present. Any pigmentation is noted, and fluorescence of
cultures is tested under ultraviolet radiation. The mode of breakdown of glucose is examined as is oxidase activity and other
characteristics. Other bacterial groups will require different tests after
initial examination of the Gram-staining reaction and cell morph-

ology. These can be found for a number of genera in the reports by Gibbs and Skinner (1966) and Gibbs and Shapton (1968).

If the polysaccharide-producing isolates are yeasts, the identification procedure obviously differs from that used for bacteria. Much use can be made of tests for the utilization of sugars and related compounds, of nitrate, and culture pigmentation. Detailed techniques for identification by such methods are provided by Beech *et al.* (1968). Serological techniques developed by Campbell and others have become increasingly important in identification of yeasts, and the application of these along with other more traditional tests in the numerical taxonomy and identification of various yeast genera is reported by Campbell (1970, 1971).

D. Selection and Further Testing

The isolates chosen for further examination will be those producing relatively large amounts of polysaccharide with interesting viscosity characteristics. Those strains which produce only small amounts of polymer or which grow poorly under all observed culture conditions should be discarded. In terms of personnel and equipment, this part of the development programme is the most demanding, as it will probably be necessary to test strains for numerous characteristics in both batch and continuous culture as well as preparing and characterizing quantities of exopolysaccharide produced.

Each of a number of variables should be tested using a basic growth medium. The variables include pH value, concentration of carbon source, concentration of organic and inorganic nitrogen sources, buffering of the media, mineral concentrations and aeration. For general aspects of formulation of laboratory culture media, the reader is referred to Calam (1969) and, for fermenter design, to Blakebrough (1969). Other relevant information is to be found in 'Methods in Microbiology' edited by J. R. Norris and D. W. Ribbons (1969 onwards). Of potential importance from the economic aspect of polysaccharide production are the use of waste carbohydrate products as carbon and energy sources, and the extent of conversion of substrate to polymer. It is unlikely that organisms producing polysaccharide in yields below 40%, and in batch times longer than 90 h, will have a great chance of commercial success.

V. THE PRODUCTS

A. Chemical and Physicochemical Properties

It will be important, at an early stage, to characterize polysaccharides both chemically and for their physical characteristics. The reader is referred to the many texts on polysaccharide analysis (Whistler and BeMiller, 1965; Coffey, 1967) for details but, until a fairly clear idea of certain basis structural features is obtained (nature of sugars, sugar ratios and acyl substituents, positions of linkage and molecular weight), it will be difficult if not impossible to ascertain what effect the various fermentation manipulations have on polymer structure. Also, for the purpose of writing patents and applying to food authorities, it is important to have as much chemical and physical data as possible. In terms of physical characteristics, the flow behaviour of aqueous solutions must be determined (see Fig. 3, pp. 350–351) as well as whether and in what way, the polysaccharides form gels

1. *Flow Behaviour*

Many polysaccharides isolated will have marked non-Newtonian properties, i.e. the apparent viscosity of solutions of the polymers will not be a constant function of the rate of shear. It will be important, therefore, to define the rate of shear, possibly employing the 'Power Law' approach described in Fig. 3 (pg. 351). In practice, capillary viscometers and rotational viscometers having no defined shear boundaries are tedious to use, and produce rather imprecise results. Several cone and plate instruments such as the Wells Brookfield, Shirley Ferranti, Haake Rotovisco, Contraves Rheomat and Hercules models operate at defined rates of shear, and the reader is advised seriously to consider this type of instrument. Another point of practical importance is that, in laboratory fermentations, volumes of culture available for resting are often small (2–10 ml) and many viscometers require several hundred ml for operation.

Having obtained a sufficient quantity of dry polysaccharide for evaluation, the following properties should be determined (test (a) on a routine basis and tests (b)–(g) in the course of rigorous product definition); (a) shear stress–shear rate relationships at standard concentrations and temperatures, applying the power law to obtain standard k and n values; (b) temperature–apparent viscosity at defined shear rate; (c) pH–apparent viscosity at defined shear rate; (d) salt–apparent viscosity at defined shear rate; (e) concentration–apparent viscosity at defined shear rate; (f) shear stability, i.e. the effect of high shear on polymer structure; (g) freeze–thaw stability, i.e. the effect of repeated freeze–thaw cycles on polymer structure.

The data obtained from these experiments will allow the polysaccharide to be assessed objectively for flow behaviour, both on a day-to-day basis (a) and more precisely (b)–(g).

2. Gel Properties

In order to ascertain whether, and in what way, the polysaccharide forms stable gels, it will be necessary to undertake a series of empirical tests based upon the gel behaviour of known polysaccharide systems. Aqueous solutions of the polysaccharide should be subjected, for example, to (a) high and low pH values, (b) high and low temperatures, (c) addition of polyvalent cations, (d) heating and cooling with other gums, e.g. guar, carob, and (e) quaternary ammonium compounds. Conditions (a) and (b) should be applied to (c), (d) and (e). Examples of gel-forming polysaccharides are given in Table 2 (pg. 331).

In an initial screen, a visual assessment of gel or precipitate formation will suffice. However, more precise determination of gel strength in terms of elasticity and rigidity will require the use of instruments such as the Instron gel tester. Eventually, biological testing will be required to ensure that the product is free from toxicity or untoward side effects. Only with all of this and further relevant information, together with the necessary costings, can a decision be taken to develop products further.

B. Biological Criteria

Two problems which may be encountered during the isolation of exopolysaccharide-forming micro-organisms may decrease the number of strains which can effectively be considered. Some polysaccharide-producing species may be human or animal pathogens and

must be excluded from consideration because of the potential danger
to operatives, especially during large-scale production, and because of
the ethical requirements for the final product; all pathogenic micro-
organisms must effectively be removed from the product.
Examples of species which cannot be used for these reasons are the
yeast *Cryptococcus neoformans* and the opportunist bacterial pathogen *Pseudomonas aeruginosa* (a potential source of microbial alginate). Another potential danger is in the use of plant pathogens.
Many bacterial species which are pathogenic for plants produce
exopolysaccharides. Such species include members of the genera
Corynbacterium, Pseudomonas, Xanthomonas and *Agrobacterium.*
Obviously, the large-scale culture of such bacteria can be contemplated only if the system for growth and harvesting is completely
enclosed and precludes contamination of the environment, and if
treatment of the final product is such that complete destruction of
the bacteria can be guaranteed. In Britain, for example, legislation
exists to control the use of plant pathogens. Some species are
regarded as legally non-indigenous to the country, and a licence must
be obtained from the Ministry of Agriculture, Fisheries and Food
before cultures of these species can be held. Other countries have
their own regulations regarding the importation or use of microbial
cultures.

As well as restrictions on the use of micro-organisms with
potential pathogenicity for Man, animals or plants, the polysaccharide produced must be free from toxic microbial components.
This is especially true if the intended use of the polysaccharide is in
the pharmaceutical, cosmetic or food industries. Toxic products of
micro-organisms may be either intracellular or extracellular. The
latter are usually proteins, if the species are bacteria, or compounds of
the aflatoxin type from moulds. If polysaccharide isolation follows
precipitation with organic solvents, some types of toxin are likely to
be removed by destruction and by dissolution in the solvent,
respectively. Much more of a problem are likely to be cell-bound
components which are difficult to remove entirely from the viscous
polysaccharide solutions. Thus, Gram-negative bacteria contain lipopolysaccharide in the cell wall. Although the bacteria themselves are
not pathogenic for Man or for animals, the lipopolysaccharide possesses
endotoxic properties. When ingested in sufficient quantity, symptoms similar to those of salmonella food poisoning are induced,

symptoms which include nausea, diarrhoea and vomiting. Obviously, the amount which must be ingested is relatively high, and is unlikely to be reached in purified preparations of exopolysaccharides. There might, however, be a danger of sensitization occurring in operatives repeatedly exposed to the material.

Regulations for the use of emulsifiers and stabilizers in food in Great Britain are at present laid down by the Ministry of Agriculture, Fisheries and Food. However, member countries of the European Economic Community are also subject to a directive on emulsifiers and stabilizers for food use adopted by the Council of Ministers on 18th June, 1974 (74/329/E.E.C.). A statutory list of permitted emulsifiers and stabilizers will be drawn up as a result of this directive, and no food sold, consigned, delivered or imported within the Economic Community shall have in it any added emulsifier which is not on the permitted list. In the new list are alginic acid, sodium, potassium, ammonium and calcium alginates, and propylene glycol alginates as specified in the Food Chemicals Codex (1972). This does not at present include bacterial alginates, but does contain xanthan gum (Food Chemicals Codex, 1972, pg. 856). Additional requirements in the case of this microbial product are the absence of viable bacteria and a limitation in the nitrogen content to 1.5%. There are also general purity criteria for all of the emulsifying agents with respect to heavy-metal ions such as arsenic, lead, copper and zinc, Similar regulations for food additives are to be found in other countries outside the E.E.C.

Specifications for xanthan gum are of particular interest in that they may well indicate the purity standards which would be required of any further microbial product to be added to the permitted list in future. Preparations for non-food use are not necessarily of such a high specification, but it is obviously desirable that viable micro-organisms are absent. The nitrogen content permitted in the specification for xanthan gum is considerably higher than that obtained in laboratory preparations of the polysaccharide from *Xanthomonas campestris* (Jeanes *et al.*, 1961) or of the polymer from the yeast *Cryptococcus laurentii* var. *flavescens* (Jeanes *et al.*, 1964). The preparations contained 0.08% and 0.23% nitrogen, respectively. The presence of relatively large amounts of nitrogenous impurities is undesirable both because it is indicative of contamination with bacterial cells, albeit dead, and because it provides nutrient material supporting growth of microbial contaminants. A product data sheet from the manufacturers of xanthan gum recommends addition of preservatives when polymer solutions are to be stored, and it seems probable that growth of contaminating micro-organisms is at the expense of contaminating nitrogenous material rather than of the polymer itself. Such considerations would apply to all microbial exopolysaccharide held in solution, the problem being greatest in cruder grades used for purposes other than food or pharmaceutical ingredients, in processes involving holding the polysaccharide solutions at ambient temperature or above for considerable periods of time.

In the U.S.A., any ingredient in food is regarded as a food additive and must be covered by a food additive regulation unless it is *generally regarded as safe* (*GRAS*) under conditions of its intended use. Thus, the so-called *GRAS* list serves as the condition of acceptability for microbial (and other) polysaccharides used as food additives. This list is monitored by the Food and Drug Administration, and their authorization at present covers the use of xanthan gum. This polysaccharide was approved for use in foods as a stabilizer, emulsifier, thickener and suspending and bodying agent, as well as a foam enhancer. Various possible uses, and a summary of the main properties of xanthan gum solutions, have been discussed by Rocks (1971). The test programme for xanthan gum lasted almost 12 years and included feeding to animals—dogs were given intakes up to 2 g/kg body weight daily (Robbins *et al.*, 1964). Toxic effects were not noted other than persistent diarrhoea in those animals fed the highest levels of polysaccharide. Nor was degradation of polysaccharide noted in the gastro-intestinal tract (Booth *et al.*, 1963). Acceptance of xanthan gum for food and other uses may well serve as a useful precedent for the development and introduction of other microbial polysaccharides.

C. Modification of Products

The polysaccharide, as isolated from culture and subsequently purified, may not be ideal for all potential applications. The uses to which it may be put can probably be extended by either chemical or enzymic modification.

1. Chemical Alteration of Polysaccharides

Bacterial alginates, xanthan gum and many other microbial exo-polysaccharides are frequently acetylated in the natural state. The presence of acetyl groups, although making the polymer more lipophilic than would otherwise be the case, may not be desirable and the simplest chemical modification of the polymer is by treatment with mild alkali. Removal of acetyl groups is likely to affect the solubility in aqueous and other solvents, but may not markedly affect the overall rheological characteristics of poly-saccharide solutions.

Dextran is modified by controlled acid hydrolysis starting with material of molecular weight considerably *greater* than required. Inclusion of an anti-oxidant is needed and sodium sulphite is used, while neutrality is maintained with calcium carbonate in a thermal degradation, or acid at pH 2 is used in the mild-acid hydrolysis. Dextran can also be sulphated to yield dextran sulphate, the process being performed with chlorsulphonic acid in pyridine. Other modifi-cations of the neutral dextran polymer are used to produce the well-known sephadex range of adsorbents. These range from cross-linked polymers of varying size to materials into which functional groups have been introduced, attached to the glucose units by ether linkages. Anion-exchange groups, functioning as either anion exchangers (DEAE-Sephadex) or cation exchangers (CM-Sephadex), are available by modification of the original polysaccharides in this way. It is also possible to introduce lipophilic activities by intro-duction of hydroxypropyl groups.

While it is not possible, or even necessary, to alter all poly-saccharides in this way, the versatility of the various dextran products indicates the great range of products which can be ob-tained.

Chemical modification of alginates is also practised when specific properties are needed. Triethanolamine alginate is soluble in water and in 75% ethanol, and forms soft films. Alginic acid amide, although soluble in hot water, forms gels on cooling. Propylene glycol alginate, in which the carboxyl groups are esterified in the free acid, is used in products such as soft drinks and beer where the acidity of the beverage would produce a gel composed of alginic acid as mentioned earlier. Further applications of this particular derivative of alginate are also found in the preparation of photographic emulsions.

2. *Enzymic Modification of Polymers*

Modification of preformed polysaccharide is sometimes found in micro-organisms which produce an enzyme that hydrolyses their own product. An example of this is seen in some hyaluronic acid-producing bacteria which also produce (later in growth) hyaluronidases. Some alginate-forming micro-organisms may also produce enzymes that modify the monosaccharides comprising the polymer (by epimerization) and enzymes that degrade the polysaccharide to yield fragments of smaller molecular weight. Obviously, such modifications are not desirable if the activities of the enzymes cannot be controlled satisfactorily. The only microbial exopolysaccharide for which enzymic modification currently plays an important part is dextran. Careful treatment of dextran with suitable enzymes can be used to produce material in a particular molecular-weight range.

As more biopolymers are introduced and an increasing variety of uses developed for them, it is probable that chemical modification or alteration of molecular weights with enzymes will find an increasingly important role.

VI. POTENTIAL FOR STRAIN IMPROVEMENT BY MUTAGENESIS

Once a strain of bacteria or yeast which produces a polysaccharide with commercially useful properties has been isolated, and initial cultural studies have been completed, further development is possible. Mutagenesis of the microbial strain can have one of five aims: (i) increased polysaccharide production; (ii) modification of the polysaccharide produced; (iii) alteration of the surface properties of the microbial cells; (iv) elimination of polysaccharase synthesis, or (v) change of species.

A. Increased Polysaccharide Production

This is dependent on a thorough knowledge of the biosynthetic mechanisms involved and, if possible, the genetic control system. Formation of the polysaccharide may be a relatively simple process in terms of the numbers of enzymes involved, as is the case for a homopolysaccharide such as dextran. If the polymer is a hetero-polysaccharide, as complex as the type synthesized by *Xanthomonas campestris* (Jeanes *et al.*, 1961) or *Escherichia coli* and *Salmonella* species (Sutherland, 1969; Lawson *et al.*, 1969), many more enzymes and much more genetic information are involved. As an example of a relatively simple system, bacterial alginate will be used and compared with the much more complex polysaccharide from *Xanthomonas campestris*. Bacterial alginate appears to be produced initially as a linear homopolymer of D-mannuronic acid residues. The homo-polymer is converted to a heteropolymer by an extracellular epi-merase isolated from *Azotobacter* species (Haug and Larsen, 1971). The product is a heteropolymer composed of D-mannuronic acid and L-guluronic acid residues distributed in an apparently non-uniform manner along the polymer molecule.

Assuming that the basic mechanism of synthesis is similar to that found in other bacterial exopolysaccharides, formation of the homo-polymer requires production of GDP-mannose and its oxidation to GDP-mannuronic acid (Pindar and Bucke, 1975). Thereafter several enzymes assemble mannuronosyl residues on a carrier lipid and extrude the finished polymer from the bacterial surface. Although relatively little research on these fundamental processes involved in alginate synthesis has been performed, it is probable that, in two alginate-producing bacterial species, namely *Azotobacter vinelandii* (C. Bucke, unpublished results) and *Pseudomonas aeruginosa* (I. W. Sutherland, unpublished results), some at the least of the enzymic processes are as indicated in Fig. 13. Thus the complete process is probably analogous to that involved in production of the capsular or slime polysaccharide of *Klebsiella aerogenes* (Troy *et al.*, 1971; Sutherland and Norval, 1970). In addition there is the unusual role of an extracellular epimerase.

From the postulated path of formation of the alginate, it is apparent that polysaccharide production may be enhanced either by increasing the specific activities of the enzymes involved, by altering control mechanisms such as feedback inhibition, or by increasing the amounts of available precursors or cofactors. Unfortunately, no

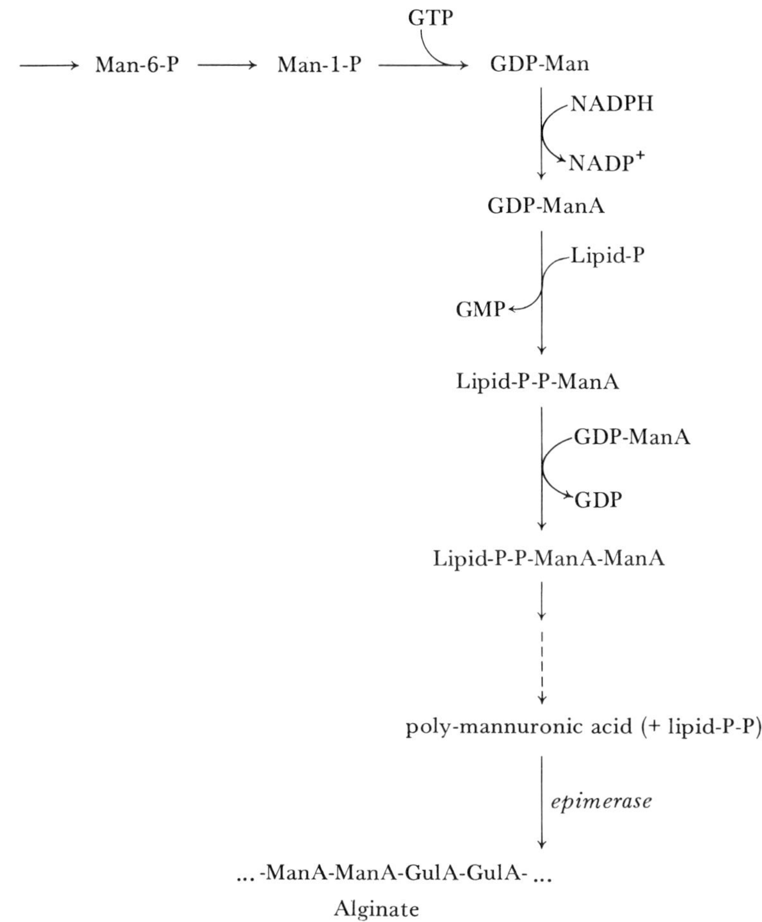

Fig. 13. Proposed biosynthetic pathway for bacterial alginate.

specific selection methods are available to obtain mutants with increased levels of the enzymes involved in alginate synthesis. After mutagenesis, expression of the mutation and plating on a suitable medium, one must observe the colonies and note those which have altered morphology, especially those which appear to produce more polysaccharide (or possibly also those producing a mucoid colony with a different viscosity from the wild type). In general, pyrophosphorylases are known to be key enzymes in the regulation of nucleoside diphosphate sugars and their derivatives (e.g. Bernstein and Robbins, 1965) and increased GDP-mannose pyrophosphorylase production might yield increased alginate. Other enzymes may be less likely to produce this effect. Alternatively, increased formation of carrier lipid (C_{55} isoprenoid alcohol phosphate; *syn.* bactoprenyl phosphate) by the bacteria may lead to increased polysaccharide production. The same carrier lipid is also involved in production of other polymers located outside the membrane of Gram-negative bacteria such as *Azotobacter* or *Pseudomonas* species. These are

$$CH_3-\underset{\underset{\displaystyle CH_3}{|}}{C}=CH-CH_2\,(CH_2-\underset{\underset{\displaystyle CH_3}{|}}{C}=CH-CH_2)_9-CH_2-\underset{\underset{\displaystyle CH_3}{|}}{C}=CH-CH_2OH$$

Fig. 14. Structure of the isoprenoid alcohol utilized as a carrier lipid in polysaccharide biosynthesis.

peptidoglycans and lipopolysaccharides. Although the bacteria *may* then produce more peptidoglycan or lipopolysaccharide, in the presence of increased amounts of carrier lipid, a more likely result is synthesis of greater quantities of exopolysaccharide; there is no constraint on this as there would be for the two structural polymers of the bacterial wall, as one must consider possible spatial effects if substantially more than normal amounts of lipopolysaccharide or peptidoglycan were formed. Unlike mutations directly affecting polysaccharide synthesis through the enzymes involved in GDP-mannose or in polysaccharide formation, selective means probably exist for isolating bacteria with the capability of producing larger amounts of bactoprenol.

Bacitracin is known to inhibit dephosphorylation of the isoprenoid alcohol pyrophosphate (Fig. 14) by binding with the prenyl phosphate in the presence of divalent cations (Stone and Strominger, 1971). Consequently, selection for bacitracin resistance should isolate mutants with higher than normal cellular levels of carrier lipid to circumvent the action of the bacitracin. Preliminary experiments with *Klebsiella aerogenes* have shown that it is possible to obtain mutants in which the bacitracin resistance has been greatly increased, and tests with these mutants indicated that the rate of transfer of hexose 1-phosphate from UDP-glucose to lipid phosphate was higher than in the original strains. Polysaccharide production in washed suspensions (i.e. non-growing cells) was also increased in the mutants. This suggests that the level of isoprenoid lipid in the cell may well be important in determining the extent of polysaccharide synthesis (Sutherland, 1977).

B. Modification of the Polysaccharide

Almost nothing is known about the enzymes which are affected when micro-organisms produce an altered exopolysaccharide. Most mutations that affect extracellular polysaccharide formation are involved in an 'all or none' relationship with the product. As most stages in biosynthesis of the final polymer are sequential, loss of one enzyme in the sequence leads to failure to form any polysaccharide. An exception to this *may* be found in acylating enzymes. Several different bacterial species in the Enterobacteriaceae synthesize a polysaccharide with the same carbohydrate structure. This polymer has been called colanic acid (Goebel, 1963). The polysaccharide isolated from different species was later shown to differ considerably in the proportion of acyl groups present (Garegg *et al.*, 1971). Although the terminal galactosyl residue on the side chain always retained a carboxyethylidene or similar substituent (Fig. 15), one

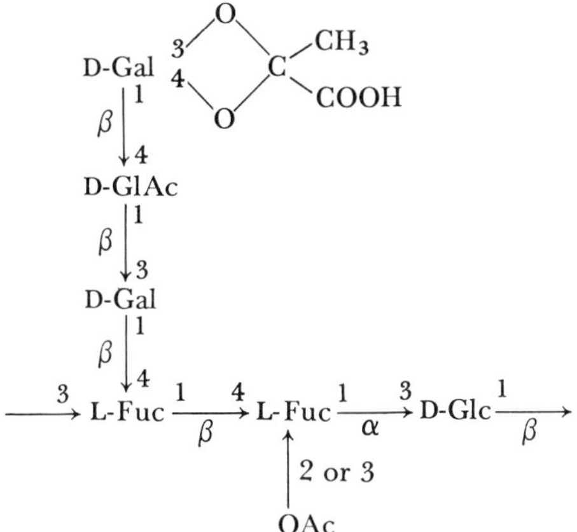

Fig. 15. Structure of colanic acid from *Escherichia coli* K12. From the results of Sutherland (1969) and Garegg *et al.* (1971).

strain produced a non-acetylated polymer, apparently through lack of the appropriate specific polysaccharide acetylase. It is by no means clear whether this is a limited phenomenon or whether it is widespread. As one can isolate serologically identical capsulate bacteria in the genus *Klebsiella*, in which different acyl residues are present on the exopolysaccharides (Sutherland, 1972), it seems likely that acyl groups can be lost or modified without loss of polysaccharide-synthesizing potential.

C. Alteration of Surface Properties

Alteration of the surface properties of the microbial cell can greatly affect the ease of harvesting extracellular polysaccharides. Most bacteria of size 2– 6 μm x 1– 2 μm would require a centrifugal force of 4– 6,000 g for adequate sedimentation. The polysaccharide increases the viscosity of the culture fluid to such an extent that very much greater centrifugal forces must be applied. Yeasts and larger bacteria are not so difficult to harvest. If the surface of the bacteria is altered, particularly if the cells become 'rough' through loss of part of the surface polymeric material, e.g. lipopolysaccharides from Gram-negative bacteria, the culture has a tendency to 'auto-agglutinate'. This means that, in unstirred cultures, flocculation of the cells occurs spontaneously; consequently, recovery of bacteria from culture fluids requires considerably less centrifugation. This is, of course, also true in the more viscous solutions formed by exopolysaccharide formation. One drawback to the production of mutants, which have lost part of their surface polymeric material, is that some of these mutants are 'leaky', that is, the mutants lose cell components, especially proteins normally found in the periplasmic space, into the growth medium, where they are liable to contaminate the product. Lysis may also occur more readily than with the wild-type organisms, either in the culture or during centrifugation to harvest the polysaccharide. The cell components released on lysis then contaminate the polysaccharide product, and some of them, such as nucleic acids, may be extremely difficult to separate or destroy.

A second approach can be used for micro-organisms which produce exopolysaccharides mainly as capsules tightly bound to the cell surface. To eliminate the problems associated with removing capsules from large quantities of cells, a search can be made for slime-forming (Sl) mutants. These are stable mutants which form slime only and not discrete capsules. As far as can be determined by chemical and enzymic techniques, the polysaccharides which have been examined from Sl mutants and the corresponding capsular material of the parent strains, possess identical compositions (Wilkinson *et al.*, 1955).

A type of mutant which was discovered in *Klebsiella aerogenes* has properties which might be of value under certain conditions for polysaccharide production. This group of conditional mutants (Norval and Sutherland, 1969) had the property of producing polysaccharide at the restrictive temperatures only after growth had ceased, or when the bacteria had been harvested from nutrient medium and transferred to buffer containing salts and a carbohydrate source but no nitrogen source. This could be used as the basis for a two-stage process. The primary stage would be used under nutrient conditions to produce a large quantity of cells. The bacteria so grown would be transferred to a secondary fermenter in which polysaccharide production could take place without further growth.

If bacteria are used as a source of polysaccharide and they are heterotrophic species, there is always the possibility that such strains may become infected with bacteriophages. This results in loss of viability and failure to produce any exopolysaccharide. To avoid problems with bacteriophages, normal aseptic techniques and filtered air supplies are essential, but there is always a risk that stock cultures may become infected. It is therefore advisable to have known phage-resistant strains available. Their isolation presents little difficulty, but one must ensure that all phages are eliminated from the resistant derivatives, that the strain has not been lysogenized, and that polysaccharide production has not been affected. Many of the phage receptors are portions of the lipopolysaccharide in the walls of Gram-negative bacteria, and phage resistance is associated with loss of the receptor (Lindberg, 1977). The phage-resistant mutants may, therefore, have lost enzymes involved in intermediary carbohydrate metabolism leading simultaneously to loss of exopolysaccharide-synthesizing capability. The type of resistant mutant required is one in which receptor synthesis only is affected and exopolysaccharide is formed as in the wild type.

D. Elimination of Polysaccharases

Few micro-organisms which synthesize exopolysaccharides have, so far, also been shown to degrade them. There are a number of exceptions. Some micro-organisms excrete exopolysaccharides at one stage of growth then degrade them through excretion of hydrolytic enzymes at some later stage in the growth cycle. Thus, *Streptococcus* species are capable of forming both hyaluronic acid and hyaluronidases; *Azotobacter vinelandii* synthesizes alginate and also small quantities of alginase; and *Streptococcus mutans* produces both insoluble dextran-type polymers and 'mutanase' which degrades them.

It is obviously unsatisfactory if a micro-organism is used to produce a polysaccharide on a commercial scale, and if it then destroys or degrades the product. Cultural conditions may be carefully controlled so that production of the polysaccharase is suppressed during growth and during product isolation. This is possible in strains in which there is a distinct lag between production of the polysaccharides and release of the hydrolytic enzymes. A much better prospect is to use mutants which lack polysaccharases. This can be accomplished by examination of microbial colonies after mutagenesis. Colonies lacking the enzymes are likely to remain mucoid under all conditions of incubation, whereas the wild-type clones, if subjected to prolonged incubation, would lose their mucoid appearance and would eventually appear similar to non-mucoid variants.

E. Inter-Strain Transfer of Genetic Material

The genetic determinants for polysaccharide synthesis can be transferred between strains of the same species. This has been observed using transformation in *Streptococcus pneumoniae* (e.g. Austrian *et al.*, 1959). and using transduction in *Escherichia coli* (Markovitz *et al.*, 1968). It may prove possible to mobilize the appropriate genetic material from a bacterial species that produces an exopolysaccharide with desirable characteristics and introduce the genetic material in a fully functional form into another bacterium. This type of transfer of genetic material, controlling a number of related functions, has been very successful in deriving nitrogen-fixing strains of *E. coli* using genetic material from *Klebsiella pneumoniae* (Dixon and Postgate, 1972). Future applications for similar transfers involving polysaccharide synthesis might perhaps include the transfer of the ability to form alginate from the pathogen *Pseudomonas aeruginosa* to other completely harmless *Pseudomonas* species. Alternatively, the ability to synthesize 'xanthan gum' might be introduced from the plant pathogenic *Xanthomonas campestris* into another species of this group of bacteria, use of which would not present problems with national regulations for control of plant diseases. One further possibility might be transfer of alginate synthesis from the slow-growing *Azotobacter vinelandii* to another faster-growing bacterial species.

Considerable study of the genetic systems of the bacteria under consideration would be needed and, at present, it is not possible to forecast the chances of success. Problems might arise with respect to the stability of the genetic information in the recipient bacteria. By analogy with the *nif* genes (controlling nitrogen fixation) in *Klebsiella pneumoniae* the original genes were probably chromosomal whereas, in *E. coli* after transfer from the donor, the genes were usually located on plasmids (extrachromosomal DNA) and consequently less stable (Cannon *et al.*, 1974).

VII. GENERAL CONCLUSIONS ON MICRO-ORGANISMS PRODUCING BIOPOLYMERS

It is still not possible to predict with certainty which microbial species will produce potentially useful biopolymers. Some species of bacteria (e.g. *X. campestris*, *A. vinelandii* or *Ps. aeruginosa*) appear to be potentially more likely to form polymers of the xanthan gum or alginate types, respectively, and could be the subject for research if these polymers have the required characteristics for a specific market or application. There are certainly no chemical characteristics common to all the gel-forming polysaccharides studied so far (Table 2, pg. 311) as not all types contain uronic acids, acyl groups or neutral monosaccharide residues.

Production of large amounts of biopolymer necessitates a suitably cheap source of carbohydrate and, in the case of dextrans and levans, a specific substrate for incorporation into the growth medium. An exception is succinoglucan. which can be produced using alternative carbon and energy sources such as ethylene glycol (see pg. 368). The efficiency of conversion of the carbohydrate source to polymer varies with the microbial species and the growth conditions used, but it can be extremely high (see pg. 356). This may not always be desirable if the viscosity of the culture fluids is so high that removal of the microbial cells is rendered exceedingly difficult, although there is evidence that enzymes may be useful in solubilizing cells (Kelco, 1975a). A balance may, therefore, have to be reached between optimal conversion of substrate to polymer and ease of recovery of the product free from contaminating micro-organisms. Associated with this problem may be the grade of purity demanded for a specific market. All polysaccharides can be recovered by precipitation with organic solvents, such as methanol and isopropanol, although the proportion of solvent needed is considerably higher for polymers with a high content of acyl groups, such as the succinoglucan. Reduction of costs is achieved by recovery of the solvent, distillation and recycling. Alternative recovery procedures can be used for alginates or succinoglucan (Harada, 1974) as both of these types of polysaccharide are precipitable in the acid form; alginates can also be prepared as their calcium salts which are insoluble in water. Xanthan gum has been recovered as an insoluble calcium salt, followed by an ion exchange to the free acid and titration to the potassium salt (Mehltretter, 1965).

It should be remembered that, even if specific polysaccharides do not have all of the qualities required, gels may be obtained via synergistic polysaccharide interaction. This is particularly true if galactomannan (from locust beans) is mixed with other polysaccharides. The interaction between galactomannan and xanthan gum has been studied by Rees (1972) using optical rotation. He showed that formation of a stiff gel at about 32°C occurred at a maximum in the optical rotation–temperature curve, consistent with a change in conformation, a property absent from solutions of either of the individual components. The number of such combinations which could be examined is obviously large, and provides considerable potential for future investigation.

Finally, while it cannot be denied that production of microbial polysaccharides is a subject of great current interest to the fermentation industry, it should be stressed that only two products, dextran and xanthan gum, have been commercialized. This situation exists because of the large number of hurdles that a potential microbial polysaccharide product must pass before ultimate commercial success is achieved. It is probably true to say that there will be no great flood of microbial gums onto the market, but rather a steady but significant introduction of products finding application in entirely new areas or in areas where the 'in use' products are not performing particularly well.

REFERENCES

American Gas Association, American Petroleum Institute and Canadian Petroleum Association (1975). Reserves of Crude Oil, Natural Gas Liquids and Natural Gas in the United States and Canada, and United States Production Capacity as of December 31 1974.

Austrian, R., Bernheimer, H. P., Smith, E. E. B. and Mills, G. T. (1959). *Journal of Experimental Medicine* 110, 585.

Baird, J. K., Longyear, V. M. C. and Ellwood, D. C. (1973). *Microbios* 8, 143.

Beech, F. W., Davenport, R. R., Goswell, R. W. and Burnett, J. K. (1968). *In* "Identification Methods for Microbiologists", (B. M. Gibbs and D. A. Shapton, eds.), part B, pp. 151-176. Academic Press, London,

Bernstein, R. L. and Robbins, T. W. (1965). *Journal of Biological Chemistry* 240, 391.

Blakebrough, N. (1969). *In* "Methods in Microbiology", (J. R. Norris and D. W. Ribbons, eds.), vol. 1, pp. 473-504. Academic Press, London.

Booth, A. N., Hendrickson, A. P. and Deeds, F. (1963). *Toxicology and Applied Pharmacology* 5, 478.

Calam, C. T. (1969). *In* "Methods in Microbiology", (J. R. Norris and D. W. Ribbons, eds.), vol. 1, pp. 567-592. Academic Press, London.

Campbell, I. (1970). *Process Biochemistry* 5, 47.

Campbell, I. (1971). *Journal of General Microbiology* 67, 223.

Cannon, F. C., Dixon, R. A., Postgate, J. R. and Primrose, S. B. (1974). *Journal of General Microbiology* 80, 227.

Carlson, D. M. and Matthews, L. W. (1966). *Biochemistry, New York* 5, 2817.

Catley, B. J. and Whelan, W. J. (1971). *Archives of Biochemistry and Biophysics* 143, 138.

Coffey, S. (1967). "Rodds Chemistry of Carbon Compounds", vol. 1(F).

Dintzis, F. R., Babcock, G. E. and Tobin, R. (1970). *Carbohydrate Research* 13, 257.

Dixon, R. A. and Postgate, J. R. (1972). *Nature, London* 237, 102.

Evans, L. R. and Linker, A. (1973). *Journal of Bacteriology* 116, 915.

Food Chemicals Codex (1972). 2nd Edition. National Academy of Sciences, Washington, D.C.

Federal Register (1969). Xanthan Gum. Section 121. 1224, pp. 5376-5377.

Fischer, F. G. and Dörfel, H. (1955). *Zeitschrift für Physiologische Chemische* **301**, 224.

Garegg, P. J., Lindberg, B., Onn, T. and Sutherland, I. W. (1971). *Acta Chemica Scandanavica* **25**, 2103.

Gibbs, B. M. and Shapton, D. A., eds. (1968). "Identification Methods for Microbiologists", part B. Academic Press, London,

Gibbs, B. M. and Skinner, F. A., eds. (1966). "Identification Methods for Microbiologists", part A. Academic Press, London.

Goebel, W. F. (1963). *Proceedings of the National Academy of Sciences of the United States of America* **49**, 464.

Glicksman, M. (1969). "Gum Technology in the Food Industry". Academic Press, New York.

Gorin, P. G. J. and Spencer, J. F. T. (1966). *Canadian Journal of Chemistry* **44**, 993.

Guggenheim, B. (1970). *Helvetica Odontica Acta, Supplement* **14**, 589.

Harada, T. (1974). *Process Biochemistry* **9**, 21.

Haug, A. (1964). "Composition and Properties of Alginates". Report No. 30. Norwegian Institute of Seaweed Research, Trondheim.

Haug, A. and Larsen, B. (1971). *Carbohydrate Research* **17**, 297.

Hehre, E. J. (1956). *Journal of Biological Chemistry* **222**, 739.

Hendrie, M. S. and Shewan, J. M. (1966). *In* "Identification Methods for Microbiologists", (B. M. Gibbs and F. A. Skinner, eds.), part A, pp. 1-8. Academic Press, London.

Imrie, F. K. E. (1973). British Patent 1331771.

Jansson, P. E., Kenne, L. and Lindberg, B. (1975). *Carbohydrate Research* **45**, 275.

Jeanes, A., Haynes, W. C., Wilham, C. A., Rankin, J. C., Melvin, E. H., Austin, M. J., Cluskey, J. E., Fisher, B. E., Tsuchiya, H. M. and Rist, C. E. (1954). *Journal of American Chemical Society* **76**, 5041.

Jeanes, A., Wilham, C. A., Tsuchiya, H. M. and Haynes, W. (1957). *Archives of Biochemistry and Biophysics* **71**, 293.

Jeanes, A., Pittsley, J. E. and Senti, F. R. (1961). *Journal of Applied Polymer Science* **5**, 519.

Jeanes, A., Pittsley, J. E. and Watson, P. R. (1964). *Journal of Applied Polymer Science* **8**, 2775.

Johnson, J., Kirkwood, S., Misaki, A., Nelson, T. E., Scarletti, J. V. and Smith, F. (1963). *Chemistry and Industry* 820-822.

Kang, K. S. and Kovacs, P. (1974). *In* "The Work Documents of the International Congress of Food Science and Technology", Madrid.

Kelco Co. (1971). Technical Bulletin DB 21.

Kelco Co. (1975a). Dutch Patent. Specification No. 7502243.

Kelco Co. (1975b). French Patent. Specification No. 7340683.

Kimura, H. (1972). Abstract. 32nd Institute of Food Technologists Meeting, Minneapolis.

Larsen, B. and Haug, A. (1971). *Carbohydrate Research* **17**, 287.

Lawrence, A. A. (1973). "Edible Gums and Related Substances". Food Technology Review No. 9. Noyes Data Corporation, New Jersey.

Lawson, C. J., McCleary, C. W., Nakada, H. I., Rees, D. A., Sutherland, I. W. and Wilkinson, J. F. (1969). *Biochemical Journal* **115**, 947.

Lawson, C. J. and Symes, K. C. (1977). *In* "Extracellular Microbial Polysaccharides", (P. A. Sandford and A. Laskin, eds.), American Chemical Society Symposium, no. 45, pp. 183-191.

Lindberg, A. A. (1977). *In* "Surface Carbohydrates of Eukaryotes", (I. W. Sutherland, ed.), pp. 289–356. Academic Press, London.

Linker, A. and Jones, R. S. (1964). *Nature, London* **204**, 187.

Linker, A. and Jones, R. S. (1966). *Journal of Biological Chemistry* **241**, 3845.

Markovitz, A., Rosenbaum, N. and Baker, B. (1968). *Journal of Bacteriology* **96**, 221.

Mehltretter, C. L. (1965). *Biotechnology and Bioengineering* **7**, 171.

Melton, L. D., Mindt, L., Rees, D. A. and Sanderson, G. R. (1976). *Carbohydrate Research* **46**, 245.

Meltzer, Y. L. (1972). "Water Soluble Polymers". Chemical Process Review No. 64. Noyes Data Corporation, New Jersey.

Mukasa, H. and Slade, H. D. (1973). *Infection and Immunity* **8**, 555.

Norris, J. R. and Ribbons, D. W., eds. (1969). "Methods in Microbiology", vol. 3A. Academic Press, London.

Norris, J. R. and Ribbons, D. W., eds. (1970). "Methods in Microbiology", vol. 3B. Academic Press, London.

Norval, M. and Sutherland, I. W. (1969). *Journal of General Microbiology* **57**, 369.

Penman, A. and Sanderson, G. R. (1972). *Carbohydrate Research* **25**, 273.

Percival, E. and McDowell, R. H. (1967). *In* "Chemistry and Enzymology of Marine Algal Polysaccharides", pp. 99–126. Academic Press, London.

Pindar, D. F. and Bucke, C. (1975). *Biochemical Journal* **152**, 617.

Rees, D. A. (1972). *Biochemical Journal* **126**, 257.

Rees, D. A. (1975). *In* "MTP International Review of Science", (W. J. Whelan, ed.), Biochemistry Series One, Vol. 5, pp. 1–42. Butterworths, London.

Robbins, D. J., Moulton, J. E. and Booth, A. N. (1964). *Food and Cosmetics Toxicology* **2**, 545.

Rocks, J. K. (1971). *Food Technology* **25**, 476.

Rogovin, S. P. (1969). U.S. Patent 3,485,719.

Rogovin, S. P., Anderson, R. F. and Cadmus, M. C. (1961). *Journal of Biochemical and Microbiological, Technology and Engineering* **3**, 51.

Sawai, T., Toriyama, K. and Yano, K. (1973). *Journal of Biochemistry, Tokyo* **75**, 105.

Sharp, J. M. (1975). "The Potential of Enhanced Recovery Processes", SPE No. 5557. 50th Annual Meeting of the Society of Petroleum Engineers of AIME, Dallas. Texas.

Sherman, P. (1975). *Chemistry in Britain* **11**, 321.

Siddiqui, I. R. (1967). *Carbohydrate Research* **4**, 284.

Sloneker, J. H. and Jeanes, A. (1962). *Canadian Journal of Chemistry* **40**, 2066.

Sloneker, J. H., Orentas, D. G. and Jeanes, A. (1964). *Canadian Journal of Chemistry* **42**, 1261.

Smith, E. E. (1970). *Federation of European Biochemical Societies Letters* **12**, 33.

Stone, K. J. and Strominger, J. L. (1971). *Proceedings of the National Academy of Sciences of the United States of America* **68**, 5107.

Sutherland, I. W. (1969). *Biochemical Journal* **115**, 935.

Sutherland, I. W. (1972). *In* "Advances in Microbial Physiology", (A. H. Rose and D. W. Tempest. eds.), vol. 8, pg. 143.

Sutherland, I. W. (1977). *In* "Extracellular Microbial Polysaccharides", (P. A. Sandford and A. Laskin, eds.), American Chemical Society Symposium no. 45, pp. 40–57.

Sutherland, I. W. and Wilkinson, J. F. (1965). *Journal of General Microbiology* **39**, 373.

Sutherland, I. W. and Norval, M. (1970). *Biochemical Journal* **120**, 567.

Troy, F. A., Frerman, F. E. and Heath, E. C. (1971). *Journal of Biological Chemistry* **246**, 118.

Umland, C. W. (1974). "Presentation for the Federal Energy Administration. Enhanced oil and Gas Recovery Symposium", Washington D.C.

Whistler, R. L. (1973a). *In* "Industrial Gums", (R. L. Whistler, ed.), 2nd edn., 807 pp. Academic Press, New York.

Whistler, R. L. (1973b). *In* "Industrial Gums", (R. L. Whistler, ed.), 2nd edn., pp. 514–520. Academic Press, New York.

Whistler, R. L. and BeMiller, J. N. (1965). *In* "Methods in Carbohydrate Chemistry", (R. L. Whistler and M. L. Wolfrom, eds.), vol. 5. Academic Press, New York.

Wilkinson, J. F. (1958). *Bacteriological Reviews* **22**, 46.

Wilkinson, J. F., Dudman, W. F. and Aspinall, G. O. (1955). *Biochemical Journal* **59**, 446.

Yuen, S. (1974). *Process Biochemistry* **9**, 7–9 & 22.

Zevenhuizen, L.P.T.M. (1968). *Carbohydrate Research* **6**, 310.

Note added in Proof

Since the original manuscript was drafted, a number of significant events, with important implications for the microbial-gum industry, have occurred. In early 1976, Tate and Lyle Ltd. and Hercules Inc. signed a Joint Development Agreement for xanthan gum and other microbial polysaccharides. As a result, a production plant for xanthan is currently under construction in Liverpool, England, designed to come on stream in mid-1979.

In the autumn of 1976, it was announced that the joint venture between Rhone-Poulenc and General Mills was not to continue, and that no further work would be undertaken on the construction of a plant for xanthan production in Iowa, U.S.A.

A number of significant publications have recently appeared. The conformation of xanthan gum has been examined (Holzwarth, 1976) on the basis of optical rotation, viscometry and potentiometric titration. On the basis of measurements of the observed, reversible, and thermally-induced transition to a denatured structure, it was suggested that the native polysaccharide possessed an ordered secondary structure stabilized by nonionic interactions. In a further study using electron microscopy (Holzwarth and Prestridge, 1977), xanthan was postulated to exist as a multistranded helix. In other studies, the conformation of xanthan was examined by X-ray crystallography (Moorhouse *et al.*, 1977) with the result that a somewhat more simple five-fold helical model was developed. Using rheological, chiroptical and nuclear magnetic resonance techniques, the association mechanism of bacterial polysaccharides, e.g. xanthan, with glactomannans, and locust-bean gum (Barnes *et al.*, 1977) was studied. Evidence was presented for the importance of conformation in the recognition between xanthomonas pathogen and its plant host (Morris *et al.*, 1977).

Finally, a recent book on the subject of extracellular microbial polysaccharides is recommended (Sandford and Laskin, 1977). Containing 22 papers on a wide range of topics, the book is a useful contribution to our knowledge of this subject.

References

Barnes, H. A., Dea, I. C. M., Morris, E. R., Price, J., Rees, D. A. and Walsh, E. J. (1977). *Carbohydrate Research* **77**, 249.

Holzwarth, G. (1976). *Biochemistry, New York* **15**, 4333.

Holzwarth, G. and Prestridge, E. B. (1977). *Science, New York* **197**, 757.

Moorhouse, R., Walkingshaw, M. D. and Arnott, S. (1977). In 'Extracellular Microbial Polysaccharides', (P. A. Sandford and A. Laskin, eds.). American Chemical Society Symposium Series no. 45.

Morris, E. R., Rees, D. A., Young, G., Walkingshaw, M. D. and Darke, A. (1977) *Journal of Molecular Biology* **110**, 1.

Sandford, P. A. and Laskin, A. eds. (1977) 'Extracellular Microbial Polysaccharides', American Chemical Society Symposium Series no. 45.

10. Production of Polyhydroxy Alcohols by Osmotolerant Yeasts

J. F. T. SPENCER AND DOROTHY M. SPENCER

Biology Department, Goldsmiths' College, University of London, England

I. INTRODUCTION

Of the polyhydroxy alcohols produced by micro-organisms, glycerol, erythritol, arabinitol and mannitol are of the greatest potential interest. All of these are produced from sugars in relatively good yields by various species of yeasts. Mannitol is available from other sources, and glycerol can be obtained as a byproduct of the petrochemical industry, but there is no chemical synthesis for erythritol and arabinitol which is comparable in simplicity and cost to the microbial process. Some other polyhydroxy alcohols which are less well known are also formed by variations of the processes for microbial synthesis of those previously mentioned. Xylitol, for instance, is formed by a strain of *Pichia farinosa*, originally classified as *Pichia miso*, if D-xylose is used as the original substrate, and two heptitols are formed as minor byproducts of this process. Very likely, a few other polyhydroxy alcohols could be formed by analogous processes, if suitable microbial species are discovered.

II. HISTORICAL

A. Organisms

Although polyhydroxy alcohols are formed by a wide range of micro-organisms, for the purposes of this chapter the discussion will be limited to those polyhydroxy alcohols produced by yeasts. Most yeasts produce traces of one or more of these compounds, but the highest yields have been obtained from species which show a relatively strong tolerance for high concentrations of salts and sugars. Many strains of osmotolerant yeasts have been isolated from various sources, and given a variety of names, but most of them have since been reclassified into a relatively few species. *Saccharomyces rouxii*, *Saccharomyces mellis*, *Pichia farinosa*, *Pichia etchellsii* and *Schizosaccharomyces octosporus* include most of the highly osmotolerant species.

The sugar-tolerant *Saccharomyces* species have mostly been isolated from fermenting honey (Lochhead and Heron, 1929), spoiled canned fruits (Phillipov, 1932), fermented soft-centred candies (Church *et al.*, 1927), sugar refinery byproducts (De Whalley and Scarr, 1947; Mossel, 1951; Scarr, 1951), dried fruits (Baker and Mrak, 1938; Mrak and McClung, 1938; Mrak *et al.*, 1942; Tanaka and Miller, 1963) and similar habitats. A second important group of species and strains has been isolated from salty environments, and constitutes salt-tolerant strains of *Saccharomyces rouxii* and *Pichia farinosa* isolated from soy sauce and miso paste, and other strains of

Pichia isolated from pickles. Onishi (1960) reported production of glycerol and arabinitol by a number of species of *Pichia*, *Hansenula*, *Debaryomyces*, *Candida* and *Torulopsis*. A few of these species formed small quantities of erythritol. Some of the yeasts isolated by Onishi were previously known species, and a number were assigned to several new species.

The yeasts isolated from honey, jams and jellies, confectionery, pickles and similar foods were investigated because of their role as spoilage agents. The species and strains, especially of *Saccharomyces rouxii*, isolated from soy sauce, miso paste and other such products, however, play an important role in the manufacturing process for these foods. It is not at present known whether the formation of polyhydroxy alcohols contributes significantly to the quality of the final product.

B. Process Development

Pasteur (1858) observed that 2.5–3.0 percent of the sugar fermented in brewing and wine making was converted to glycerol. Studies of the phenomenon were at first limited to development of processes for recovery of glycerol from distillation residues, but subsequently the discovery that the yield of glycerol could be increased considerably by modification of the conditions of fermentation (Neuberg *et al.*, 1949) led to the development of several large-scale processes for production of this polyhydroxy alcohol by fermentation of various sugars by brewing or baking yeasts. These processes, developed by Connstein and Ludecke (1919) in Germany, Cocking and Lilly (1922) in England, and by Eoff *et al.* (1919) in the United States, were based on the interruption of the normal alcoholic fermentation, using strains of *Saccharomyces cerevisiae*, by addition of 'steering' agents such as bisulphites or alkalis, so that the amount of ethanol formed was considerably decreased, and that of glycerol was increased. The process yielded ethanol and acetate or acetaldehyde (as the bisulphite complex) as byproducts, and about 25% of the original sugar present was converted to glycerol. The presence of relatively high concentrations of sulphites, bisulphites or alkalis interfered with recovery of the glycerol, and for this and other reasons the process never became economic.

Subsequent research on the microbiological production of glycerol was concerned mainly with minor improvements and other variations in the alkali and bisulphite processes, such as the use of different reagents to replace bisulphites in formation of the complex with acetaldehyde (Egorov *et al.*, 1957, Sala and De Ayala, 1959). Since none of these led to a commercially viable process, nor revealed any new basic principles, they are nearly all of minor historical interest only. These investigations have been discussed briefly in an earlier review (Spencer, 1968). However, a variant of the bisulphite process, developed at the United States Department of Agriculture Forest Products Laboratory in Madison, Wisconsin, is worthy of attention (Harris and Hajny, 1960). The concentration of bisulphite in the culture was controlled continuously by automatic titration, and the yeast was recovered from the spent mash by centrifugation and returned to the culture vessel. Sulphur dioxide was also recovered and re-used. The process could be operated continuously, so that yields on the basis of sugar metabolized were: acetaldehyde 11%, ethanol 17% and glycerol 25%. The final concentration of glycerol in the culture was 5%. No new methods for solving the problems of glycerol recovery were developed, unfortunately.

A possible solution to the problems raised by the presence of steering agents in the final culture, namely slower and uncertain fermentations, increased susceptibility of the culture to contamination, and difficulty in recovery of the glycerol, was suggested by Nickerson and Carroll (1945). They found that a yeast strain, then classified as *Zygosaccharomyces acidifaciens*, apparently converted about 22% of the sugar metabolized into glycerol, in the absence of bisulphites, alkalis or other steering agents. The yeast used by Nickerson and Carroll (1945) was eventually reclassified in the genus *Saccharomyces*, when the term *Zygosaccharomyces* was discarded by the Dutch taxonomists, and has been currently classified as *Saccharomyces baillii* (Van der Walt, 1970). Investigations by other workers suggested that part of the product was probably arabinitol (J. F. T. Spencer, unpublished data). Under the essentially micro-aerophilic conditions of the work by Nickerson and Carroll (1945), the fermentation took about ten days and utilization of sugar was incomplete. Figard (1951) used the same yeast strain in an attempt to improve yields of glycerol obtained from starch hydrolysates, but reverted to the use of sulphites as steering agents. He obtained little improvement in yields, and the fermentation was slow and incomplete.

Other strains of sugar-tolerant yeasts were used in studies begun in 1951 at the Prairie Regional Laboratory of the National Research Council of Canada at Saskatoon in Saskatchewan (Spencer and Sallans, 1956; Spencer *et al.*, 1957). The yeasts used were isolated from flowers, fermenting honey, and dried fruits, and were mostly classified as strains of *Saccharomyces rouxii* and *Torulopsis magnoliae* (L. J. Wickerham, personal communication). Strains of the former species produced glycerol and arabinitol in ratios depending on the conditions of culture, and the latter, glycerol and erythritol. Later, strains of *Endomycopsis capsularis* and a new species, *Trichosporonoides oedocephalis* (Haskins and Spencer, 1967), were isolated and found to produce polyhydroxy alcohols. *Endomycopsis capsularis* produced good yields of arabinitol (J. F. T. Spencer, unpublished data) and *Trichosporonoides oedocephalis* formed similar quantities of erythritol.

The essential differences in the process from the 'steered' fermentations were that they employed aerobic rather than micro-aerophilic or anaerobic conditions of growth, a relatively high content of nitrogenous compounds and a considerably higher sugar content in the medium, an improved rate of conversion of sugar to product, and a much improved yield of polyhydroxy alcohols. Under optimum conditions of aeration, polyhydroxy alcohol formation was at a maximum and ethanol production was eliminated.

The results obtained at the Prairie Regional Laboratory were verified by investigators at the United States Department of Agriculture Forest Products Laboratory in Madison, Wisconsin. They tested a number of strains of '*Zygosaccharomyces*', most of which have been reclassified as *Saccharomyces rouxii*, and obtained results which were essentially the same as those from the Prairie Regional Laboratory (Peterson *et al.*, 1958). They also isolated other strains and species of yeasts and yeast-like organisms which produced various polyhydroxy alcohols. They found a strain of *Torulopsis magnoliae* (T_2b) which produced good yields of glycerol (Hajny *et al.*, 1960), one of *Endomycopsis chodatii* (now classed as a synonym of *Endomycopsis burtonii*; Kreger-van Rij, 1970) which produced arabinitol (Hajny, 1964), and a fungus which they did not identify which produced erythritol (Hajny *et al.*, 1964). The last species may have been the same or similar to the one isolated from honeycomb in California and named *Trichosporonoides oedocephalis* by the investigators at the Prairie Regional Laboratory.

Recently, production of erythritol from *n*-alkanes by *Candida zeylanoides*, and mannitol by *Candida lipolytica*, has been reported (Hattori and Suzuki, 1974a, b; Tabuchi and Hara, 1973). Good yields of mannitol are also obtainable from various *Aspergillus* species. In addition, Onishi and Suzuki (1970) have demonstrated formation of erythritol from glucose or glycerol by the yeast *Trigonopsis variabilis*.

III. THE PROCESS

Formation of polyhydroxy alcohols is an integral part of the normal growth processes of the various yeasts which produce them, and the requirements for growth determine the requirements for polyol production. Yields of the different polyhydroxy alcohols, however, can be influenced by the conditions of growth and production. Of these, the composition of the medium, especially the choice of carbon and nitrogen sources, the level of aeration, and the temperature have the greatest effect on yields and rate of fermentation.

A. Composition of the Medium

1. Carbon Sources

a. *Sugars.* The nature of the carbon source depends on the yeast species, and the sugars it normally uses, as far as production of polyols by osmotolerant yeasts is concerned. *Saccharomyces rouxii* uses glucose, maltose and sometimes galactose, trehalose and sucrose, and produces glycerol and arabinitol during growth on these substrates. *Torulopsis magnoliae* assimilates glucose, galactose, raffinose and sucrose, but not maltose or trehalose. *Pichia farinosa*, however, uses a wider range of common sugars, including as well sorbose, cellobiose, xylose, ribose, and sometimes lactose and L-arabinose. *Endomycopsis burtonii* (*E. chodatii*) uses all of the sugars employed in the current scheme used for identification except lactose, melibiose, D-arabinose and L-rhamnose (Lodder, 1970). *Trichosporon-*

oides oedocephalis ferments and assimilates glucose, galactose, maltose and sucrose (Haskins and Spencer, 1967). Most of these sugars should be useable for production of polyhydroxy alcohols, under proper conditions.

In practice, only a few of the possible sugars have been tested as carbon sources for formation of polyhydroxy alcohols. Maltose can be converted to glycerol and arabinitol by *Sacch. rouxii*, though the yields are lower than if glucose is used (J. F. T. Spencer, unpublished data). *Hansenula subpelliculosa* and *Torulopsis magnoliae* will form glycerol from sucrose (Charles and Graham, 1961; Hajny *et al.*, 1964), *E. burtonii* will convert sucrose to arabinitol (Hajny, 1964), and the yeast-like fungus tested by Hajny and his coworkers (1964) will convert it to erythritol. The behaviour of *Pichia farinosa* (*Pi. miso*) was similar to that of *Sacch. rouxii* in that it would form the same polyols from glucose, fructose and mannose, but not from galactose or sucrose. The latter pair of sugars were metabolized, according to Onishi (1960), to some extent but no appreciable quantity of product was formed. It should be noted that, according to the standard description of *Pi. farinosa* (Kreger-van Rij, 1970), Onishi's strain of *Pi. farinosa* is a variant which uses sucrose; the type strain does not.

The concentration of sugars and salts affects the yields and the ratios of the products formed (Spencer *et al.*, 1957; Onishi, 1960). With *Sacch. rouxii*, increasing the concentration of glucose raised the yield of glycerol about 50% without bringing about much change in arabinitol yield (Table 1). Increasing the sugar concentration above 30% increased the rate of glucose utilization, but lowered the percentage conversion to polyhydroxy alcohols. The Wisconsin workers found that the same principles applied to their results, as well, for production of glycerol by *T. magnoliae* (Hajny *et al.*, 1960; Table 1). Yields increased at first, as the glucose concentration was raised, but decreased after an optimum concentration was passed. The optimum concentration of glucose was not fixed, but was related to the aeration rate so that, at the optimum availability of oxygen, the optimum sugar concentration for maximum glycerol yield could be increased considerably. Hajny *et al.* (1960) obtained a conversion of glucose to glycerol of 42.5% and a final glycerol concentration of 7.9%. Onishi (1960), likewise, found an optimum concentration of sugar for production of arabinitol and glycerol by

Table I

Effect of nitrogen source, sugar concentration and aeration on yields of polyhydroxy alcohols from yeasts

Yields of products (% of glucose consumed)

Nitrogen source (%, w/v)	Sugar con-centration (%, w/v)	Aeration rate (mmoles/l/h)	Ethanol	Glycerol			Erythritol (unidentified fungus)	Arabinitol	
				Saccharomyces rouxii	*Torulopsis*[b] *magnoliae*	*Pichia farinosa*		*Saccharomyces rouxii*	*Endomycopsis burtonii*
Yeast extract									
0.1			0.6^a		35.8	49.0^a			
0.25					42.4		37.5^d		
0.5	10	42			38.4		38.4^d		32.1^e
		200			40.9				32.3^e
	20	42		39	27.7			61	36.2^e
		200			43.0				
	30	42		103	15.1			51^c	34.8^e
		200			35.6				
1.0	10	42			24.5		33.1^d		
		200							21.8^e
	20	42							32.7^e
		200							18.2^e
1.5	30	42							33.4^e
		200					9.1^d		
4.0			64.1^a			6.1^a			
Urea, corn-steep liquor; 0.35 and 0.8%	20	5^c	58	10					
		12^c	28	30					
		38^c	2	62					

[a]Data from Onishi *et al.* (1961); [b]Data from Hajny *et al.* (1960); [c]Data from Spencer *et al.* (1957); [d] Data from Hajny *et al.* (1964); [e]Data from Hajny *et al.* (1964);

his strain of *Pi. farinosa*, at approximately 29%. If he raised the sugar concentration above 48%, the yield fell considerably and only about 30% of the sugar was metabolized.

He found in addition that the presence of high concentrations of sodium chloride (18%) in the medium affected the ability of *Sacch. rouxii* to metabolize various sugars. In the presence of sodium chloride, his strains utilized glucose, but not galactose, maltose or sucrose, which they normally did to a greater or less extent. Likewise *Torulopsis etchelssii* and *Hansenula subpelliculosa* did not utilize galactose and maltose, respectively, when sodium chloride was included in the medium. A number of other yeast species which Onishi tested, however, were not affected in this way by high concentrations of sodium chloride (Onishi, 1961).

b. *Hydrocarbons*. The most recent advances in production of polyhydroxy alcohols by yeasts have come from a study of the species which use *n*-alkanes (Tabuchi and Hara, 1973; Hattori and Suzuki, 1974a, b). *Candida lipolytica*, which is normally used in the conversion of hydrocarbons into single-cell protein, produces mannitol from *n*-alkanes. More interesting is the production of erythritol by *Candida zeylanoides*. Hattori and Suzuki (1974a, b) used strain KY6166, a glycerol-requiring auxotroph which has been shown to produce citric acid from *n*-alkanes. If the pH value was lowered to 3.0 or below, citric acid production was suppressed and erythritol formation began. If the phosphate concentration was increased to 0.1–0.2% (as KH_2PO_4), mannitol was produced instead of erythritol. Hattori and Suzuki (1974a, b) used a medium containing *n*-alkanes in the C_{12}–C_{15} range as well as mineral salts, and obtained yields of 180 mg of erythritol per ml after 160 h in aerated culture. Production began after cell growth had ceased, that is after about 20 hours.

Candida zeylanoides, *C. lipolytica* and other hydrocarbon-utilizing yeasts do not normally produce polyhydroxy alcohols from carbohydrate substrates. Both species are strongly aerobic, fermentation of sugars being absent or very weak.

2. *Nitrogen Sources*

Most of the yeasts which are commonly used in the production of polyhydroxy alcohols use ammonium salts only as a source of nitrogen, *Torulopsis magnoliae* (Van Uden and Vidal-Leiria, 1970) and *Trichosporon oedocephalis* (Haskins and Spencer, 1967) being the only ones which can assimilate nitrate as a source of inorganic nitrogen. Most yeasts, however, can use a variety of amino acids and other organic nitrogen compounds, and complex nitrogen sources are usually used in media for polyhydroxy alcohol production. Yeast extract and corn-steep liquor are satisfactory sources of nitrogen, and supply an adequate quantity of growth factors as well.

Yeast extract, with various supplements, has been used extensively in studies in production of various polyols (Table 1, pg. 400). Concentrations of yeast extract used have ranged from 0.5 to 4.0%, depending on the conditions of the experiments, and the researchers have not always agreed upon the optimum concentration. The Wisconsin workers generally found decreased yields of all three polyols as the yeast extract concentration was increased beyond 0.5%, while the investigators at the Prairie Regional Laboratory obtained maximum yields at concentrations of about 1%. In some cases, the effects of availability of oxygen and increased cell growth were not separated from those of increased nitrogen concentration, so that lower yields of polyol may have been brought about by conversion of an appreciable fraction of the sugar to ethanol under conditions of diminished oxygen supply, i.e. the increased cell populations, resulting from an increased concentration of nitrogen, depleted the amount of dissolved oxygen in the medium actually available to the cells (Campbell and Hegbom, 1957). The importance of this effect could be seen in data obtained by Onishi *et al.* (1961) where 0.1 and 4.0% yeast extract was used (Table 1, pg. 400). Conversion of glucose to ethanol increased from 0.6 to 64%, and conversion to polyols decreased from 49 to 6%. In work at the Prairie Regional Laboratory, the aeration rate was normally adjusted so that ethanol production was suppressed and maximum conversion of sugar to polyhydroxy alcohols was obtained (Table 1, pg. 400).

Most workers used small concentrations of urea as a supplement to the yeast extract, obtaining slightly improved yields thereby. When corn-steep liquor was substituted for yeast extract, as a more economic nitrogen source, an increased concentration of urea was required. Corn-steep liquor alone was not a satisfactory nitrogen source. The pH value of the fermentation was considerably lower than when yeast extract was used, but adjustment of the pH value with calcium carbonate did not improve the yield. However, when the concentration of corn-steep liquor was decreased to 0.75% and that of urea was increased from 0.1 to 0.35%, yields of polyols comparable to those obtained with yeast extract as a nitrogen source were obtained (Spencer *et al.*, 1957).

A number of ammonium salts were tried as supplementary nitrogen sources, without notable success (Spencer *et al.*, 1957). The presence of ammonium sulphate, chloride or phosphate in the medium usually resulted in lowered yields, while ammonium lactate or tartrate gave yields about as good as those obtained with urea. Onishi (1960) obtained some improvement of yields by using buffered medium with ammonium sulphate or chloride, so that increased acid formation as the ammonium ion was used up may have been partly responsible for the decreased yields obtained with these compounds. However, a complete explanation for the beneficial effect of urea is not known.

Other nitrogen sources can be successfully used. Hajny (1964)

Table 2

Effect of salt concentration on yields of polyhydroxy alcohols

Organism	Salt	Concentration (M)	Total polyols (as glycerol)	Erythritol	Glycerol
			Yields (% of glucose consumed)		
Unidentified fungus[a]	KCl				
		0		38.7	0.0
		1		22.2	13.0
		2		15.8	28.0
		3		17.7	31.6
Pichia farinosa[b]	NaCl				
		0	22.9		
		0.5	29.9		
		1.0	21.4		
		2.0	31.5		
		3.0	23.6		
	Glucose				
		1.7	45.7		

[a] Data from Hajny *et al.* (1964); [b] Data from Onishi (1963).

used a medium containing blackstrap molasses, glucose and urea for arabinitol production by *E. burtonii*, and obtained moderately good conversions of sugar to product. He and his coworkers (Hajny *et al.*, 1964) were also able to use malt, malt sprouts and distiller's dried solubles as nitrogen sources. Onishi (1960) used beef extract and polypeptone as well as yeast extract or corn-steep liquor for production of a mixture of polyhydroxy alcohols by *Pi. farinosa*, with satisfactory results. Polypeptone especially gave good conversion of glucose to polyols. Casamino acids were also a satisfactory nitrogen source.

In the process used for production of mannitol and erythritol by *C. lipolytica* and *C. zeylanoides*, only inorganic nitrogen sources were used (Hattori and Suzuki, 1974a, b). Growth was satisfactory, as the maximum cell yield was obtained after about 20 hours.

3. *Sodium and Potassium Chlorides*

These salts are important for their effect on the osmotic tension of the medium, rather than for any direct involvement in metabolism of the organism. In the manufacture of soy sauce and other fermented foods, where *Sacch. rouxii* plays a key role, concentrations of sodium chloride up to 18% are normally used, and the yeast strains are acclimatized to it. The effects of high concentrations of salt on production of polyhydroxy alcohols, however, are sometimes deleterious (Table 2). Hajny *et al.* (1964) found a considerable decrease in glucose utilization as the concentration of potassium chloride was increased from zero to 3M, and at the same time erythritol yield decreased, and glycerol, which was not formed in the potassium chloride-free medium, appeared in increasing yield. Onishi (1963) likewise found a shift in the nature of the products formed by *Pi. farinosa*, in the presence of 18% NaCl, from arabinitol. In the absence of sodium chloride, the yeast produced both polyols but, as the salt concentration was increased, the concentration of arabinitol decreased until at 6% NaCl (i.e. 1 M) it had virtually ceased to be produced and the sole product formed was glycerol. Other yeast species tested showed the same shift in the nature of the products formed as salt concentration was increased.

Onishi (1960) also observed that the presence of high concentrations of salt enabled the yeasts to grow at significantly higher temperatures. This phenomenon is similar to the observation by Lochhead and Heron (1929), in which the yeasts were found to be obligately osmophilic at 37°C but did not require high concentrations of sugar at 30°C. The same effect of protection against high temperatures by elevated sugar concentrations was observed by the investigators at the Prairie Regional Laboratory (Spencer *et al.*, 1957).

4. *Phosphate and Other Mineral Nutrients; Growth Factors and Amino Acids*

The major effect of these nutrients now appears to be to increase cell growth. Some early studies indicated that changes in phosphate concentration might influence yields of ethanol and glycerol directly (Spencer and Shu, 1957), but further studies indicated that the effect was due to increased cell growth and consequent exhaustion of oxygen from under-aerated medium. This results in a decrease in oxidative metabolism and an increase in fermentation, with a consequent decrease in glycerol and increase in ethanol yield. Changes in phosphate concentration did not significantly affect the yields of polyhydroxy alcohols other than glycerol. The amounts of other nutrient salts in the media usually used for polyol production are sufficient for normal cell growth and maximum yields of products.

The yeast species and strains used for polyhydroxy alcohol production are known to require various growth factors, including biotin (some strains of *Sacch. rouxii*; Lochhead and Landerkin, 1942) and thiamin (for erythritol production by Hajny's organism; Hajny *et al.*, 1964). However, addition of amino acids or growth factors to media had no effect on yields of glycerol or arabinitol, so the requirements of these organisms for these compounds are apparently fully met without adding them to the medium (Peterson *et al.*, 1958; Hajny, 1964). Most of the yeasts used can utilize many amino acids as sole nitrogen sources for growth (LaRue and Spencer, 1967), and could likely use them for polyhydroxy alcohol production. Some of the D-isomers as well as the L-isomers can be utilized by most strains.

B. Aeration

Saccharomyces rouxii, like *Sacch. cerevisiae*, exhibits the Pasteur effect as the level of aeration is varied so that, at low levels of aeration, conversion of glucose (or other fermentable sugar) to ethanol is relatively great, and decreases as the level of aeration is

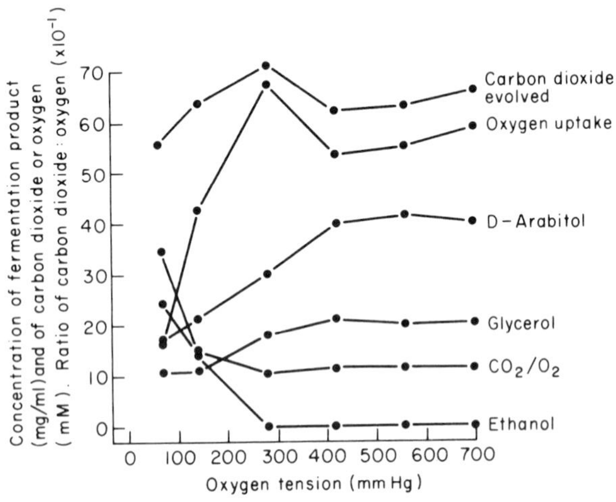

Fig. 1. Effect of oxygen tension on oxygen uptake and production of fermentation products by *Saccharomyces rouxii* in a medium lacking phosphate. From Spencer and Shu (1957).

increased. Besides this, it was early observed at the Prairie Regional Laboratory that, as the yield of ethanol was decreased, the yield of glycerol increased (Spencer *et al.*, 1957; Table 1, pg. 400). Early indications were that the level of aeration did not affect the yield of arabinitol, but subsequent experiments provided some evidence that arabinitol yields also increased as aeration levels were raised (Fig. 1; Spencer and Shu, 1957).

In general, the results obtained by the Wisconsin workers confirmed those from the Prairie Regional Laboratory. In some of the early work, the yield of ethanol obtained was high enough to suggest that the aeration rates, and thus the levels of dissolved oxygen, were inadequate for maximum yields of polyols. Their later results were obtained using aeration rates at the other end of the optimum range, but the yields were approximately equivalent to those obtained at the Prairie Regional Laboratory. Hajny *et al.* (1964) also obtained yields of glycerol with an optimum level of aeration somewhere between 42 and 200 mmoles O_2/l/h. They obtained similar results in studies of arabinitol production by *E. burtonii* (Hajny, 1964; Table 1, pg. 400).

Onishi obtained further confirmation of the effect, using *Sacch. rouxii* and *Pi. farinosa.* Glycerol production was lowered as the aeration level was decreased (by increasing the volume of medium in shaken flasks). At the lowest level of aeration, obtained either by increasing the volume of medium in the flask to 200 ml or using static cultures, glycerol and arabinitol production by *Pi. farinosa* was greatly decreased, and erythritol production was suppressed. Further experiments using a Waldhof fermentor demonstrated convincingly that yields were a function of the degree of dispersion of the air in the medium rather than of the air flow rate (Onishi and Saito, 1962).

The rate of oxygen supply per unit of biomass, which is the critical factor determining the percentage conversion of glucose to polyhydroxy alcohols, is determined by the density of the yeast population in relation to the rate of solution of oxygen in the medium. This in turn is a function of other nutritional factors in the culuture, especially the nature and concentration of the nitrogen source and the carbon/nitrogen ratio in the medium. Campbell and Hegbom (1957) called attention to a similar effect in their studies of factors affecting cell growth in *Sacch. fragilis.*

C. Temperature

Early experiments on the effect of temperature on production of polyols by *Sacch. rouxii* indicated that the ratio of the concentration of glycerol to arabinitol was increased considerably by increasing the temperature of the fermentation from 30° to 37°C (Spencer *et al.,* 1957). These results were not observed in later experiments. On the contrary, increases in yields of total polyols were the result of increased production of arabinitol. There was a significant increase in the rate of glucose utilization at the higher temperatures, as would be expected. Similar results were obtained by Peterson's group at the University of Wisconsin (Peterson *et al.,* 1958; Hajny *et al.,* 1960;

Hajny, 1964). The optimum temperatures for polyol production by
E. burtonii, T. magnoliae and various strains of '*Zygosaccharomyces*'
lay between 30°C and 35°C, with slightly higher yields and faster
rates of sugar utilization at the higher temperatures. During production of erythritol by the organism used by Hajny, the yield of
erythritol was decreased at 35°C (Hajny *et al.*, 1964), though the
rate of sugar utilization was increased and the yield of glycerol was
increased at the expense of erythritol. The relationship of these
phenomena to the observation by Lochhead and Heron (1929) that
these yeasts became obligate osmophiles at 37°C is not known.

D. Product Recovery

Recovery of polyols from aqueous solutions, which is essentially the
problem encountered in separating them from fermentation media,
presents no technical difficulties (Roxburgh *et al.*, 1956). Essentially,
the method consists of removing yeast cells by settling or centrifugation in a rough primary clarification step after acidification of the
broth. Most of the water is then removed by evaporation, preferably
under vacuum. Hot ethanol is then added, the gums are filtered off,
and the solution is decolourized with activated charcoal. Erythritol
and arabinitol are only slightly soluble in cold mixtures of glycerol
and ethanol, and they crystallize from the solution on cooling. After
the solids are removed by filtration, alcohol is recovered from the
liquid by distillation at atmospheric pressure, and glycerol is vacuum
distilled in the conventional manner and recovered in fair yield.
Some arabinitol is lost because of its solubility in glycerol–alcohol
mixtures, which increases significantly if even small amounts of
water are present. It is necessary to ensure that as much water as
possible is removed before the crystallization step.

If an organism is used which produces only arabinitol or erythritol, with little or no glycerol being formed, the problem becomes one
of recovering these products essentially from alcohol–water mixtures. Losses are, of course, much lower if all or nearly all of the
water is removed first. Further purification of the product, whatever
the process, may be desirable, and can usually be done satisfactorily
by recrystallization from ethanol after dissolving the crude crystals in
a minimum amount of water.

IV. PHYSIOLOGY AND BIOCHEMISTRY OF POLYHYDROXY ALCOHOL FORMATION

A. General

Sugar- and salt-tolerant yeasts occur in several genera (Onishi, 1963), and there is nothing species-specific about this characteristic. Early work on the physiology of these micro-organisms was done on isolates from spoiled foods, especially honey, and the primary interest of the investigators in these organisms was in their role as spoilage agents (Fabian, 1928; Fabian and Quinet, 1928). Of particular interest was the observation that, in honey, the yeasts were limited in their growth primarily by the moisture content of the substrate. In properly ripened honey, the moisture content was too low, and consequently the sugar content was too high to permit growth of the yeasts. However, honey exposed to a moist atmosphere took up a minimum of approximately 26% and a maximum of 33% water. The critical water content permitting growth of the yeasts was about 21%, which is well below the concentration which is reached when honey is allowed to take up moisture from air.

The yeasts isolated by earlier investigators were probably most strains of *Sacch. rouxii*, and the metabolic properties reported are essentially those of this species, namely fermentation of glucose and sucrose only, formation of approximately equimolar amounts of ethanol and carbon dioxide, formation of small quantities of volatile acids (Marvin *et al.*, 1931) and improved fermentation in the presence of oxygen (Krumbholz, 1936; Borries, 1934). *Saccharomyces rouxii* is tolerant of wide ranges of acidity and sugar and salt concentration (English, 1954).

The osmotolerant yeasts studied by Lochhead and his coworkers varied in their requirements for growth factors (Lochhead and Farrell, 1931; Farrell and Lochhead, 1931; Lochhead and Landerkin, 1942). All of the strains required biotin, but varied in their requirements for pantothenic acid, inositol and thiamin. Nickerson and Thimann (1941, 1943) found that a factor from *Aspergillus niger* stimulated conjugation in these haploid species of *Saccharomyces*. Riboflavin and α-oxoglutaric acid produced a similar stimulatory effect, but the *A. niger* factor was not definitely identified with them. Interrelationships between temperature and osmotic pressure, which were referred to previously (see pg. 405), were observed by three of the principal workers in this field. Lochhead and Heron (1929) found that their strains required high concentrations of sugar for growth at 37°C but that, at 30°C, growth occurred at all concentrations of sugar tested. Onishi (1963) observed the same effect in his study of salt-tolerant yeasts from soy sauce mashes, i.e. at elevated temperatures a high (18%) concentration of sodium chloride was necessary for yeast growth. Workers at the Prairie Regional Laboratory (Spencer *et al.*, 1957) found that growth and polyhydroxy alcohol production occurred, at somewhat decreased rates and yields, at temperatures above 40°C, as long as the sugar concentration was greater than 30%. Growth and polyol production stopped at these temperatures when the sugar concentration fell significantly below 30%. However, S. E. Windisch (personal communication) isolated a number of strains of *Sacch. rouxii* which did not show a temperature-concentration effect. The reason for the difference in Windisch's results from those of the other investigators is not known. Nor is the nature of the effect understood. There is an analogy to the behaviour of temperature-sensitive osmotic remedial mutants of *Sacch. cerevisiae* (Hawthorne and Friis, 1964) in which a block in metabolism at the restrictive temperature is removed and growth allowed to proceed by the use of increased concentrations of potassium chloride in the medium. In this case, an altered protein is probably formed by the mutant, which is not functional at the restrictive temperature unless it is stabilized by the presence of increased concentrations of salts. In the osmotolerant yeasts, the same mechanisms may be responsible, i.e. the enzyme or enzymes which are inactivated by elevated temperatures in dilute media are stabilized by high concentrations of sugar or salts as in the *Sacch. cerevisiae* mutants.

Onishi (1963) has contributed most of the significant information concerning the nature of salt tolerance in yeasts isolated from soy sauce and miso fermentations. He found that yeasts which were freshly isolated from soy mashes sometimes grew more slowly on ordinary media, especially at high temperatures. The limiting osmotic pressure was considerably lower in media containing sodium chloride rather than sugars, and the toxicity of various salts decreased through the series: LiCl, $CaCl_2$, Na_2SO_4, $MgSO_4$, NaCl and KCl. This series is slightly different from that observed at the Prairie Regional Laboratory (J. F. T. Spencer, unpublished data) where magnesium salts appeared to be less toxic than sodium chloride. The yeasts used by the latter investigators, however, were isolated from honeycomb and dried fruits and similar sources, and had not been previously acclimatized to high concentrations of salts.

Onishi (1963) found an increase in permeability of cells grown in media containing sodium chloride, and that such cells required a metabolizable substrate for survival in a phosphate buffer containing 18% sodium chloride. He also searched for enzymes which required or tolerated high concentrations of salt, but none was found. Some or all of the enzymes required for glycerol production were sensitive to high concentrations of sodium chloride, since glycerol formation in cell-free extracts was suppressed by high concentrations of salt. Onishi (1963) suggested that, since there were apparently no halo-tolerant enzymes in the cell-free extracts, osmotolerance in yeasts depended on an exclusion mechanism, but he offered no further evidence for this.

Brown's (1974) recent work is a significant advance in the understanding of the nature of osmotolerance in yeasts. Of two possible explanations for the phenomenon, the presence of halo-tolerant enzymes is apparently ruled out by the evidence given previously, and if an exclusion mechanism is responsible for the effect it cannot be one which results in the cell contents being more dilute than the surrounding medium, since there is no mechanism to prevent water loss from the cell to the more concentrated solution outside the cell membrane. Brown has confirmed that NAD^+-specific isocitrate dehydrogenase from *Sacch. rouxii* has the same kinetics, water relationships and electrophoretic properties as the same enzyme from *Sacch. cerevisiae*, and has also shown that cells of *Sacch. rouxii* contain high concentrations of arabinitol while *Sacch. cerevisiae* cells do not. Brown suggests that arabinitol functions as a compatible solute, which protects intracellular enzymes against inhibition or inactivation at biologically low levels of water activity. Further study may show whether the high concentration of intra-cellular arabinitol is essentially part of an active exclusion mechanism, as suggested by Onishi (1963), and which makes it possible for the cell actively to exclude sodium chloride and other salts which inhibit or inactivate any sensitive enzymes, without at the same time suffering any water loss.

B. Enzymic Processes and Regulatory Mechanisms in Polyhydroxy Alcohol Formation

1. *The Pasteur Effect*

The Pasteur effect is possibly the best known and least understood phenomenon in yeast metabolism (Sols *et al.*, 1971). Its importance in polyhydroxy alcohol formation stems from its involvement in the regulation of the balance between ethanol and glycerol production. As cultural conditions become more aerobic, the yield of ethanol is decreased and that of glycerol increases. A number of mechanisms for the Pasteur effect have been suggested, most of which have been too simplistic; competition between the fermentative and respiratory pathways for inorganic phosphate, ADP, or both (Johnson, 1941) was favoured for some time but, though plausible, it is probably not the complete answer. Sols *et al.* (1971) suggest a mechanism in which regulation of the balance between the fermentative pathways depends upon allosteric feedback mechanisms affecting several enzymes: (1) NAD^+-dependent isocitrate dehydrogenase, which catalyses the first irreversible step on the pathway of citrate oxidation through the tricarboxylic acid cycle; (2) phosphofructokinase, catalysing the first irreversible step of the common glycolytic pathway from glucose 1-phosphate. Action of this enzyme is controlled by the content of ATP following anaerobiosis, and of ATP and citrate during aerobiosis, (3) enzymes involved in glucose transport across the cell membrane.

The scheme as presented has many attractive features. It has not been studied in the osmotolerant yeasts, and it does not seem at first sight to account for the concomitant increase in production of glycerol and other polyols with increasing aeration and decreasing ethanol formation. Since this is not observed in *Sacch. cerevisiae* to any noticeable extent, investigators of the Pasteur effect have not attempted to account for what may be described as the other half of the phenomenon. Investigation of the Pasteur effect in *Sacch. rouxii* might yield more positive answers than have so far been obtained using *Sacch. cerevisiae*.

2. *Polyol Dehydrogenases*

Yeasts have been shown to contain numerous polyol dehydrogenases, some of which are necessary in the final stages of formation of the various polyhydroxy alcohols in osmotolerant yeasts. Polyol dehydrogenases catalyse the last reaction or reactions on the pathways leading to formation of the different polyols, and they usually reduce one or the other of the corresponding ketoses. Weimberg (1962) outlined a scheme for the formation of arabinitol as follows:

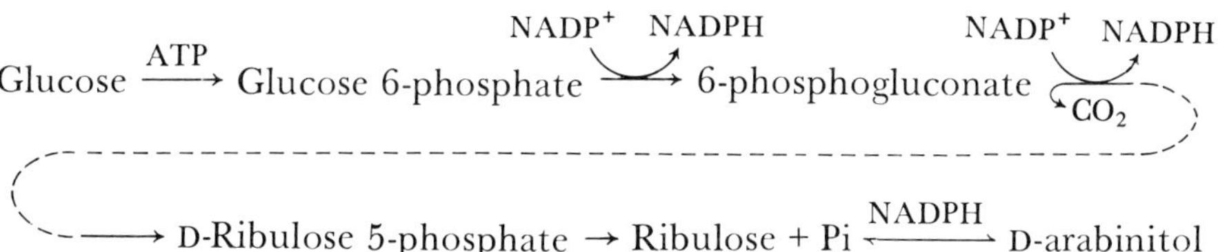

This scheme accounts for only half of the NADPH formed. However, the remaining NADPH can be accounted for in the production of arabinitol *via* transketolase-transaldolase-catalysed reactions, which are postulated to explain the distribution of radio-activity in arabinitol formed from C-1- and C-2-labelled glucose.

While Weimberg's (1962) scheme postulated a polyol dehydrogenase which catalysed a reversible oxidation-reduction reaction producing arabinitol from glucose, other workers have observed somewhat different reactions. For instance, xylulose rather than ribulose was reduced to arabinitol by a reductase from *Sacch. rouxii* (Blakley and Spencer, 1962). Crude cell-free extract reduced both xylulose and ribulose, however, so that both enzymes were present in the intact yeast. Brewer's yeast possesses dehydrogenases which oxidize mannitol, sorbitol, dulcitol and erythritol (Wilhelmsen, 1961). Onishi and Saito (1962) isolated an enzyme which catalysed the general reaction:

$$\text{Polyalcohol} + \text{NAD}^+ \rightleftharpoons \text{Ketose} + \text{NADH} + \text{H}^+$$

The enzyme catalysed oxidation of arabinitol to xylulose or reduction of xylulose to arabinitol. Onishi's preparation differed from that used by Blakley and Spencer (1962) in catalysing oxidation of several polyalcohols including mannitol and sorbitol to fructose, dulcitol to (probably) tagatose, ribitol to ribulose, xylitol to xylulose, erythritol to erythrulose and glycerol to dihydroxyacetone. The optimum pH value of this preparation was about 7.0, higher than the enzyme used by Blakley and Spencer (1962), and required Fe^{2+} ions. It may have consisted of a mixture of enzymes.

Polyol dehydrogenases have been found in a number of other micro-organisms. *Penicillium chryosgenum* has an enzyme system which converts L-arabinose to L-arabinitol and thence to L-xylulose (Chiang and Knight, 1961). L-Ribulose was also formed in small quantities, probably by a reaction catalysed by a separate enzyme. *Aerobacter aerogenes*, likewise, metabolizes L-arabinitol by a pathway which included phosphorylated intermediates:

L-Arabinitol $\longrightarrow$ L-Xylulose $\longrightarrow$ L-Xylulose 5-phosphate $\longrightarrow$

L-Ribulose 5-phosphate $\longrightarrow$ D-xylulose 5-phosphate

(Fossit *et al.*, 1964). Xylitol was oxidized directly to D-xylulose and then converted to D-xylulose 5-phosphate. *Aerobacter aerogenes* has also a ribitol dehydrogenase which oxidizes only ribitol but synthesis of which is induced by ribitol and arabinitol. Under anaerobic conditions, oxidation of ribitol was limited by the availability of hydrogen acceptors. *Cellvibrio polyoltrophicus* has two NAD^+-linked polyol dehydrogenases, inducible by mannitol, sorbitol and arabinitol, and one of which oxidizes sorbitol to fructose, and the other arabinitol to xylulose and mannitol to fructose (Scolnick and Lin, 1962). Many bacteria and yeasts both form and utilize a variety of polyhydroxy alcohols, and polyol dehydrogenases would be expected to play an essential part in this part of their metabolism. Touster and Shaw (1962) have made a comprehensive review of the literature concerning the biochemistry of the acyclic polyols, including their metabolism in plants and animals, and the polyol dehydrogenases of mammalian tissues.

3. *Phosphatases*

Since formation of polyols apparently proceeds via a number of phosphorylated intermediates, a variety of phosphatases are necessary for the final step of the process, leading to conversion of polyol phosphates to the corresponding polyhydroxy alcohols. Some of these phosphatases have been detected in *Sacch. mellis* (Weimberg and Orton, 1963) and *Sacch. rouxii* (Ingram and Wood, 1965). The former workers found an acid phosphomono-esterase having broad specificity and an inorganic pyrophosphatase with an optimum pH value of 7.5 in *Sacch. mellis*. Formation of the former enzymes was inhibited by the presence of phosphate in the culture medium, while the latter enzymes were not. The role of these enzymes in formation of polyhydroxy alcohols is not known. Ingram and Wood (1965) likewise demonstrated the presence of an acid phosphatase in *Sacch. rouxii*, but could not find a specific D-ribulose 5-phosphatase. The enzymes these workers found, however, are sufficient to meet the needs of the last reaction on the pathways of polyol formation.

C. Metabolic Pathways in Biosynthesis of Polyhydroxy Alcohols

1. Glycerol

Glycerol is formed via the Embden–Meyerhof pathway (Fig. 2) in brewer's and baker's yeasts, other fungi and other organisms, as well as in *Sacch. rouxii* and other osmotolerant yeasts. Briefly, formation of glycerol results from reduction and dephosphorylation of dihydroxy-acetone phosphate, formed by scission of fructose 1,6-diphosphate by aldolase. The pathway is well known, and the more interesting aspect of glycerol formation is the mechanism which controls the relative yields of glycerol and ethanol. Under normal conditions of growth, most of the glucose is converted to ethanol (in *Sacch. cerevisiae*). During the process, NAD^+ is first reduced to NADH, then re-oxidized during reduction of acetaldehyde in the final stage of ethanol formation. A little of the NADH is diverted and used up in reduction of dihydroxyacetone phosphate to glycerol phosphate, which is then dephosphorylated yielding glycerol. The problem in processes for the production of glycerol by fermentation is to influence the reactions of the Embden–Meyerhof pathway in such a way as to suppress formation of ethanol and divert more of the NADH formed earlier to reduction of dihydroxyacetone phosphate to glycerol phosphate.

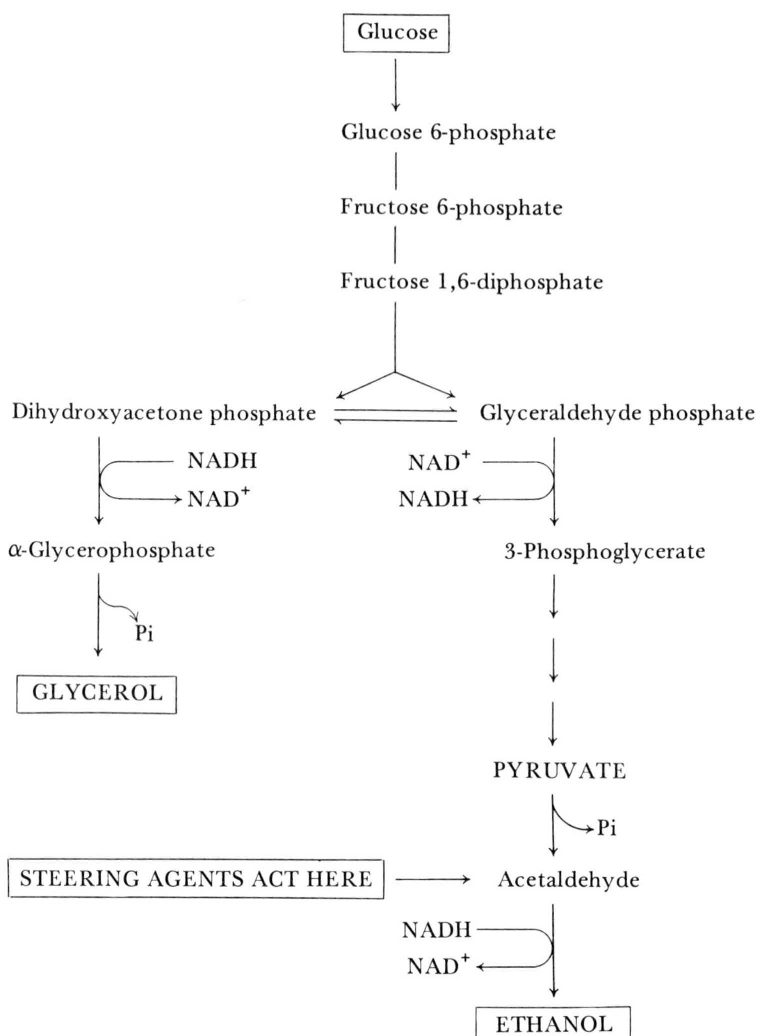

Fig. 2. Formation of glycerol and ethanol by anaerobic dissimilation of glucose *via* the Embden–Meyerhof pathway, showing the point of action of steering agents.

In the 'steered' fermentation processes, this is achieved by the use of sulphites, bisulphites or other acceptors, or alkalis. Acetaldehyde binds to bisulphite to form a complex which cannot react with NADH, which in turn then reacts with dihydroxyacetone phosphate yielding glycerol. Other reagents such as dimedon, which bind acetaldehyde before it can be reduced by NADH, act in a similar manner. The control is not complete, as ethanol is still formed in lower yield, but a much larger fraction of the glucose metabolized is converted to glycerol.

The control mechanism in the alkaline process depends on the occurrence of a Canizzaro reaction in which one molecule of acetaldehyde is oxidized to acetate, while a second is simultaneously reduced to ethanol. For some time, this reaction was assumed to be non-enzymic, but it does not occur in the absence of living yeast cells and so represents an alteration induced by alkalis in the normal metabolism (Freeman and Donald, 1957). Thus, one molecule of acetaldehyde functions as a hydrogen donor resulting in the removal of another molecule of acetaldehyde from the normal reaction system, so that again there is a surplus of NADH, which can take part

Table 3

Distribution of labelled carbon in products from fermentation of labelled glucose by *Saccharomyces rouxii* and *Torulopsis magnoliae*

	^{14}C as % of total in product		
Products	[1-^{14}C]-Glucose	[2-^{14}C]-Glucose	[6-^{14}C]-Glucose
Glycerol			
$-CH_2OH$	99.2	5.0	99.0
$-CHOH$	0.8	95.0	1.0
Ethanol			
$-CH_3$	99.2	6.0	
$-CH_2OH$	0.8	94.0	
D-Arabinitol			
C-1	57.3	59.4	
C-2	1.2	21.4	
C-3	1.1	1.2	
C-4	0.6	13.5	
C-5	39.8	4.5	
Erythritol			
C-1	24	44	1
C-2	17	10	20
C-3		30	0
C-4	59	10	80

Erythritol was formed by *Torulopsis magnoliae*. The percentage of total label from [6-^{14}C]-glucose in the erythritol formed was much greater than in that from [1-^{14}C]-glucose and [2-^{14}C]-glucose, so that even more of the erythritol was formed directly from the former than is apparent from the table (Spencer and Gorin, 1960).

in the reduction of dihydroxyacetone phosphate leading to increased yields of glycerol.

Steering agents are not necessary in processes for polyhydroxy alcohol production using osmotolerant yeasts, and the controlling factors are not known, except that they appear to be related to the respiratory metabolism. An adequate, but not excessive, supply of oxygen is necessary for maximum yields of products, and growth of the organisms under conditions of suboptimum aeration results in increased quantities of ethanol being formed and a considerable decrease in the yield of glycerol. The metabolic pathway is the same as in baker's yeast (Table 3, Fig. 2; Spencer *et al.*, 1956). Glycerol formed from $[1\text{-}^{14}C]$-glucose is labelled in the $-CH_2OH$ carbons, and from $[2\text{-}^{14}C]$-glucose, in the $-CHOH$ carbons. The labelling in the ethanol formed is also in accordance with the glucose being metabolized to these products via the Embden–Meyerhof pathway.

2. *Other Polyhydroxy Alcohols*

At least two mechanisms are involved in formation of D-arabinitol (Spencer *et al.*, 1956). It was at first assumed that the main pathway would be decarboxylation of 6-phosphogluconate via the hexose monophosphate (Warburg–Dickens) pathway (Fig. 3). The specific activity of $^{14}CO_2$ formed from $[1\text{-}^{14}C]$-glucose was approximately four times that of the $^{14}CO_2$ from $[2\text{-}^{14}C]$-glucose, which is in accordance with this idea. However, D-arabinitol formed from $[1\text{-}^{14}C]$-glucose is labelled in C-1 and C-5 (Table 3), which cannot be explained if the hexose monophosphate pathway is the sole one involved in its formation. Likewise, D-arabinitol formed from $[2\text{-}^{14}C]$-glucose is labelled in C-1, C-2 and C-4. Labelling in C-1 would be expected, but the label in the other two positions cannot be explained by a simple decarboxylation of 6-phosphogluconate.

These patterns of labelling can, however, be explained by assuming operation of the enzyme transketolase in this system. Transketolase transfers two-carbon units from ketoses or ketose phosphates to a variety of acceptors. In this case, the acceptor is one of the triose phosphate fragments formed by action of aldolase on fructose 1,6-diphosphate, and the result is the formation of 5-carbon chains with labelling in carbons 1 and 5 from $[1\text{-}^{14}C]$-glucose, and in C-2 and C-4 from $[2\text{-}^{14}C]$-glucose. The label in C-5 arises from that in the triose phosphate which originates from C-1 of the glucose, and is probably less than that in C-1 of the arabinitol because some of the 2-carbon units are transferred by transketolase to unlabelled triose phosphate arising from carbons 4, 5 and 6 of glucose. The considerable excess of ^{14}C in C-1 of arabinitol formed from $[2\text{-}^{14}C]$-glucose probably gives a rough measure of the amount of arabinitol formed

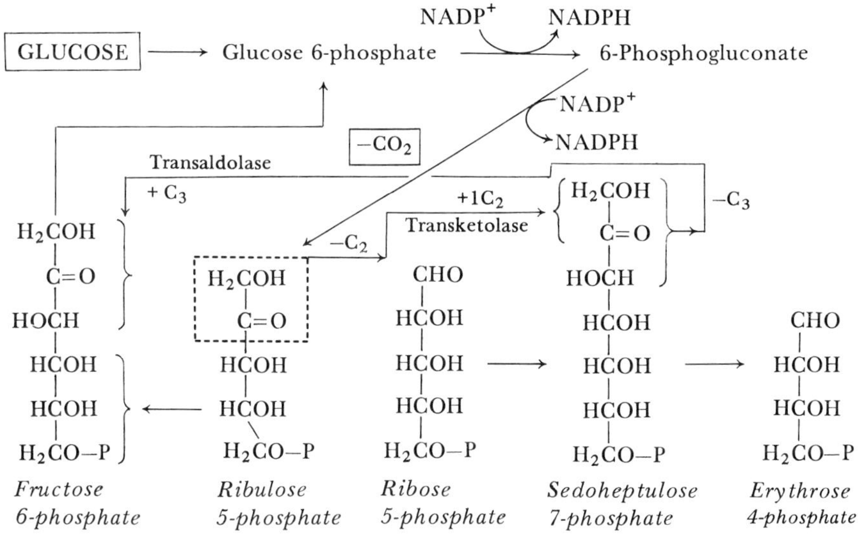

Fig. 3. Pentose phosphate pathway for aerobic dissimilation of glucose, and formation of C_4 and C_5 chains. P indicates a phosphate residue.

via the hexose monophosphate pathway, since this is the only direct pathway which leads to labelling of C-1 in arabinitol formed from $[2\text{-}^{14}C]$-glucose. A number of later reactions in the hexose monophosphate pathway involving pentose phosphates lead to further reshuffling of the label from either $[1\text{-}^{14}C]$- or $[2\text{-}^{14}C]$-glucose, but these probably play no more than a minor role in the formation of D-arabinitol.

Similarly, erythritol is probably formed principally by the action of transketolase on fructose 6-phosphate, which leaves a 4-carbon fragment that is probably reduced and dephosphorylated. If $[6\text{-}^{14}C]$-glucose is used as substrate, most of the label is in C-4 of the resulting erythritol (Spencer and Gorin, 1960). The other reactions of the hexose monophosphate pathway play a small role in the labelling of other carbon atoms of erythritol, but the major source of this polyol is the reaction catalysed by transketolase. These reactions are summarized in Fig. 4.

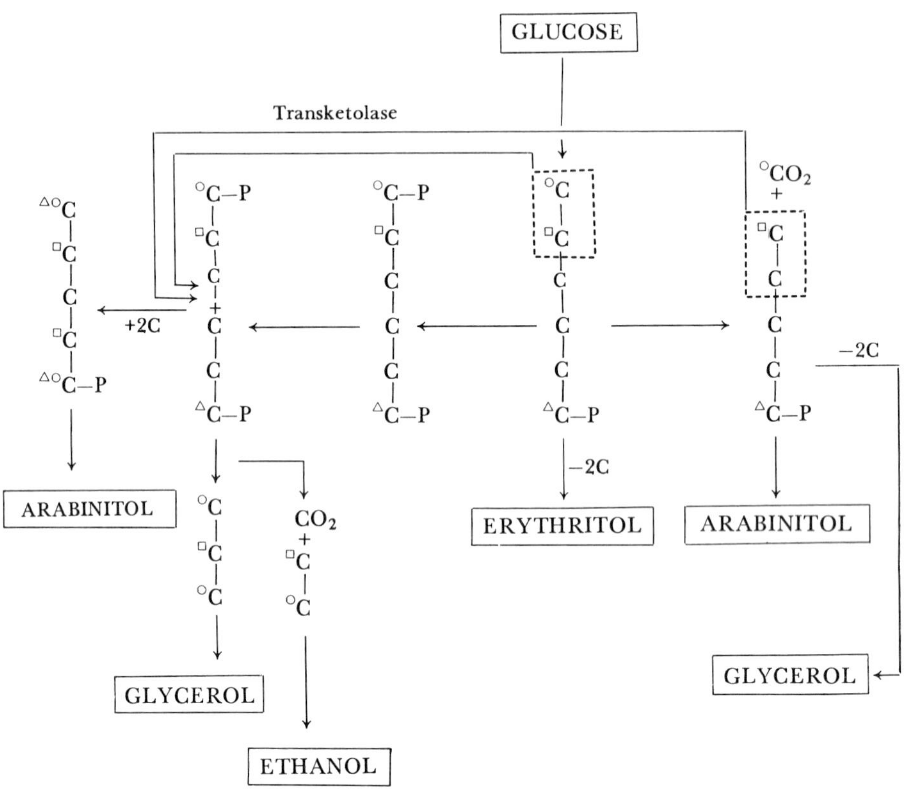

Fig. 4. Distribution of label in potential carbon skeletons of arabinitol, erythritol, glycerol and ethanol formed from [1-^{14}C]-, [2-^{14}C]- and [6-^{14}C]-glucose by *Saccharomyces rouxii* and *Torulopsis magnoliae*. P indicates a phosphate residue. The carbon atoms are tagged with different symbols to help the reader to follow the fate of the carbon atoms.

Finally, the heptitols found by Onishi to be formed from D-xylose by *Pi. farinosa* (Onishi and Perry, 1965) probably are formed *via* the action of transketolase in transferring 2-carbon units to pentose or pentose phosphate. The high yields of xylitol obtained at the same time probably arose by direct reduction of D-xylose or, possibly, pentose phosphate. Transketolase, then, seems to play a considerable part in the formation of polyhydroxy alcohols by osmophilic yeasts.

As far as is known, investigations of the metabolic pathways involved in the formation of polyhydroxy alcohols from hydrocarbons have not been made. However, enough is known of the mechanisms of hydrocarbon catabolism in micro-organisms that possible pathways can be suggested. One or both of the terminal carbons of an *n*-alkane can be oxidized to the corresponding carboxylic acid. The carbon chain can then be decreased in length by β-oxidation yielding 2-carbon units which can be introduced into the rest of the metabolic pathways of the organism, for growth or for production of other metabolites. Erythritol or mannitol could be synthesized by condensation of two or three acetyl-CoA units, for instance. In *Candida zeylanoides*, where citric acid production is increased by a decrease in the pH value to below 3.0 (Hattori and Suzuki, 1974a), there is evidently a diversion of 2-carbon units from citrate synthesis to erythritol at the lower pH value, but the mechanism for this is not known.

The major unsolved problem in polyhydric alcohol formation is that of metabolic control, of which very little is known. Control of glycerol formation in unsteered fermentations using osmotolerant yeasts is connected with the Pasteur effect, and probably raises questions concerning the validity of the mechanisms currently suggested for this effect. Nothing is known of the mechanisms which determine which of the polyhydroxy alcohols will be produced by any particular yeast or other fungal species. Traces of most possible polyols are usually found in the culture medium, no matter which species is used, but it is seldom that more than one or at the most two are produced in significant yields. So far, no explanation has been offered for the change in *Candida zeylanoides* from citric acid to erythritol production at a pH value below 3. Brown's (1974) work has suggested a logical reason for production of polyhydroxy alcohols by osmotolerant yeasts, but does not explain why any specific one should be formed. Further studies along these lines may provide an answer.

V. SUMMARY

Polyhydroxy alcohols are produced in considerable quantities by many micro-organisms, especially by sugar-tolerant and salt-tolerant yeasts and yeast-like organisms. These fungi occur naturally in association with flowers, fruits and insects such as domestic bees and bumblebees, and as spoilage agents of jams, jellies, honey, confectioneries and some types of stored fruits. A few are found in salt meats and in similar habitats. Some species are used in the production of a number of types of fermented foods having a high content of salt which are in common use in Japan, Indonesia and other countries in that area of the World.

The metabolic pathways involved in formation of the carbon skeletons of the various polyhydroxy alcohols have been determined, and are all well known. Biosynthesis of all of the polyols can be assumed to proceed by combinations and modifications of the Embden–Meyerhof and pentose phosphate pathways. The remaining problems concern the nature of the control systems which determine which of the various products are formed by the different species and under different cultural conditions.

A function for the different polyols in the osmotolerant yeasts has now been suggested. They appear to serve as compatible solutes, and play an essential role in preventing inhibition or inactivation of enzymes in conditions of low levels of water activity, i.e. during growth of the yeasts in media, natural or artificial, containing high concentrations of salts or sugars.

Actual and potential uses for polyhydroxy alcohols are extremely numerous. At present there are no current uses for erythritol or arabinitol, though the latter compound may have some potential use as an alternative sweetener for special purposes. It has a very pleasant 'cool' taste, but problems with digestive upsets have been reported, and these would have to be overcome before arabinitol could be used in foods. Mannitol and especially glycerol have a wide variety of commercial uses, too numerous for individual mention. Glycerol, like ethanol, has for some time been a product of the petrochemical industry. However, recent increases in the price of crude oil may reduce the price differential between the petrochemical and the fermentation product, so that glycerol produced by fermentation may become economically feasible, and constitute a significant fraction of glycerol used in industry.

REFERENCES

Baker, E. E. and Mrak, E. M. (1938). *Journal of Bacteriology* **36**, 317.

Blakley, E. R. and Spencer, J. F. T. (1962). *Canadian Journal of Biochemistry and Physiology* **40**, 1737.

Borries, G. (1934). *Zeitschrift für Untersuchung der Lebensmittel* **67**, 65. (*Chemical Abstracts* **28**, 2806).

Brown, A. D. (1974). *Journal of Bacteriology* **118**, 769.

Campbell, L. A. and Hegbom, S. S. (1957). *Canadian Journal of Microbiology* **3**, 599.

Charles, J. and Graham, J. (1961). British Patent 870,622.

Chiang, C. and Knight, S. G. (1961). *Biochimica et Biophysica Acta* **46**, 271.

Church, M. B., Paine, H. S. and Hamilton, J. (1927). *Industrial and Engineering Chemistry* **19**, 353.

Cocking, A. T. and Lilly, C. H. (1922). United States Patent 1,425,838.

Connstein, W. and Ludecke, K. (1919). *Chemische Berichte* **52**, 1385.

De Whalley, H. C. S. and Scarr, M. P. (1947). *Chemistry and Industry* pg. 351.

Egorov, A. S., Visnevskaya, G. L. and Skirstymonski, A. I. (1957). U.S.S.R. Patent 104,880. (*Chemical Abstracts* **51**, 10001).

English, M. P. (1954). *Journal of General Microbiology* **10**, 328.

Eoff, J. R., Lindner, W. V. and Beyer, G. F. (1919). *Industrial and Engineering Chemistry* **11**, 842.

Fabian, F. W. (1928). *Fruit Products Journal and American Vinegar Industry* **8**, 18. (*Chemical Abstracts* **23**, 450).

Fabian, F. W. and Quinet, R. I. (1928). *Michigan Agricultural Experiment Station, Technical Bulletin* **92**, 1.

Farrell, L. and Lochhead, A. G. (1931). *Canadian Journal of Research* **5**, 539.

Figard, P. H. (1951). *Iowa State College Journal of Science* **25**, 208.

Fossit, D., Mortlock, R. P., Anderson, R. L. and Wood, W. A. (1964). *Journal of Biological Chemistry* **239**, 2110.

Freeman, G. G. and Donald, G. M. S. (1957). *Applied Microbiology* **5**, 197, 211, 216.

Hajny, G. J. (1964). *Applied Microbiology* **12**, 87.

Hajny, G. J., Hendershot, W. F. and Peterson, W. H. (1960). *Applied Microbiology* **8**, 5.

Hajny, G. J., Smith, J. H. and Garver, J. C. (1964). *Applied Microbiology* **12**, 240.

Harris, J. F. and Hajny, G. J. (1960). *Biotechnology and Bioengineering* **2**, 9.

Haskins, R. H. and Spencer, J. F. T. (1967). *Canadian Journal of Botany* **45**, 515.

Hattori, K. and Suzuki, T. (1974a). *Agricultural and Biological Chemistry* **38**, 581.

Hattori, K. and Suzuki, T. (1974b). *Agricultural and Biological Chemistry* **38**, 1203.

Hawthorne, D. C. and Friis, J. (1964). *Genetics, Princeton* **50**, 829.

Ingram, J. M. and Wood, W. A. (1965). *Journal of Bacteriology* **89**, 1186.

Johnson, M. J. (1941). *Science, New York* **94**, 200.

Kreger-van Rij, N. J. W. (1970). *In* "The Yeasts, A Taxonomic Study", (J. Lodder, ed.), pp. 166–208. North Holland Publishing Company, Ltd., Amsterdam.

Krumbholz, G. (1936). *Obst-und-Gemuse-verwertungsindustrie* **23**, 70, 72, 85, 96, 113. (*Chemical Abstracts* **31**, 6365).

LaRue, T. A. and Spencer, J. F. T. (1967). *Canadian Journal of Microbiology* **13**, 777.

Lochhead, A. G. and Farrell, L. (1931). *Canadian Journal of Research* **5**, 529.

Lochhead, A. G. and Heron, D. A. (1929). *Dominion Department of Agriculture Bulletin* **116** (new series), 1.

Lockhead, A. G. and Landerkin, G. B. (1942). *Journal of Bacteriology* **44**, 343.

Lodder, J. (1970). "The Yeasts, A Taxonomic Study", 2nd edition. North Holland Publishing Company Ltd., Amsterdam.

Marvin, G. E., Peterson, W. H., Fred, E. B. and Wilson, H. F. (1931). *Journal of Agricultural Research* **43**, 121.

Mrak, E. M. and McClung, L. S. (1938). *Journal of Bacteriology* **36**, 316.

Mrak, E. M., Phaff, H. J. and Vaughn, R. H. (1942). *Journal of Bacteriology* **43**, 689.

Mossel, D. A. A. (1951). *Antonie van Leeuwenhoek* **17**, 146.

Neuberg, C. *In* "Industrial Microbiology", (S. C. Prescott and C. G. Dunn, eds.), 2nd edition. McGraw-Hill Book Company, Inc., New York.

Nickerson, W. J. and Carroll, W. R. (1945). *Archives of Biochemistry* **7**, 257.

Nickerson, W. J. and Thimann, K. V. (1941). *American Journal of Botany* **28**, 617.

Onishi, H. (1960). *Bulletin of the Agricultural Society of Japan* **24**, 131, 226.

Onishi, H. (1961). *Agricultural and Biological Chemistry* **25**, 341.

Onishi, H. (1963). *In* "Advances in Food Research", (E. M. Mrak and G. F. Steward, eds.), vol. 12, pg. 53, Academic Press, New York.

Onishi, H. and Perry, M. B. (1965). *Canadian Journal of Microbiology* **11**, 929.

Onishi, H. and Saito, T. (1962). *Agricultural and Biological Chemistry* **26**, 804.

Onishi, H. and Suzuki, T. (1970). *Journal of Fermentation Technology* **48**, 568.

Onishi, H., Saito, N. and Koshiyama, I. (1961). *Agricultural and Biological Chemistry* **25**, 124.

Pasteur, L. (1858). *Comptes Rendues Hebdomadaire des Séances de l'Academie des Séances* **46**, 857.

Peterson, W. H., Hendershot, W. F. and Hajny, G. J. (1958). *Applied Microbiology* **6**, 349.

Phillipov, G. S. (1932). *Schriften des Zentralen Biochemischen Forschungs-Instituts der Nahrungs-und Genussmittelindustrie* U.S.S.R. **2**, 26. (*Chemical Abstracts* **27**, 4597).

Roxburgh, J. M., Spencer, J. F. T. and Sallans, H. R. (1956). *Canadian Journal of Technology* **34**, 248.

Sala, J. P. and De Ayala, E. B. (1959). French Patent 1,178,479. (*Chemical Abstracts* **55**, 12766.

Scarr, M. P. (1951). *Journal of General Microbiology* **5**, 704.

Scolnick, E. M. and Lin, E. C. C. (1962). *Journal of Bacteriology* **84**, 631.

Sols, A., Gancedo, C. and De LaFuente, G. (1971). *In* "The Yeasts", (A. H. Rose and J. S. Harrison, eds.), vol. 2, pp. 271–307. Academic Press, London.

Spencer, J. F. T. (1968). *In* "Progress in Industrial Microbiology", (D. J. D. Hockenhall), vol. 7, pg. 1. Churchill-Livingstone, London.

Spencer, J. F. T. and Gorin, P. A. J. (1960). *Canadian Journal of Biochemistry and Physiology* **38**, 157.

Spencer, J. F. T. and Sallans, H. R. (1956). *Canadian Journal of Microbiology* 2, 72.

Spencer, J. F. T. and Shu, P. (1957). *Canadian Journal of Microbiology* 3, 559.

Spencer, J. F. T., Neish, A. C., Blackwood, A. C. and Sallans, H. R. (1956). *Canadian Journal of Biochemistry and Physiology* 34, 495.

Spencer, J. F. T., Roxburgh, J. M. and Sallans, H. R. (1957). *Journal of Agricultural and Food Chemistry* 5, 64.

Tabuchi, T. and Hara, S. (1973). *Nippon Nogeikagaku Kaishi* 47, 485.

Tanaka, H. and Miller, M. W. (1963). *Hilgardia* 34, 167, 171.

Touster, O. and Shaw, D. R. D. (1962). *Physiological Reviews* 42, 181.

Van der Walt, J. P. (1970). *In* "The Yeasts, A Taxonomic Study", (J. Lodder, ed.), pp. 555-724. North Holland Press, Amsterdam.

Van Uden, N. and Vidal-Leiria, M. (1970). *In* "The Yeasts, A Taxonomic Study", (J. Lodder, ed.), pp. 1235-1308. North Holland Publishing Company, Amsterdam.

Weimberg, R. (1962). *Biochemical and Biophysical Research Communications* 8, 442.

Weimberg, R. and Orton, L. L. (1963). *Journal of Bacteriology* 86, 805.

Wilhelmsen. J. B. (1961). *Enzymologia* 23, 259.

AUTHOR INDEX

Numbers in italic are those pages on which References are listed

A

Abe, R., 249, *260*
Abe, S., 203, 205, 206, *206*, *207*, 212, 218, 227, *256*, *260*, *261*
Abolins, T., 239, *256*
Acres, G. J. K., 104, *111*
Adams, F., 123, *183*
Adiga, P. R., 53, *111*
Aida, K., 226, *258*
Akita, S., 70, *114*, 210, *257*
Akiya, T., 202, 203, *206*, *207*
Akiyama, S., 72, *111*
Albright, F., 93, *111*
Alentyeva, E. S., 289, *296*
Alikhanian, S. I., 241, *256*
Allen, L. A., 273, 289, *296*
Allen, R. H., 93, *111*
Allgeier, R. J., 125, *183*
Al-Sohaily, I. A., 274, 281, 295, *299*
Al-Sultan, A. S., 274, 281, 295, *299*
Amann, P. F., 87, *114*
Ambekar, G. R., 102, *111*
Amelung, H., 59, *111*
American Gas Association, 355, *389*
Ames, B. N., 250, 251, *256*
Anderson, D. G., 321, 323, *325*
Anderson, J. G., 68, *111*
Anderson, L. G., 275, 279, *297*
Anderson, R. F., 88, 107, *116*, *117*, 318, 319, *324*, *325*, 352, *391*
Anderson, R. L., 414, *423*

Andreevskaya, V. D., 272, 275, 291, 295, *299*, *301*
Andrewes, A. G., 321, *325*
Anthony, D. S., 89, *113*
Anton, D. N., 251, *259*
Arai, Y., 244, 250, *256*
Araki, K., 223, 235, 239, 240, 241, 249, 250, 251, *256*, *258*, *259*
Arnold, M., 319, *324*
Arnstein, H. R. V., 93, *111*
Arpai, J., 82, *111*
Asai, T., 131, 132, 133, 170, 176, 183
Aspinall, G. O., 385, *392*
Atkinson, D. E., 16, *29*
Ault, R. G., 137, 141, *183*, *184*
Austin, M. J., 346, 347, *390*
Austrian, R., 386, *389*

B

Baba, S., 66, *117*
Babcock, G. E., 358, *389*
Babij, T., 291, *296*
Baird, J. K., 349, *389*
Baker, B., 386, *391*
Baker, D. L., 104, *111*
Baker, E. E., 394, *423*
Baldwin, I. L., 51, *113*
Baldwin, N. Y., 211, *256*
Bantz, A. C., 110, *118*

Murata, K., 231, *258*
Murray, E. G. D., 132, *183*
Murtaugh, J. J., 104, *116*

N

Naegeli, C., 132, *185*
Nagami, I., 239, *256*
Naganuma, F., 249, *257*
Naganuma, T., 284, *301*
Nägeli, K. W., 2, *30*
Naguib, K., 273, 274, 281, 289, 295, 299
Naim, M. S., 85, *116*
Nakada, H. I., 380, *390*
Nakagawa, M., 241, 242, 244, *260*
Nakahara, T., 89, 91, 92, *113, 117, 118*
Nakajima, J., 249, *258*
Nakamori, S., 226, 232, *258, 260*
Nakamura, I., 83, 84, *114*
Nakamura, T., 226, 250, *258*
Nakanishi, T., 66, 72, *114*, 219, 255, *258*
Nakao, Y., 72, *111*, 215, 218, *257, 258*
Nakatsu, I., 214, *257*
Nakayama, K., 188, 205, 206, *207, 208*, 211, 219, 223, 225, 226, 227, 228, 229, 230, 238, 240, 241, 242, 244, 250, 251, *256, 257, 259, 260, 261*
Nakayama, T., 318, *325*
Nakazawa, H., 239, 241, *259, 260*
Nanji, Y. Y., 180, *183*
Nara, T., 188, 203, 204, 205, 206, *207, 208*, 229, 252, *259*
Negrete, M., 273, *300*
Neilson, N. E., 57, *116*
Neish, A. C., 54, *117*, 418, *425*
Nelson, G. E. N., 79, 106, 109, *114, 115, 116*, 318, 319, 320, 321, *324*
Nelson, S. J., 291, *298*
Nelson, T. E., 367, *390*
Neuberg, C., 21, *30*, 172, *185*, 395, *424*
Neuhard, J., 188, *208*
Newton, R., 137, 141, *183*
Nichol, G. B., 125, 160, *183, 184*
Nicholas, D. J. D., 53, *116*
Nicholson, D. E., 10, *29*

Nickerson, W. J., 369, 409, *424*
Nielsen, N., 279, *299*
Nierlich, D. P., 195, *208*
Niethammer, A., 278, *296*
Niewiadomski, H., 277, 288, 291, *300*
Nilsson, N. G., 279, *299*
Ninet, L., 318, 319, 320, *325*
Ninimura, N., 249, *260*
Nishimura, Y., 215, *259*
Nishio, Y., 214, *257*
Nixon, I. S., 273, 282, *296*
Nobukuni, T., 218, *260*
Nogami, I., 198, 202, *208*
Nonomura, S., 94, *116*
Nord, E. F., 172, *185*
Norris, J. R., 371, 372, *391*
Norval, M., 380, 385, *391*
Nour el Dein, M. S., 85, *117*
Noury and van der Lande, N. V., 102, *116*
Nowakowska-Waszczuk, A., 85, *116*
Nubel, R. C., 80, *116*

O

Obraztsova, N. V., 272, 275, 291, 295, *299, 301*
Ochs, I. L., 123, *185*
O'Donovan, G. A., 188, *208*
Ogata, K., 188, *208*, 292, *299*
Ogston, A. G., 57, *116*
Ohmori, S., 160, *184*
Ohsawa, H., 201, *207*
Ohuchi, S., 199, *207*
Oishi, K., 241, *259*
Oka, T., 205, *208*
Okachi, R., 206, *207*
Okada, H., 218, *260*
Okazaki, H., 217, *257*
Okazaki, Y., 215, *257*
Oki, T., 215, *259*
Okii, M., 242, *261*
Okuhara, M., 292, *299*
Okumura, S., 214, 222, 223, 226, 229, 239, 244, 245, 249, 254, 255, *257, 258, 259, 260, 261*
O'Leary, W. M., 287, *299*
Olson, J. A., 59, *116*
Onda, T., 244, *258*

SUBJECT INDEX

A

B

C

M

Macrocyctis pyrifyra, alginates from, problems, 339

Magnesium
Aspergillus niger requirements, 53
in itaconic acid production, 79, 83

Malate dehydrogenase in glutamic acid biosynthesis from glucose, 217

Malawi, vinegar production, 164

Malaya, seed oil production, 268

Malaysia, seed oil production, 266

Malbranchea pulchella
fat production, 274, 285
fatty-acyl compositions of lipids from, 276

L-Maleic acid, *Rhizopus arrhizus* production, 89

Malic acid, 27
in vinegar analysis, 179

DL-Malic acid, regulations, 90

L-Malic acid, 90, 91

Malt vinegar
analytical values, 179
definition, 126, 127
history, 124, 125
imports, 161
raw materials, treatment, 134
regulations, 129

Maltol, production from kojic acid, 95

Manganese
Aspergillus niger requirements, 53
in citric acid production, 52, 64

Manganese ions in alanine production, 252

Manganese salts in salvage synthesis of purine nucleotides, 205, 206

Mannitol
Candida lipolytica production, 404
production, from alkanes, 401
by *Candida lipolytica*, 398
by osmotolerant yeasts, 393

D-Mannitol, oxidation by *Acetobacter suboxydans*, 175

Margarine, 264
manufacture, 43

Mashing in malt vinegar manufacture, 135

Mechanical agitation in acetone-butyl alcohol fermentation, 39

Media
for *Acetobacter* spp. 164
for *Aspergillus niger* gluconic acid production, 100
for β-carotene production, 319
citric acid production, 50
itaconic acid production, 78
for itaconic acid submerged culture, 79
polyhydroxy alcohol production by osmotolerant yeasts and, 398–405

Medicine, vinegar as, 123

Mesogloia vermiculata, alginate production, composition, 336

Mesotartaric acid from epoxysuccinic acid, 93

Metabolic pathways, 6
primary, 7–13

Methane
microbial fat production from, 295
production in sewage works, 27

Methanol
in citric acid production, 23, 66
from hydrocarbons, 72
in fumaric acid production, 88
glutamic acid production from, 212
microbial fat production from, 295
synthesis from hydrogen from acetone-butyl alcohol fermentation, 43

Methionine, 15
in lysine production, 220

L-Methionine, production, 210

Methylcellulose, World production, 333

Methylene blue in oxidation of alcohols and aldehydes by *Acetobacter* spp., 174

Microbial alginates (*see also* Alginates)
biosynthesis, 381
commercial production, 345
production, 342, 359–367
uronic acids in, structure, 362

Microbial metabolism,
manipulation, 21, 22
primary products, 1–30
regulation, 5–20
secondary products, 3–5

Microbacterium ammoniaphilum, L-glutamic acid production, 212

T

U

V